CATHODOLUMINESCENCE

Edited by **Naoki Yamamoto**

Cathodoluminescence
Edited by Naoki Yamamoto

Published by InTech
Janeza Trdine 9, 51000 Rijeka, Croatia

Publishing Process Manager Danijela Duric

Technical Editor Teodora Smiljanic

Cover Designer InTech Design Team

Additional hard copies can be obtained from orders@intechopen.com

Cathodoluminescence, Edited by Naoki Yamamoto
p. cm.
ISBN 978-953-51-0362-2

For EU product safety concerns:
IN TECH d.o.o., Prolaz Marije Krucifikse Kozulić 3, 51000 Rijeka, Croatia,
info@intechopen.com or visit our website at intechopen.com.

Meet the editor

Dr. Naoki Yamamoto is an Associate Professor of Graduate School of Condensed Matter Physics (Physics Department) at Tokyo Institute of Technology. Dr. Yamamoto is particularly known for his long experience of cathodoluminescence (CL) technique using a transmission electron microscope and its application to characterization of defects in semiconductors and semiconductor quantum structures such as quantum dots and wires. His interest is also directed to light emission induced by an electron beam, i.e., transition radiation, Cherenkov radiation and Smith-Purcell radiation. Dr. Yamamoto has recently applied CL technique to the study of localized surface plasmons in metal nano-particles and surface plasmon polaritons in plasmonic crystals.

Contents

Preface XI

Part 1 Fundamental Techniques 1

Chapter 1 **Cathodoluminescence Studies of Electron Injection Effects in Wide-Band-Gap Semiconductors 3**
Casey Schwarz, Leonid Chernyak and Elena Flitsiyan

Chapter 2 **What is the Pulsed Cathodoluminescence? 31**
Vladimir Solomonov and Alfiya Spirina

Chapter 3 **Pulsed Cathodoluminescence of Natural and Synthetic Diamonds Excited by Nanosecond and Subnanosecond Electron Beams 51**
E.I. Lipatov, V.M. Lisitsyn, V.I. Oleshko, E.F. Polisadova, V.F. Tarasenko and E.H. Baksht

Chapter 4 **Study of Defects by Cathodoluminescence Measurements 71**
A. Djemel and R-J. Tarento

Part 2 Application to Semiconductors 87

Chapter 5 **Cathodoluminescence Properties of Zno Thin Films 89**
M. Addou, J. Ebothé, A. El Hichou, A. Bougrine, J.L. Bubendorff, M. Troyon, Z. Sofiani, M. EL Jouad, K. Bahedi and M. Lamrani

Chapter 6 **Multicolor Luminescence from Semiconductor Nanocrystal Composites Tunable in an Electric Field 121**
G.N. Panin

Chapter 7 **Cathodoluminescence from Amorphous and Nanocrystalline Nitride Thin Films Doped with Rare Earth and Transition Metals 161**
Muhammad Maqbool, Wojciech M. Jadwisienczak and Martin E. Kordesch

Part 3 Application to Oxides and Minerals 207

Chapter 8 **Peculiarity of the Cathodoluminescence of lpha- Alumina Prepared by Calcination of Gibbsite Powder or Generated by Oxidation of a Metallic FeCrAl Alloy 209**
Djelloul Abdelkader and Boumaza Abdecharif

Chapter 9 **Cathodoluminescence Properties of SiO_2: Ce^{3+},Tb^{3+}, SiO_2:Ce^{3+}, Pr^{3+} and SiO_2: PbS 233**
Odireleng M. Ntwaeaborwa, Gugu H. Mhlongo, Shreyas S. Pitale, Mokhotswa S. Dhlamini, Robin E. Kroon and Hendrik C. Swart

Chapter 10 **Cathodo- and Photo- Luminescence of Silicon Rich Oxide Films Obtained by LPCVD 253**
Rosa López-Estopier, Mariano Aceves-Mijares and Ciro Falcony

Chapter 11 **Textural Characterization of Sedimentary Zircon and Its Implication 273**
Yixiong Qian, Wen Su, Zhong Li and Shoutao Peng

Part 4 Related Luminescence 283

Chapter 12 **Novel Problems in the Solid State Cathodoluminescence of Organic Materials 285**
Zheng Xu and Suling Zhao

Chapter 13 **A Lighting Mechanism for Flat Electron Emission Lamp 305**
Jung-Yu Li, Ming-Chung Liu, Yi-Ping Lin, Shih-Pu Chen, Tai-Chiung Hsieh, Po-Hung Wang, Chang-Lin Chiang, Ming-Shan Jeng, Li-Ling Lee, Hui-Kai Zeng and Jenh-Yih Juang

Preface

Cathodoluminescence (CL) is a non-destructive technique to characterize optical and electronic properties of nanostructures in many kinds of materials. Major subject is to investigate basic parameters in semiconductors, impurities in oxides and phase determination of minerals. CL is similar to photoluminescence (PL) technique, but the excitation by high energy electrons can produce all the transition to the higher energy excitation states and induce light emission in the wide energy range concerned with versatile types of electronic transitions in materials. This book involves 13 chapters to present the basics in the CL technique and applications to particles, thin films and nanostructures in semiconductors, oxides and minerals.

The first section presents fundamental techniques to measure basic parameters in semiconductors such as carrier concentration, diffusion length and life time. Prof. Flitsiyan shows how to get those parameters as a function of temperature and beam irradiation using the CL related techniques. Pulsed CL is introduced by Prof. Solomonov to be used for determination of concentration in binary phase. Prof. Tarasenko applied the pulsed CL to study spectral-kinetic characterization of the natural and synthetic diamonds. CL becomes a powerful imaging technique to realize a high spatial resolution when it combines with a scanning electron microscope (SEM) and transmission electron microscope (TEM). Prof. Djemel demonstrates the CL imaging of defects in semiconductors by using SEM.

The application of CL is widely spread over many types of materials. Wide gap semiconductors are recently intensively studied because of its advantage for optoelectronic and power devices operated at high temperatures. Prof. Addou presents the study on CL properties of impurity doped ZnO thin films. Prof. Panin shows CL spectra from ZnO nanorods, nanoparticles and other nano-structures, and discusses the mechanism of electric field tunable luminescence. Prof. Kordesch presents CL peaks originating from impurities of rare earth and transition metals doped in AlN, GaN and other nitride thin films.

Excited electrons are bounded to atomically localized centers such as impurities and vacancies in insulators. Peculiarity of CL due to point defects in α-alumina prepared by different procedures is reported by Prof. Boubaza. Prof. Ntwaeaborwa reviews CL properties of phosphor materials prepared by incorporating Ce, Tb and other ions in amorphous SiO_2. CL and PL study of silicon rich oxide films are reviewed by Dr. Lopez-Estopier. Prof. Yixiong presents characterization of internal structures of zircon in geological rocks.

Some of the light emissions generated in devices using different excitation procedures are closely related to CL. Prof. Xu presents a new kind of electroluminescence in organic materials in the layered device structures. Prof. Li reviews several types of light emission in a new lighting device.

I would like to thank all the contributors for spending their time and energy to complete their chapters. I hope that this book provides a unique opportunity for a reader to gain knowledge about various aspects of CL technique.

Naoki Yamamoto
Physics Department, Tokyo Institute of Technology,
Japan

Part 1

Fundamental Techniques

1

Cathodoluminescence Studies of Electron Injection Effects in Wide-Band-Gap Semiconductors

Casey Schwarz, Leonid Chernyak and Elena Flitsiyan*
Physics Department, University of Central Florida, Orlando, FL, USA

1. Introduction

Recent developments in doping and growth of ZnO stimulated a renewal of interest in this material from the point of view of its applications in optoelectronic devices. As a direct wide bandgap semiconductor ($E_g \approx 3.35$ eV at room temperature) with high exciton binding energy (60 meV, compared to 25 meV in GaN), ZnO is a superior candidate for minority-carrier-based devices, such as light emitting diodes, laser diodes, and transparent p-n junctions.

In the present state of the art, the development of the full potential of ZnO applications hinges in part on the availability of quality, highly conductive materials of both n- and p-type. Similar to GaN, achieving n-type conductivity in ZnO does not present a problem, since even nominally undoped material is generally n-type, due to the electrical activity of native defects, such as zinc interstitials, zinc antisites, and oxygen vacancies [1,2], as well as hydrogen impurities [3].

On the other hand, p-type conductivity with sufficiently high carrier concentrations appears to be much more elusive, mainly due to high ionization energies of potential acceptors, such as nitrogen, phosphorus, and arsenic. This problem is compounded by high concentrations of native and unintentional donors, which act as compensating centers, thus further reducing the free carrier concentration. It has been shown that the background donor concentration can be lowered if Mg is incorporated into the ZnO lattice, since each percent of Mg increases the bandgap of ZnO by 0.02 eV, suppressing the ionization of shallow donors [4]. The conversion to p-type can then be obtained by heavily doping the resultant material with phosphorus followed by annealing in O_2 atmosphere [5].

Recently, successful p-type doping of ZnO also has been attained using phosphorus [6,7], nitrogen [8], arsenic [9], and antimony [10] yielding net hole concentrations up to 10^{18} cm^{-3}. Even with the advent of new technology enabling the production of viable p-type materials, the performance of bipolar devices is fundamentally limited by the transport properties of minority carriers. In direct band gap semiconductors, including ZnO, minority carrier

*Corresponding author

diffusion length is generally several orders of magnitude lower than in indirect gap materials such as silicon or germanium. In order to noticeably increase minority carrier diffusion length by reducing scattering by the dislocation walls, the edge threading dislocation density must be reduced by at least two orders of magnitude from a typical value of about 10^9 cm^{-2} in epitaxial ZnO layers [11-14].

Investigation of minority carrier diffusion lengths and lifetimes in both n- and p-type ZnO is an issue of practical importance, since it has direct implications on the performance of bipolar devices. Moreover, considering possible applications of these devices in high-temperature electronics, the insight into the temperature dependence of minority carrier properties is also of great value. Nonetheless, to the best of our knowledge, the reports on this subject are rather scarce. It is therefore the goal of this work to summarize the available information on the subject of minority carrier transport in ZnO and related compounds, focusing on its temperature dependence and the dynamics of non-equilibrium carrier recombination. This discussion will be preceded by a brief summary of the role of minority carrier transport in the performance of bipolar devices, as well as by the review of techniques of choice for measurement of the minority carrier diffusion length.

2. Role of minority carrier diffusion length in bipolar device performance

In general, when non-equilibrium carriers are generated in a material due to external excitation in the absence of an electric field, they diffuse over a certain distance before undergoing recombination. This parameter, namely the average distance traveled in a particular direction between generation and recombination, is characterized by the diffusion length, L. The diffusion length is related to the carrier lifetime, τ, (i.e., the time between generation and recombination of non-equilibrium carriers) through carrier diffusivity, D:

$$L = \sqrt{D\tau} \tag{1}$$

Diffusivity, or diffusion coefficient, is determined in turn by the mobility of the carriers, μ, according to the Einstein relation:

$$D = \frac{k_B T}{q}\mu \tag{2}$$

where k_B is the Boltzmann's constant, T is absolute temperature, and q is the fundamental charge.

The diffusion process is driven by concentration gradients; since external excitation has a much larger impact on the concentration of *minority* carriers than that of majority ones (because generation density is usually much lower than the majority carrier density), it is the minority carriers that are more susceptible to diffusion. The electron diffusion current density is proportional to the gradient of the electron density, n, by the following relation;

$$J_n = eD_n \frac{dn}{dx} \tag{3}$$

Where e is the charge of the electron and D_n is the electron diffusion coefficient. Relating the diffusion coefficient to temperature we get:

$$D = D_0 \exp\left(\frac{-E_a}{k_B T}\right) \tag{4}$$

With E_a being the activation energy of the process. Using equation (1) we arrive at the equation relating the diffusion length to temperature by:

$$L_n = L_0 \exp\left(\frac{-E_a}{2k_B T}\right) \tag{5}$$

CL peak intensity, I_{CL}, of Near Band Edge (NBE) luminescence decays systematically with decreasing temperature. Since CL intensity is proportional to the rate of recombination, as the intensity decreases it is shown that the number of recombination events also decreases. Also, since the intensity is inversely proportional to the lifetime of carriers in the band, the decay of I_{CL} indicates the increase of τ with temperature. This follows in an exponential relationship represented in this equation;

$$I_{CL} = \frac{A}{\left[1 + B\exp\left(-\frac{\Delta E_{a,T}}{k_B T}\right)\right]} \tag{6}$$

Where A and B are scaling factors, $\Delta E_{a,T}$ is the thermal activation energy, k_B is the Boltzmann constant and T is temperature.

Diffusion of minority carriers is a process that is fundamental to the operation of bipolar photovoltaic devices, with minority carrier diffusion length being the central parameter defining the device performance. In the presence of a p-n junction or a Schottky barrier, the non-equilibrium minority carriers generated by external excitation (e.g., light incident on a photodiode) within a few diffusion lengths of the space-charge region can be collected by the built-in field and thus contribute to the current flow across the device. The greater the diffusion length of the carriers, the more current can be collected, leading to the higher efficiency of the device. In photodiodes, it is usually only one side of the p-n junction that contributes to photocurrent. If the light is absorbed in the p-region of the junction, the quantum efficiency, η, can be represented as follows:

$$\eta = (1-r)\left[1 - \frac{\exp(-aW)}{1 + aL_n}\right] \tag{7}$$

where r and a are the reflection and absorption coefficients, respectively, W is the width of the space-charge region, and L_n is the diffusion length of minority electrons.

Quantum efficiency is directly related to the spectral responsivity, R, of a photodiode:

$$R(E) = \frac{I_{ph}}{P_{op}} = \frac{q\eta}{E} \tag{8}$$

where I_{ph} is total photocurrent, P_{op} is optical power incident on the device, q is the fundamental charge, and E is the energy of the incident photons. The relationship between minority carrier diffusion length and responsivity of Schottky photodiodes has been examined in great detail in Ref. [15]. Schottky photodiodes are among the simplest photovoltaic devices, where the non-equilibrium minority carriers, generated in the bulk of the semiconductor due to light absorption, are collected by the built-in field of the Schottky barrier deposited on the surface of the semiconductor. In order for the carriers to contribute to device current, they have to be generated within a few diffusion lengths of the collector.

For incident energies greater than the bandgap of the absorber material, non-equilibrium electron-hole pairs are generated only in the thin layer next to the surface of incidence, with the maximum depth of $1/\alpha$. This value is on the order of 100 nm in ZnO [16,17] and is generally much smaller than the thickness of the absorber layer. Considering a front-illuminated configuration (in which the incident light passes through the semitransparent Schottky contact), if L is greater than the generation depth, most of the non-equilibrium minority carriers are collected by the built-in field of the space-charge region. In this case, the internal quantum efficiency of the device approaches 100% and the responsivity is independent of the diffusion length value. Below this threshold, the responsivity decreases with L, provided that the width of the space-charge region is smaller than the generation depth.

If the energy of incident light is below the bandgap, light penetration depth is large (several micrometers), and a fair portion of the non-equilibrium carriers is generated in the neutral region of the semiconductor due to the ionization of the mid-gap levels. Since only the carriers within a few diffusion lengths of the space-charge region contribute to photocurrent, the responsivity at below-bandgap energies is limited by the diffusion length (unless the diffusion length exceeds the thickness of the absorber layer, in which case the latter is the limiting factor) [15].

3. Methods for determination of minority carrier lifetime and diffusion length

Although as of the date of this writing minority carrier transport in ZnO remains, with a few exceptions, essentially unexplored, however, the measurement of minority carrier diffusion length is a well-established subject. This section reviews three of the most widely used techniques, namely Electron Beam Induced Current (EBIC), Time-Resolved Photoluminescence (TRPL), and Time-Dependent Cathodoluminescence (CL) Measurements.

3.1 Electron beam induced current technique

Due to a unique combination of convenience and reliability, Electron Beam Induced Current (EBIC) method is among the most popular techniques for minority carrier diffusion length measurements. It requires comparatively simple sample preparation and is used *in-situ* in a scanning electron microscope (SEM). Fig. 1 shows a typical measurement configuration known as planar-collector geometry [18,19]. As a charge collection technique, EBIC method employs a Schottky barrier or a p-n junction to collect the current resulting from the non-equilibrium minority carriers generated by the beam of the SEM. As the beam is moved away from the barrier/junction in a line-scan mode, the current decays as fewer and fewer minority carriers are able to diffuse to the space-charge region.

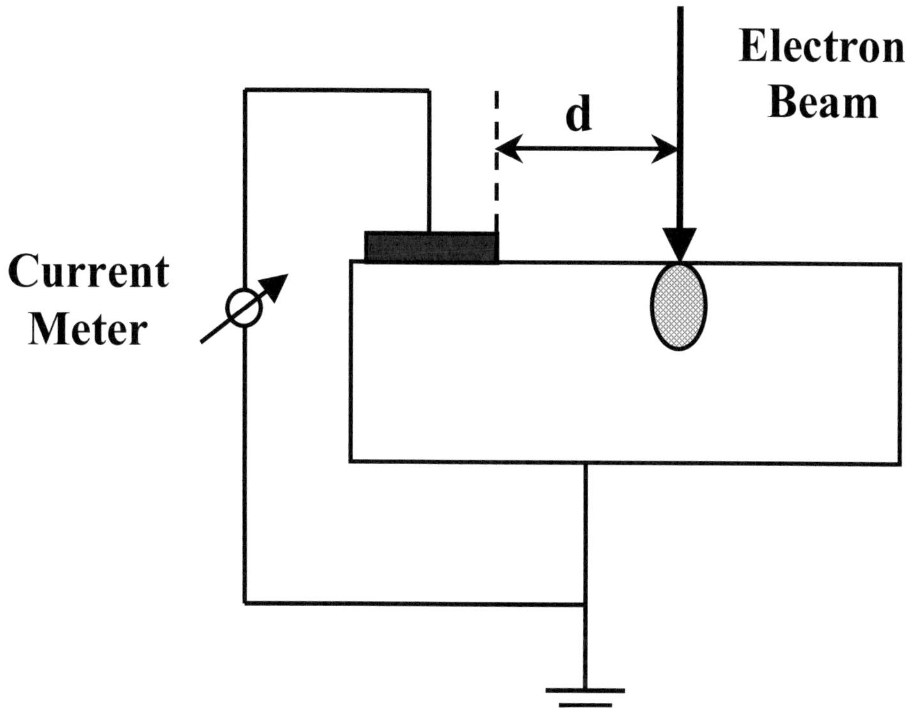

Fig. 1. Experimental setup used for EBIC measurements in planar-collector configuration. The dashed ellipse represents the generation volume; *d* is the variable distance between electron beam and the Schottky contact, represented by a solid rectangle.

The mathematical model for EBICurrent [20], shows that the decay of current, I_{EBIC}, can be described by the following expression:

$$I_{EBIC} = Ad^{\alpha} \exp\left(-\frac{d}{L}\right) \tag{9}$$

where *A* is a scaling constant, *d* is beam-to-junction distance, and α is an exponent related to the surface recombination velocity, v_s.

The diffusion length is usually extracted by rearranging the terms of equation (9):

$$\ln\left(Id^{-\alpha}\right) = -\frac{d}{L} + \ln(A) \tag{10}$$

which yields a linear relationship between $ln\left(Id^{-\alpha}\right)$ and *d* with a slope equal to $-\frac{1}{L}$. This approach is taken to be accurate for $d > 2L$. It should be noted that in Ref. [20], the authors analyzed only the two asymptotic cases, namely $v_s = 0$ and $v_s = \infty$, and found that $\alpha = -1/2$ for the former and -3/2 for the latter. Later, Chan *et. al.* [21] demonstrated that this approach can be applied to materials with arbitrary surface recombination velocity by selecting α such that a linear relationship between $ln(Id^{-\alpha})$ and *d* is obtained. However, even if *any* value of α is used (such that $-3/2 \leq \alpha \leq -1/2$), the result for the diffusion length changes by less than 20% [22] this is due to the fact that the exponential term dominating the value of I_{EBIC} is independent of α.

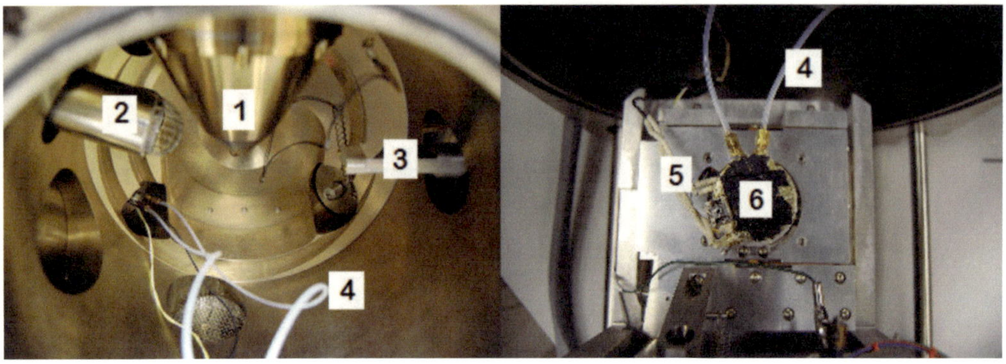

Fig. 2. Inside the vacuum chamber of the SEM. (Left) The electron gun (1), secondary electron collector (2), light guide (3) and (4) plastic hoses through which liquid nitrogen is passed through in order to cool the sample. (Right) This is the sample stage (6) with the computer controlled heater and platinum resistance thermometer (5) connected to it.

3.2 Time-dependent cathodolumenscence (CL) measurements

Cathodoluminescence is an optical and electrical phenomenon in which a luminescent material (semiconductor) emits light upon the impact of an electron beam produced from an electron gun. The high energy electron beam impacted onto a semiconductor will result in the promotion of electrons from the valence band to the conduction band. This movement leaves behind a hole, when the electron and hole recombine a photon may be emitted. These emitted photons can then be collected and analyzed by an optical system.

Minority carrier lifetime is related to the lifetime of the non-equilibrium carriers in the conduction band. CL measurements were used to detect changes in carrier lifetimes due to varying sample temperature as well as electron irradiation.

3.3 Experimental setup

CL measurements were also performed in situ inside the Philips XL30 SEM. The temperatures varies in the sample were 25°C to 125°C. The Hamamatsu photomultiplier tube is sensitive to wavelengths ranging from 185 to 850 nm. Slit size was kept at 4.5 mm. The CL experimental set up is integrated with the SEM. The monocromator is on the right hand side of the SEM and houses the mirrors and the diffraction grating system. Located on the right of the monochromator is the Hamamatsu photomultiplier tube. For each specific wavelength a PC is used to record the intensity of light by counts.

On the light guide there is a parabolic mirror which collects the emitted light from the samples. A hole is cut in the middle of the mirror to allow for the electron beam to reach the sample when the mirror is over the sample. The mirror is positioned a specific working distance, usually a few millimeters, from that sample which is optimized with Back Scattered Electrons (BSE) for maximum intensity so that its focal length coincides with the sample. The focus size is normally in tens of microns.

The light collected from this focal point comes from a parallel beam through the hollow waveguide tube and focused onto the entrance slits of the monochromator. The electron beam may be used in spot or line scan mode. New areas are sampled after each temperature measurement is made to avoid contamination.

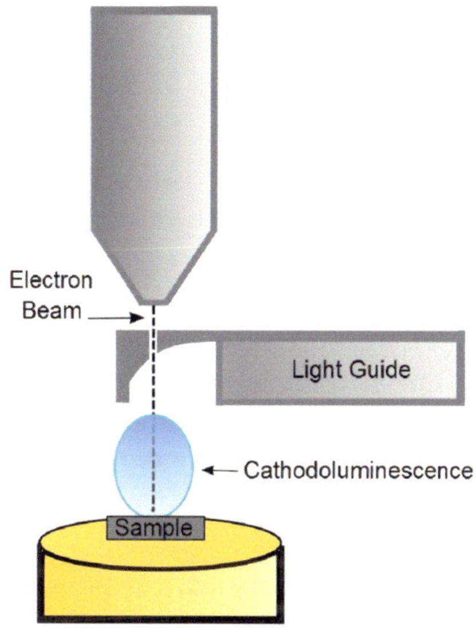

Fig. 3. Diagram of the Cathodoluminescence method including the sample, light guide and electron beam.

Determining the thermal activation energy by use of the CL measurements is similar to the way in which the diffusion length is used to calculate the thermal activation energy using the EBIC technique. Diffusion is due to the existence of a spatial variation of the carrier concentration inside a semiconductor. The carriers will move from areas of high concentration to low concentration, this is known as the diffusion current, equation (3).

When non-equilibrium carriers are generated in a material due to an external excitation in the absence of an electric field, they will diffuse over a certain distance, in a particular direction, before recombining. The average distance that the carrier has gone before succumbing to recombination is defined by L, the diffusion length, which is related to carrier lifetime ,τ, by equation (1). Carrier lifetime refers to the time between generation and recombination of non-equilibrium carriers by the Einstein relationship connecting the mobility μ and the diffusion current D_n by equation (2).

3.4 Time-resolved photoluminescence technique

Time-Resolved Photoluminescence (TRPL) technique is an indirect method that relies upon measuring the minority carrier lifetime in order to determine the diffusion length according to equations (1) and (2). It provides a time-efficient way for characterizing the transport of non-equilibrium minority carriers and is most useful for materials with good luminescence properties. An advantage of this method is that it also provides valuable insight into the nature of the recombination mechanisms governing minority carrier lifetime. The detailed description of the TRPL measurements and analysis is provided in Ref. [23].

TRPL measurements are based on recording the transient decay of near-bandgap (NBE) photoluminescence excited by a short pulse of light, such as from a pulsed laser source. If the concentration of non-equilibrium minority carriers decreases exponentially from its initial value ρ_0, the intensity of the photoluminescence, I_{PL}, also follows an exponential decay:

$$I_{PL} = \frac{\rho_0}{\tau_R}\exp\left(-\frac{t}{\tau}\right) \tag{11}$$

where τ_R is the radiative recombination lifetime. In most cases, however, the photoluminescence intensity is affected not only by lifetime of minority carriers in the band, but also by the diffusion rate out of the absorber region, since the escape of the carriers from the active region is in this case equivalent to the reduction in their lifetime. Therefore, the lifetime obtained from the TRPL measurements is often less than the true minority carrier lifetime. In order to obtain an accurate measurement, it is necessary to introduce a confinement structure to reduce the influence of diffusion. In such a confinement structure the semiconductor of interest is "sandwiched" between two layers with wider bandgap and like doping type, so that the wide bandgap layers act as minority carrier mirrors.

4. Temperature dependence of minority carrier diffusion length and lifetime

4.1 Studies in n-type ZnO

Because of its intrinsic thermal stability, ZnO is a good candidate for high-temperature optoelectronic devices. However, the subject of the temperature dependence of minority carrier transport properties has not yet been adequately addressed in the literature. This section reviews the results obtained from variable-temperature diffusion length measurements performed on n-type ZnO [24].

The samples under investigation were weakly n-type ZnO substrates with electron concentration of ~ 10^{14} cm^{-3} and mobility of ~ 150 cm^2/Vs. Secondary Ion Mass Spectroscopy (SIMS) measurements revealed the Li concentration of about 3×10^{16} cm^{-3} [25]. The Schottky contacts for EBIC measurements were deposited on the non-polar a-plane of ZnO crystal by electron beam evaporation of 80 nm-thick Au layer and subsequent lift-off. ZnO:N samples were grown using molecular beam epitaxy by SVT Associates. Hall effect measurements revealed hole concentration of 4.5×10^{17} cm^{-3} and mobility of ~1 cm^2/V·s.

The studies of L as a function of temperature were carried out using EBIC method in a planar-collector configuration with a Schottky barrier (Sec. 3.1). At each temperature, several measurements were taken by scanning the beam of the SEM along a line perpendicular to the edge of the Schottky contact and recording the exponential decay of current. The recorded data were fitted with equation (9) using $\alpha = -1/2$. This value corresponds to zero surface recombination velocity which, given the excellent luminescence properties of this sample and a good fit to the experimental results, is a reasonably good approximation. It should be noted that each EBIC line-scan was recorded on a previously unexposed area in order to avoid the influence of electron irradiation on the value of diffusion length (cf. Sec. 5 below).

Table 1 summarizes the results of EBIC measurements performed on one of the bulk ZnO samples and shows that the diffusion length of minority holes in n-ZnO increases with increasing temperature, T. The increase of L with T is not unique to this semiconductor. Similar trends were previously observed in GaAs [26] and later in GaN epitaxial layers [27]. In all cases, this increase was exponential with temperature and was modeled with the expression (5).

Temperature (°C)	Diffusion Length (μm)	CL Intensity (10^3 counts)
25°C	0.438 ± 0.022	72.1 ± 3.7
50°C	0.472 ± 0.060	54.4 ± 3.8
75°C	0.493 ± 0.028	49.2 ± 2.4
100°C	0.520 ± 0.074	44.6 ± 4.7
125°C	0.547 ± 0.086	38.5 ± 6.8
E_A (eV)	0.045 ± 0.002	0.058 ± 0.007

Table 1. Temperature dependence of minority carrier diffusion length and cathodoluminescence intensity of the near-band-edge peak in n-ZnO. After Ref. [24].

With the fit using equation (5), the experimental results for n-ZnO yields activation energy of 45 ± 2 meV [24]. This energy represents carrier de-localization energy, since it determines the increase of the diffusion length due to reduction of recombination efficiency [26]. The smaller is the activation energy, the more efficient is the de-trapping of captured carriers at a fixed temperature (see discussion below).

The role of increasing carrier lifetime is also supported by the results of cathodoluminescence (CL) measurements, carried out *in-situ* in SEM, which are presented in Fig. 4. The inset of Fig. 4 shows a cathodoluminescence spectrum in the vicinity of the NBE transition at 383 nm (3.24 eV). This feature in bulk ZnO has been attributed to the transition from the conduction band to a deep acceptor level [28]. It was observed that the peak intensity, I_{CL}, of NBE luminescence decays systematically with increasing temperature, providing direct evidence that the number of recombination events decreases. Because the intensity of the NBE luminescence is inversely proportional to the lifetime of carriers in the band, the decay of I_{CL} indicates the increase of τ with temperature. The decay proceeds exponentially according to the equation (6) [29]. Based on the fit shown in Fig. 4, the activation energy was determined to be 58 ± 7 meV. This value is in excellent agreement with that obtained by photoluminescence measurements in Li-doped ZnO films [30]. It is also consistent with the results of the variable-temperature EBIC measurements, which suggests that the same underlying process is responsible for both the increase in the diffusion length and the CL intensity decay. This process is outlined below.

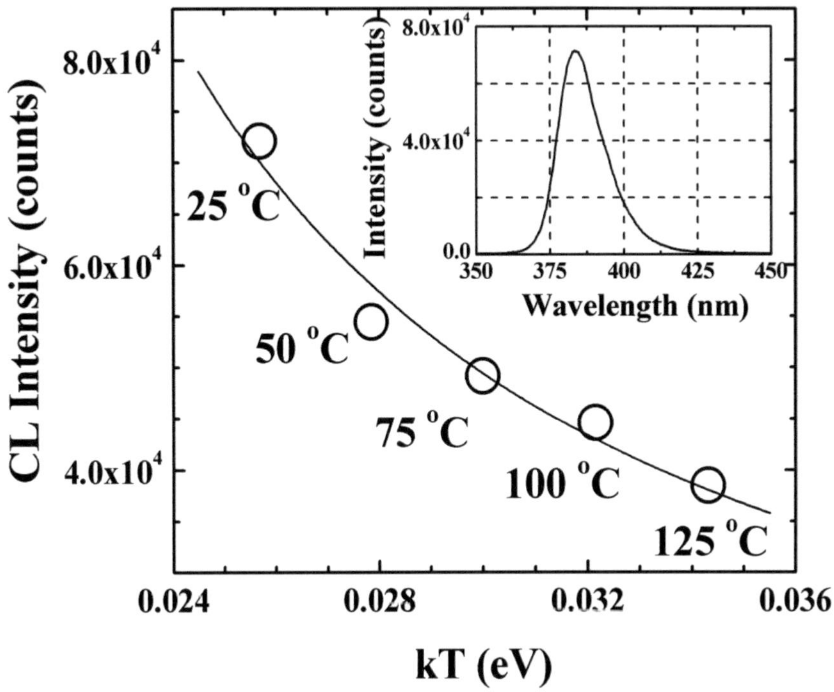

Fig. 4. Experimentally obtained values for the peak NBE CL intensity in ZnO:Li as a function of temperature (open circles) and the fit (solid line; equation (10)), yielding activation energy of 58 ± 7 meV. **Inset:** CL spectrum on n-ZnO showing the NBE transition at 25 °C. After Ref. [24].

The increase in minority hole lifetime in the valence band is likely associated with a smaller recombination capture cross-section for this carrier at elevated temperatures. In GaAs, for example, detailed analysis for temperature dependence of capture cross-section indicates an order of magnitude decrease of recombination efficiency, measured in terms of an "effective capture radius", in the temperature range from 100 to 300 K [26]. Non-equilibrium electron-hole pairs are generated by the beam of the SEM and subsequently annihilate by recombining with each other. Since the hole capture cross-section is inversely proportional to temperature [26,31], the frequency of the recombination events (and, hence, the CL intensity) decreases as the temperature is raised. This means that non-equilibrium holes exist in the valence band for longer periods of time and, consequently, diffuse longer distances before undergoing recombination. Note that carrier diffusivity, D, is also a temperature-dependent quantity and, therefore, can affect the diffusion length (cf. equation (1)). On the other hand, it has been demonstrated for n-ZnO that the mobility, μ, of majority carriers decreases in the temperature range of our experiments by about a factor of 2 [32]. Assuming that the mobility of the minority carriers exhibits the same behavior [27] and combining equations (1) and (2), it is clear that the value of the diffusion length is dominated by the growing lifetime of minority holes. From the Einstein relation (equation (2)), the above-referenced difference in mobility translates to about a 30% decrease in diffusivity at 125 °C as compared to 25 °C. Based on a 30% difference in diffusivity and using experimentally obtained values of diffusion length, we conclude that the lifetime of minority holes at 125 °C is nearly 2.5 times greater than at room temperature.

Preliminary results indicate that temperature-induced increase in carrier lifetime also occurs in epitaxial ZnO. Nitrogen-doped ZnO samples grown using molecular beam epitaxy were provided by SVT Associates. Hall effect measurements revealed hole concentration of $4.5x10^{17}$ cm^{-3} and mobility of ~1 $cm^2/V \cdot s$. Fig. 5 shows the decay of NBE CL intensity fitted with equation (6). The measurements yielded activation energy of 118 ± 12 meV. This value is comparable to the activation energy of the nitrogen acceptor in ZnO [34,35], which indicates possible non-equilibrium carrier trapping on nitrogen-related deep levels.

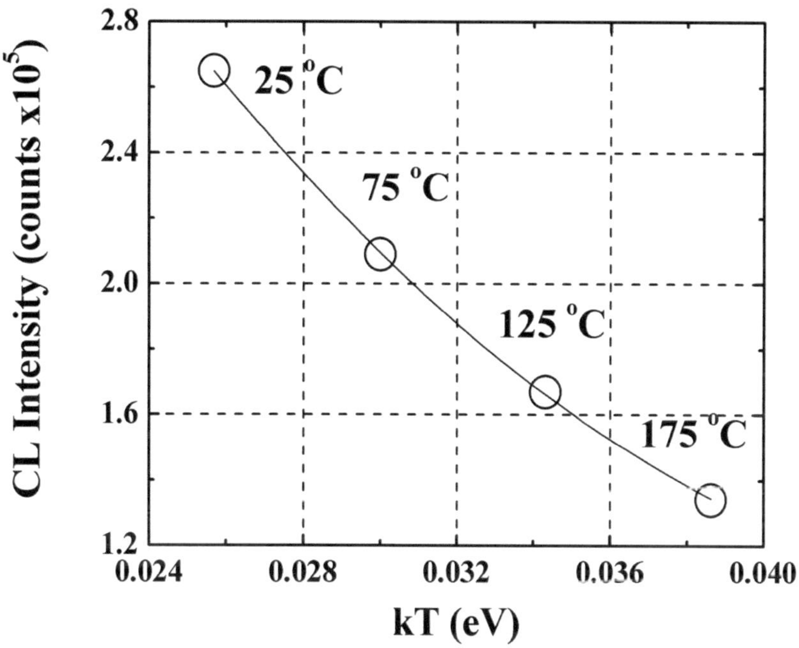

Fig. 5. Maximum CL intensity of the NBE transition in p-ZnO:N as a function of temperature (open circles). Solid line shows the fit with equation (10), resulting in activation energy of 118 ± 12 meV. After Ref. [33].

4.2 Studies in p-type ZnO doped with antimony

The possibility of p-type doping with larger radii atoms, such as antimony, has been explored in Refs. [10,36]. The studies demonstrated that despite the large size mismatch, which in principle should inhibit the substitution of this impurity on the oxygen site, effective p-type doping with hole concentrations up to 10^{20} cm^{-3} can be achieved [36]. These findings prompted the first-principles investigation by Limpijumnong *et al.*, who suggested that the role of acceptors in size-mismatched impurity doped ZnO is performed by a complex of the impurity with two zinc vacancies Sb_{Zn}-$2V_{Zn}$, the ionization energy of which is several-fold lower than that of a substitutional configuration and is consistent with the independent experimental observations [37]. Despite the encouraging predictions, however, very few attempts at achieving p-type conductivity in antimony-doped ZnO have been effective. Aoki *et al.* reported surprisingly high hole concentrations of up to 5×10^{20} cm^{-3} in ZnO:Sb films prepared by excimer laser doping [36]. Some of the authors also obtained p-type ZnO:Sb by molecular beam epitaxy (MBE) [10].

The characteristics of an acceptor level in Sb-doped, p-type ZnO were studied using cathodoluminescence spectroscopy as a function of hole concentration. Variable-temperature CL measurements allowed for the estimation of the activation energy of a Sb-related acceptor from temperature-induced decay of CL intensity. The experiments were performed on ZnO:Sb layers grown on Si (100) substrates by an electron cyclotron resonance (ECR)-assisted MBE. The detailed growth procedures are available in Ref. [10]. Hall Effect measurements revealed strong p-type conductivity, with hole concentrations up to 1.3×10^{18} cm^{-3} and mobility up to 28.0 cm^2/V s at room temperature (Table 2).

Sample Number	Hole Concentration (cm^{-3})	Carrier Mobility (cm^2/V s)
1	1.3×10^{17}	28.0
2	6.0×10^{17}	25.9
3	8.2×10^{17}	23.3
4	1.3×10^{18}	20.0

Table 2. Room-temperature electronic properties of Sb-doped p-type ZnO films. After Ref. [39].

CL measurements were conducted *in-situ* in the Philips XL30 scanning electron microscope (SEM) integrated with Gatan MonoCL cathodoluminescence system. The SEM is also fitted with a hot stage and an external temperature controller (Gatan) allowing for temperature-dependent experiments. The decay of near-band-edge (NBE) luminescence intensity was monitored as a function of temperature in the range from 25 to 175 ^{o}C. Accelerating voltage of 10 kV was used. Note that each measurement was taken in a previously unexposed area to avoid the potential influence of electron irradiation [24,38].

The investigation of the luminescence properties of Sb-doped ZnO was started with the acquisition of room-temperature cathodoluminescence spectra shown in Fig. 6.

The inset of Fig. 6 reveals that the CL spectra of all three samples are dominated by the NBE band, which generally contains the band-to-band transition as well as the transition from the conduction band to a deep, neutral acceptor level (e, A^0) [28,38]. Since acceptor levels form a band in the forbidden gap, the red shift of the NBE peak with increasing carrier concentration (i.e., higher doping levels) is consistent with the (e, A^0) emission and may indicate the broadening of the Sb-related acceptor band [40,41]. Another observation that can be made from Fig. 6 is the systematic decay in intensity of the NBE luminescence with increasing doping level. This decrease may be attributed to the reduction in radiative recombination rates as more disorder is introduced into the ZnO lattice by large-radius Sb atoms. The increasing trend in the values of the full width at half-maximum (FWHM) of the NBE spectra provide further evidence for the impact of the size-mismatched dopant - FWHM values were determined to be about 16.1, 19.4, 23.5, and 21.7 nm (corresponding to 136, 163, 196, and 178 meV) for samples 1, 2, 3, and 4, respectively. Note that while FWHM of NBE transitions in CL spectra tends to be greater than the width of photoluminescence (PL) peaks, the above values are comparable to those obtained for (e, A^0) transitions in CL spectra of other ZnO and GaN materials [24,38,42].

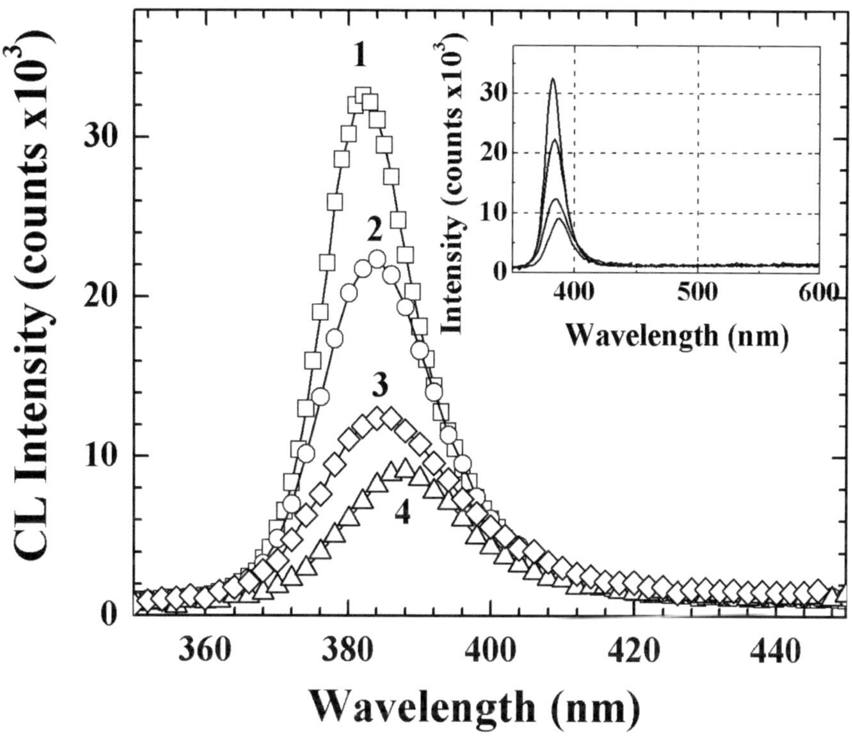

Fig. 6. NBE cathodoluminescence spectra of samples 1-4 taken at room temperature. The peaks are at 382, 384, 385, and 387 nm, respectively. Inset: broad-range CL spectra of the same samples. After. Ref. [39].

The intensity of NBE luminescence was also monitored as a function of temperature. It was observed that the intensity decays with sample temperature, T, in agreement with expression (6) [29]. From equation (6), it can be deduced that the inverse intensity, 1/I, should exhibit an exponential dependence on 1/kT. This is shown in the inset of Fig. 7 on the example of sample 1. Note that the intensity in this and subsequent figures was normalized with respect to its room-temperature value for each of the samples. The activation energies, E_A, were obtained from the slopes of Arrhenius plot shown in Fig. 7. In case of a (e, A^0) transition, E_A is related to the ionization energy of acceptors: the lower the value of the activation energy, the more likely is the ionization of the acceptor by a valence band electron ($A^0 + e \rightarrow A^-$); since an ionized level does not participate in recombination via the (e, A^0) route, the rate of these transitions (i.e., the intensity of the luminescence) decreases with E_A at any given temperature. Conversely, for a constant E_A, the intensity decays with increasing temperature as more and more acceptors are ionized.

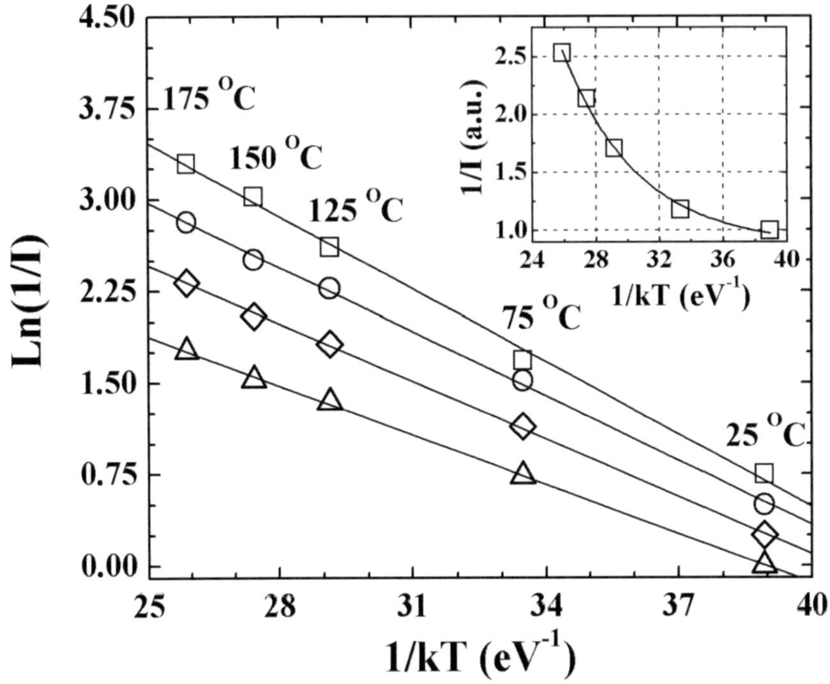

Fig. 7. Arrhenius plot showing the decay of normalized NBE luminescence intensity with increasing temperature for sample 1 (open squares), sample 2 (open circles), sample 3 (open diamonds), and sample 4 (open triangles). The linear fits (solid lines) yielded activation energies of 212 ± 28, 175 ± 20, 158 ± 22 and 135 ± 15 meV for samples 1, 2, 3, and 4, respectively. The data were vertically offset for clarity. Inset: exponential decrease of CL intensity for sample 1 (open squares) and the fit (solid line). After Ref. [39].

It is apparent from Fig. 7 that the activation energy shows a systematic dependence on the carrier concentration. The values of E_A are 212 ± 28, 175 ± 20, 158 ± 22, and 135 ± 15 meV for samples 1, 2, 3, and 4, respectively. These values are in reasonable agreement with the ionization energy of a Sb_{Zn}-$2V_{Zn}$ complex predicted by Limpijumnong *et al.* to have a value of about 160 meV [37]. Furthermore, the decay of activation energy with carrier density, *p*, follows a common pattern observed previously in other semiconductors [41, 42, 43] and is described by an equation of the type;

$$E_A(N_A^-) = E_A(0) - \alpha(N_A^-)^{1/3} \tag{12}$$

where N_A^- is the concentration of ionized acceptors, EA(0) is the ionization energy at very low doping levels, and α is a constant accounting for geometrical factors as well as for the properties of the material. Fig. 8 demonstrates that equation (12) provides a reasonable fit to the experimentally obtained activation energies under the approximation that N_A^- - N_D^+ = p, where N_D^+ is the density of ionized shallow donors (due to compensation, the p-type conductivity is determined by the difference between the concentrations of ionized donors and acceptors).

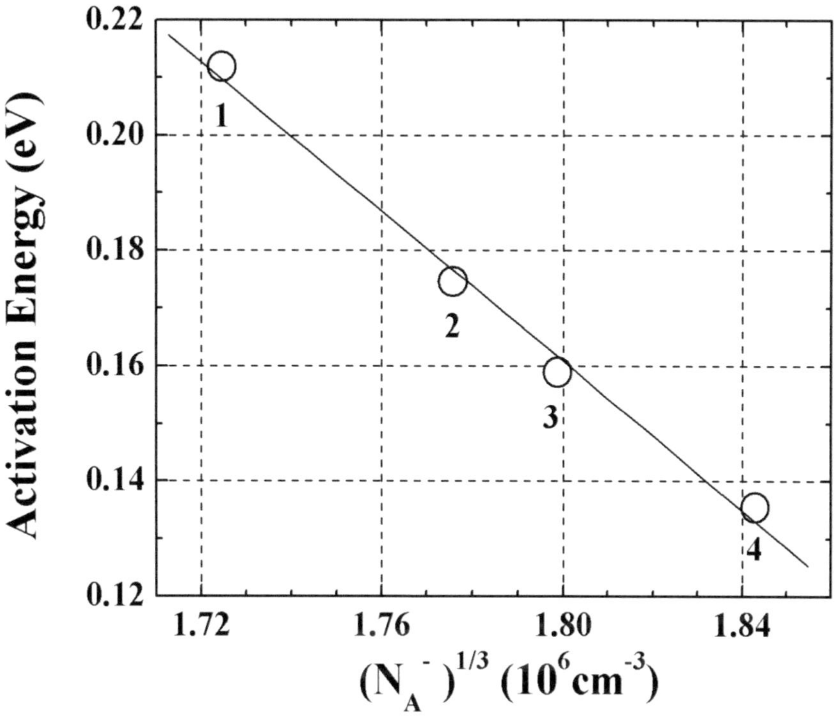

Fig. 8. Decrease of activation energy as a function of ionized acceptor concentration. After Ref. [39].

The value of α was found to be equal to 6.4 × 10 ⁷, which is comparable to that in p-GaN and p-Si [41,43]. N_D^+ can be estimated from the electron concentration in undoped, n-type ZnO samples grown by the same method and is about 5 × 10^{18} cm^{-3} [10]. We note that this is a rough estimate and does not account for the earlier observation that background donor concentration in Sb-doped samples may be different from that in undoped ZnO films due to the creation of Zn vacancies induced by Sb doping. The concentration of Zn vacancies was shown to depend on Sb doping level [10], which would in turn lead to the variations in shallow donor density among the samples under investigation.

It should be noted that earlier PL measurements performed on sample 4 showed consistent activation energy of 140 meV [10]. Furthermore, temperature-dependent measurements of hole concentration in sample 4 shown in Fig. 9 suggest that the temperature dependence of luminescence intensity is associated with acceptors. The dependence of p on temperature can be modeled with a charge-balance equation of the following form:

$$p + N_D^+ = \frac{N_A^-}{1 + \frac{p}{\phi}} \tag{13}$$

where N_D^+ and N_A^- were estimated as described above and $\phi = AT^{3/2}\exp(-E_A / kT)$, with A being a factor accounting for the degeneracy of acceptor states and the density of states in the valence band [32]. The fit of the data shown in Fig. 9 revealed activation energy of about 100 meV [10], which is in reasonable agreement with activation energy obtained by CL and is most likely related to the Sb_{Zn}-$2V_{Zn}$ complex.

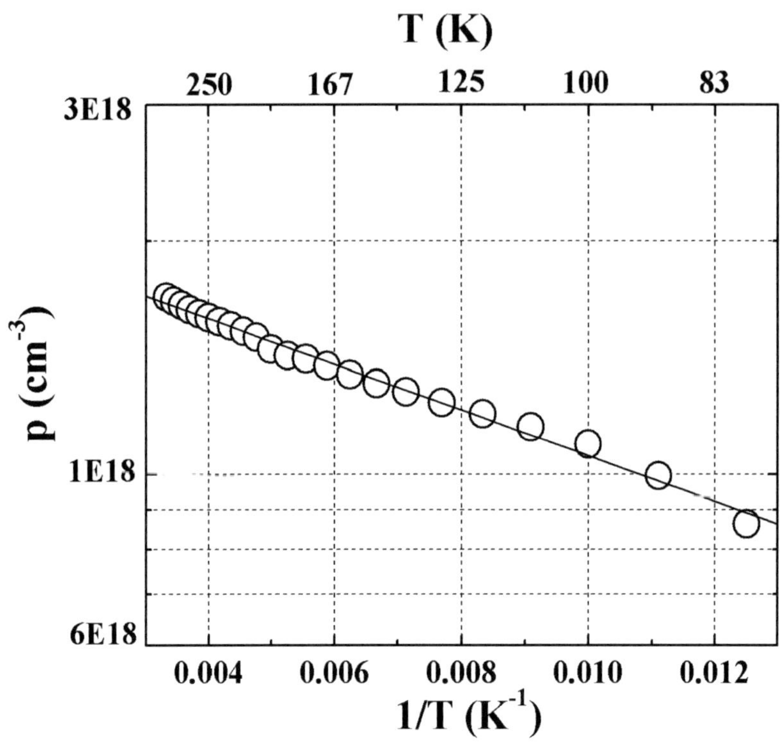

Fig. 9. Temperature dependence of hole concentration in sample 4. The fit yielded activation energy of about 100 meV. After Ref. [39].

Although the existence of other Sb-related acceptors cannot be categorically excluded, their involvement in the temperature-induced CL intensity decay is highly unlikely. The substitutional defect (Sb_O) as well as the single-vacancy complex (Sb_{Zn}-V_{Zn}) are predicted to have ionization energies about an order of magnitude greater than those obtained experimentally, while other defects can also be ruled out based on their electrical behavior and/or high formation energies [37].

The phenomenon of variation of the dopant activation energy with carrier concentration in semiconductors has been attributed to a number of causes. Among these are the formation of the band-tail states that extend into the forbidden gap, the broadening of the acceptor band in the gap, and the reduction of binding energy due to Coulomb interaction between the holes in the valence band and the ionized acceptor states [40-42]. The variable temperature cathodoluminescence studies of Sb-doped p-type ZnO allowed estimating the activation energy of the Sb-related acceptor in the range of 135-212 meV. The activation energy was found to be strongly dependent upon the hole concentration. While the nature of the acceptor cannot be determined conclusively, evidence suggests that it is a Sb_{Zn}-$2V_{Zn}$ complex proposed by Limpijumnong *et al* [37].

5. Studies of minority carrier recombination

Deep carrier traps have pronounced implications on minority carrier transport and often give rise to such undesirable phenomena as radiation-induced optical metastability, persistent photoconductivity and optical quenching of photocurrent [44-46]. On the other hand, it has been demonstrated that capture of minority carriers by deep metastable traps is associated with the increase of minority carrier diffusion length and lifetime [24,38,47]. Furthermore, in GaN this increase was shown to result in a significant (several-fold) improvement of photoresponse in agreement with equation (7) [22,47].

Extensive studies aimed at achieving p-type conductivity in ZnO reveal that most potential acceptors, such as nitrogen, phosphorus, and arsenic, tend to form acceptor levels far from the valence band maximum [48]. Since the ionization fraction of such acceptors is low (due to their high activation energy), there is a large concentration of neutral states that may act as traps for non-equilibrium electrons. The effects of electron trapping on the diffusion length and lifetime of minority carriers can be probed by subjecting the material of interest to the excitation by the electron beam of the scanning electron microscope. The remainder of this review deals with the influence of electron irradiation on minority carrier diffusion length (Sec. 5) and lifetime (Sec. 5.2).

5.1 Influence of electron trapping on minority carrier diffusion length

The measurements of diffusion length as a function of beam irradiation duration were carried out on bulk n-ZnO doped with lithium [24], molecular beam epitaxy (MBE) p-ZnO doped with nitrogen [49], phosphorus-doped ZnMgO grown by pulsed laser deposition (PLD) [38], and on Sb-doped epitaxial ZnO layers MBE grown on a Si substrate [39].

As was already mentioned, bulk ZnO samples (Tokyo Denpa Co.) were weakly n-type, showing electron concentrations of ~ 10^{14} cm^{-3} and mobility of ~ 150 cm^2/Vs at room temperature. Secondary Ion Mass Spectroscopy (SIMS) measurements revealed the Li concentration of about $3x10^{16}$ cm^{-3} [25] (Li is often added to ZnO to increase the resistivity of initially n-type samples).

Phosphorus-doped $Zn_{0.9}Mg_{0.1}O$ layers were fabricated using PLD. Capacitance-Voltage profiling of similar films, grown using the same procedure, resulted in net acceptor concentration of ~ $2x10^{18}$ cm^{-3} after annealing. Pt/Au (200/800 Å) layers were deposited on phosphorus-doped $Zn_{0.9}Mg_{0.1}O$ films by electron beam evaporation and patterned by lift-off with contact diameters ranging from 50 to 375 μm. Circular electrode pairs with significantly different surface areas were employed for the EBIC measurements. The electrodes create an asymmetric rectifying junction, based on back to back Schottky diodes, with the larger area electrode being pseudo-Ohmic.

EBIC experiments were conducted at room temperature *in-situ* in a Philips XL30 SEM using a planar-collector configuration with a Schottky barrier (Sec. 3.1) to monitor the changes in minority carrier diffusion length as a function of time. The results of EBIC experiments are presented in Fig. 10. Note that the results shown in the left panel of Fig. 10 and those discussed in Sec. 4 were obtained from different bulk ZnO samples, which may offer an explanation for the significant difference in the initial, room temperature values of diffusion length. Additionally, while the large diffusion length value of bulk ZnO can be attributed to the higher quality of the bulk material (compared to the epitaxial layers), the diffusion length of ~ 2 μm in PLD-grown ZnMgO is rather surprising because of polycrystalline nature of the layers. It appears that in the latter sample, the transport of minority carriers is not significantly influenced by scattering from the grain boundaries.

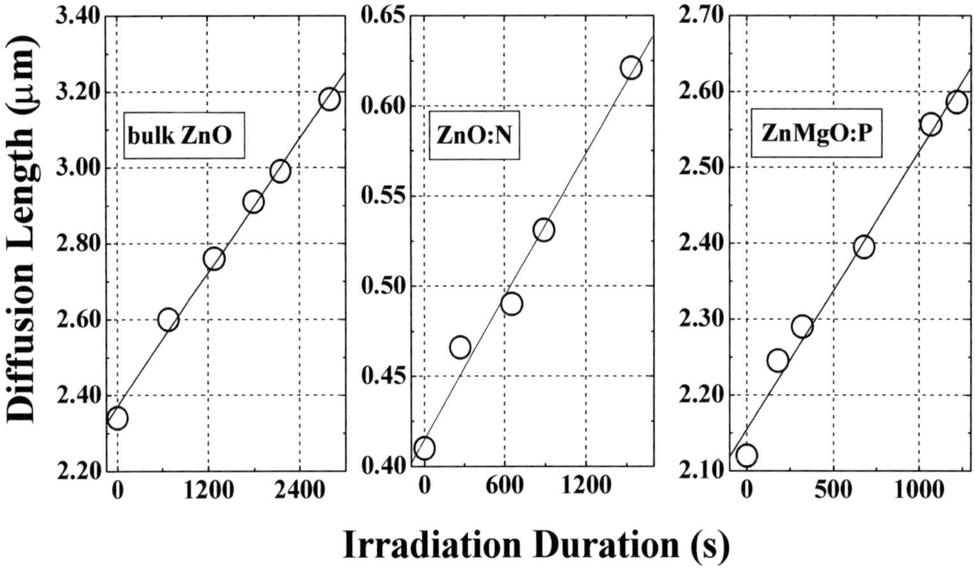

Fig. 10. Experimental dependence of minority carrier diffusion length on duration of electron beam irradiation at room temperature (open circles) and the linear fit (solid line). After Refs. [24, 38, 49].

The experiments revealed that diffusion length follows a general trend in all samples studied. Fig. 10 shows that irradiation by the electron beam clearly results in a significant increase of the carrier diffusion length, and that this increase is linear with respect to the duration of electron irradiation (t). Up to 50% increase of diffusion length was achieved (cf. Fig. 10 (center)), with the rates of diffusion length increase ranging from ~13% to ~30% per 1000 s. This appears to be a common occurrence in wide bandgap semiconductors doped with species that create deep acceptor levels, as similar observations were made in (Al)GaN doped with Mg, Mn, Fe, and C [39,50,51]. It is also noteworthy that similar experiments conducted on bulk ZnO *without* any intentional dopants did no show any significant changes in minority carrier diffusion length [38].

The observed increase of L is ascribed to charging of the deep, neutral acceptor states by the electrons generated by the SEM beam, since trapping of non-equilibrium electrons prevents these levels from participating in recombination [38,39]. Therefore, the difference in the rates of diffusion length increase is likely explained, at least in part, by the difference in the concentrations of these deep centers [52].

EBIC measurements in Sb-doped ZnO were performed on the samples by moving the electron beam of the SEM from the edge of the Schottky barrier (created on the top surface of ZnO:Sb samples by Ti/Au evaporation followed by lift off) outwards (line-scan) and recording an exponential decay of induced current.

After a single EBIC line-scan was completed (12 seconds), the excitation of the sample was continued by moving the electron beam back and forth along the same line for the total time of ~2200 seconds. EBIC measurements were periodically repeated to extract the minority carrier diffusion length [52], L, as a function of the duration of electron beam irradiation, t.

The effects of electron injection for Sb-doped 0.2 μm-thick p-type ZnO epitaxial layers ($p = 1.3x10^{17}$ cm^{-3}; $\mu = 28$ cm^2/Vs at room temperature) grown on Si substrate by MBE are summarized in Fig. 11. The activation energy for the e-beam injection-induced increase of L, $\Delta E_{A,I} = 219 \pm 8$ meV, was obtained from the graphs in Fig. 11b,c using the following equation:

$$R = R_0 \exp\left(\frac{\Delta E_{A,I}}{kT}\right)\exp\left(-\frac{\Delta E_{A,T}}{2kT}\right) \tag{14}$$

where R_0 is a scaling constant, T is temperature, k is the Boltzmann's constant, $\Delta E_{A,I}$ is the activation energy of electron irradiation effect, and $\Delta E_{A,T}$ is the activation energy of thermally-induced increase of L presented in Fig. 11a ($\Delta E_{A,T} = 184 \pm 10$ meV). The value of $\Delta E_{A,I}$ is in reasonable agreement with that for a Sb_{Zn}-$2V_{Zn}$ acceptor complex, predicted by Limpijumnong *et al.* [37] (see Ref. [53] for details).

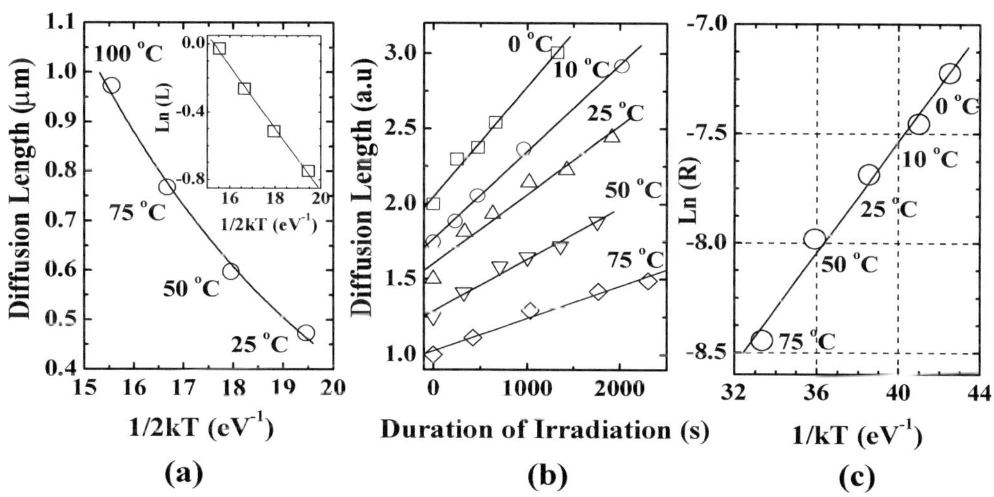

Fig. 11. a) Diffusion length of minority electrons as a function of temperature (open circles) and the fit (solid line). Inset: Arrhenius plot of the same data yielding activation energy of 184 ± 10 meV. b) Electron beam irradiation-induced increase of minority electron diffusion length at different temperatures. The values of the diffusion length were vertically offset for clarity and are not intended to illustrate the temperature dependence. c) Rate of irradiation-induced increase of diffusion length as a function of temperature (open circles).The fit with equation 5 (solid line) gives activation energy of 219 ± 8 meV. After Ref. [53].

It can also be seen from Fig. 11 that the rate, R, of the diffusion length increase is reduced with increasing temperature. The increase of the diffusion length due to trapping is counteracted by the release of the trapped electrons that occurs if the carriers gain sufficient energy to escape the trap. As the temperature is raised, the likelihood of de-trapping increases, which dampens the irradiation-induced growth of the diffusion length.

The saturation and relaxation of irradiation-induced change of diffusion length was studied at room temperature [53]. It was demonstrated that L reached its maximum value after about 50 min of continuous exposure to the electron beam. Further monitoring revealed that irradiation-induced increase persists for at least one week. Annealing the sample at 175 °C for about 30 minutes resulted in a decrease of the diffusion length to about 1 μm. This behavior further supports the involvement of deep electron traps in the phenomenon of interest, since temperature-induced de- trapping of carriers (cf. Fig. 11,c) re-activates the original recombination route (cf. Fig. 11,a), thus reducing carrier lifetime and diffusion length.

5.2 Optical studies of the effects of electron trapping on minority carrier lifetime

According to equation (1), electron irradiation-induced increase in minority carrier diffusion length discussed above is associated with the increase of lifetime of non-equilibrium carriers. Experimental evidence for this dependence was obtained from the cathodoluminescence measurements performed on the same samples. Time-dependent CL measurements were conducted *in-situ* in SEM using setup described in section 3.2. This setup allows combining periodic acquisition of CL spectra with continuous excitation of the sample by scanning the beam over the same location. For temperature-dependent CL measurements, the sample temperature was varied *in-situ* using specially designed hot stage and an external temperature controller (Gatan). At each temperature, the electron beam irradiation and CL measurements were conducted at a different location.

5.2.1 Optical studies of $Zn_{0.9}Mg_{0.1}O$ doped with phosphorus

Fig. 12 shows a series of NBE transitions in p-type $Zn_{0.9}Mg_{0.1}O$:P recorded under continuous electron irradiation and numbered in order of increasing irradiation duration. The edge of this spectrum at ~ 355 nm (see also inset of Fig. 12) is in agreement with 10% Mg content in ZnO lattice, since each atomic percent of Mg is known to increase the ZnO band gap by 0.02 eV [4]. The observed CL spectrum is attributed to the band-to-band as well as band-to-impurity (P-acceptor) optical transitions. The inset of Fig. 12 shows a wider range spectrum featuring a broad band, which is likely defect related [54].

While no changes were observed in the broad band CL (cf. inset of Fig. 12), the near band-edge luminescence in Fig. 12 exhibits a continuous decay with increasing duration of electron beam irradiation. This demonstrates that exposure to the electron beam results in the increase of carrier lifetime (τ), since I_{CL} is proportional to $1/\tau$. Similar phenomena were also observed in GaN, where the decay of NBE CL intensity occurred concomitantly with increasing diffusion length [38,50,51].

To characterize the intensity decay, we relate it to the diffusion length, L, which is known to vary linearly with duration of irradiation (cf. Fig. 13). Since L is proportional to $\tau^{1/2}$ (equation (1)), the inverse square root of normalized (with respect to the initial maximum value) intensity must also be proportional to L, and consequently, would be expected to change linearly with duration of electron irradiation. Fig. 13 shows that this is indeed the case, indicating that the observed increase of the diffusion length is attributable to the growing lifetime of non-equilibrium carriers.

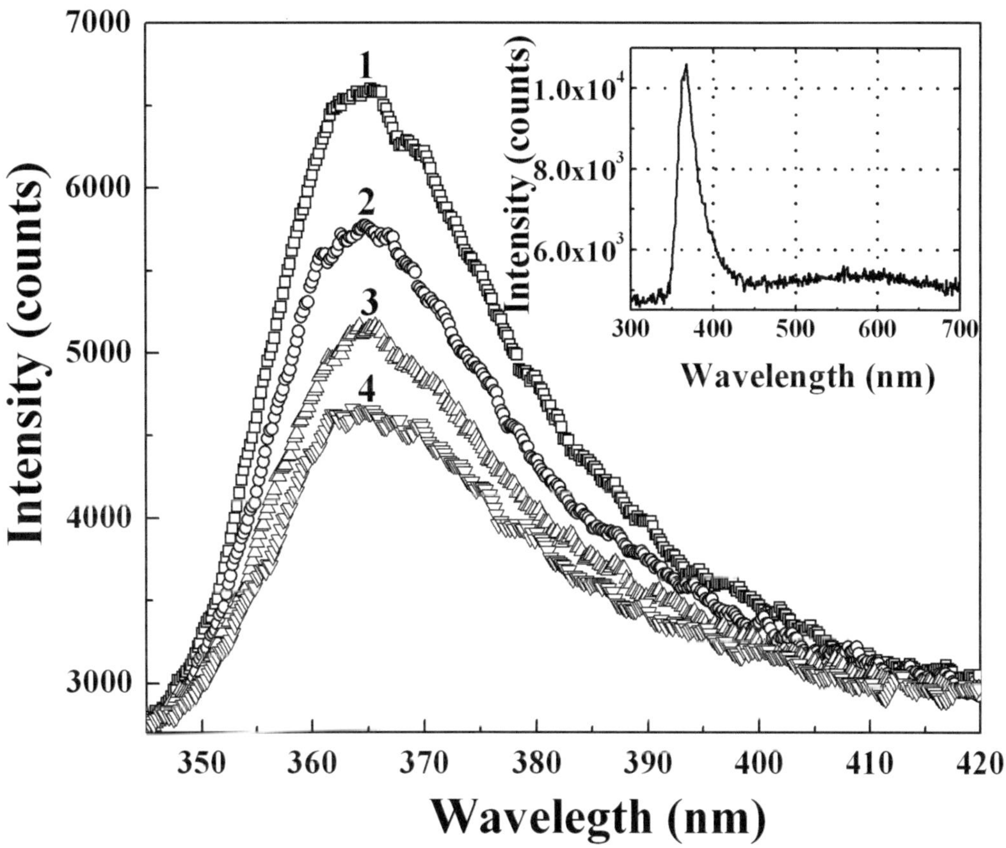

Fig. 12. Room temperature CL spectra of $Zn_{0.9}Mg_{0.1}O$:P measured in the same location at different times of electron injection. **1** is a pre-injection spectrum; **2**, **3**, and **4** correspond to duration of electron injection of 359, 793, and 1163 s, respectively. **Inset:** pre-injection broad-range CL spectrum taken in a different location than measurements in figure 12. After Ref. [38].

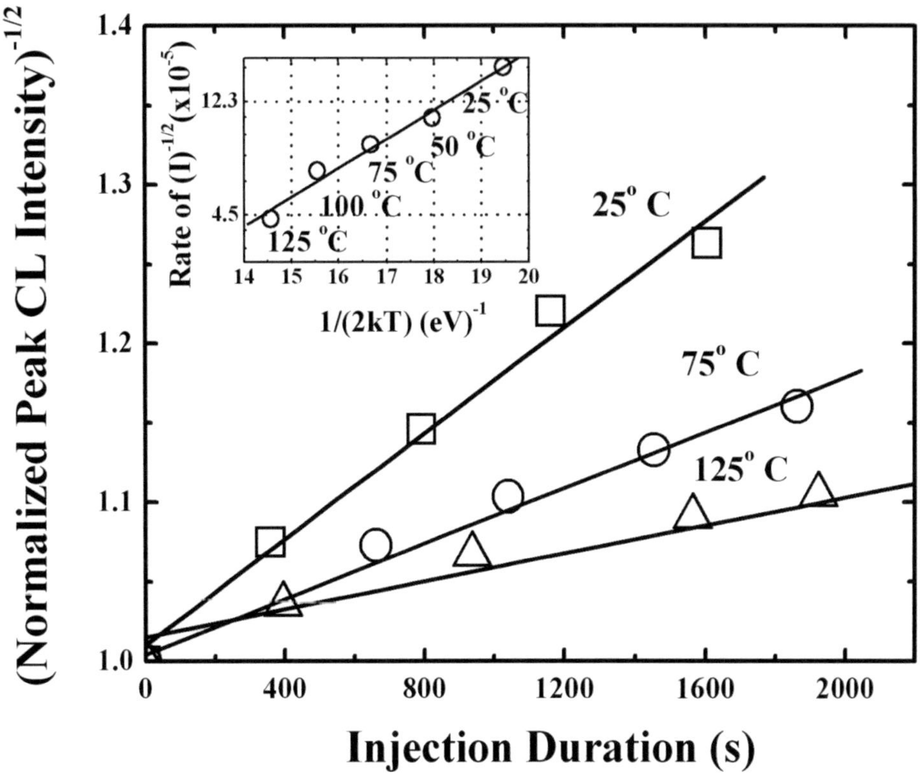

Fig. 13. Variable temperature dependence for the square root of inverse normalized intensity on duration of electron injection in $Zn_{0.9}Mg_{0.1}O$:P. The rate at every temperature is obtained from the slope of a linear fit. **Inset:** temperature dependence for the rate of the square root of inverse normalized intensity (open circles) and the fit. The slope of the graph yields ΔE_A = 256 ± 20 meV. After Ref. [38].

In CL, the temperature of the samples was varied from 25 °C to 125 °C. At each temperature, injection was performed on a site previously not exposed to electron beam. As is apparent from the inset of Fig. 13, the rate of the decrease in CL intensity (described by the slope, R, of the linear dependence of $I_{CL}^{-1/2}$ on t) diminishes with growing temperature. This suggests that while electron irradiation results in an increase of carrier lifetime, there exists another, thermally activated process that contributes to its decay.

Taking into account both injection-induced effect on R ($\propto \exp(\Delta E_A/kT)$) and its temperature dependence ($\propto \exp(\Delta E_A/-2kT)$) [26], and assuming that activation energies are similar in both cases, the temperature dependence of R can be described as follows:

$$R = R_0 \exp(\frac{\Delta E_A}{2kT}) \quad (15)$$

where R_0 is a scaling constant and ΔE_A is the activation energy for the overall process. Fitting the experimental results with this expression (inset of Fig.13) yielded activation energy of 256 ± 20 meV. This activation energy is in good agreement with that for the phosphorus acceptor obtained based on the simple hydrogenic model. The model assumes phosphorus substitution on the oxygen site and predicts the activation energy of about 250-300 meV [55]. The experimentally obtained value of the activation energy, combined with the fact that no electron irradiation effects were observed in undoped ZnO [38], suggests that carrier trapping on phosphorus acceptor levels plays a crucial role in this phenomenon.

5.2.2 Optical studies of bulk ZnO

Similarly to ZnMgO, the electron irradiation-induced increase of minority carrier diffusion length in ZnO:Li (Sec. 5.1) was found to correlate with the increase of minority carrier lifetime. Fig. 14 shows a series of room temperature NBE spectra collected under continuous electron beam excitation, in which the intensity of the NBE transition can be seen to fall steadily with increasing t. The peak of this emission occurs at 383 nm (3.24 eV) and has been assigned to the transition from the conduction band to a deep acceptor level, (e,A0) [28]. The inverse square root of maximum CL intensity (I_{CL}-1/2), which is proportional to $\sqrt{\tau}$ and, therefore, to L, changes linearly with duration of irradiation (cf. left inset of Fig. 14), which is consistent with the results of EBIC measurements and indicates that the increase in L occurs due to the irradiation-induced growth of carrier lifetime.

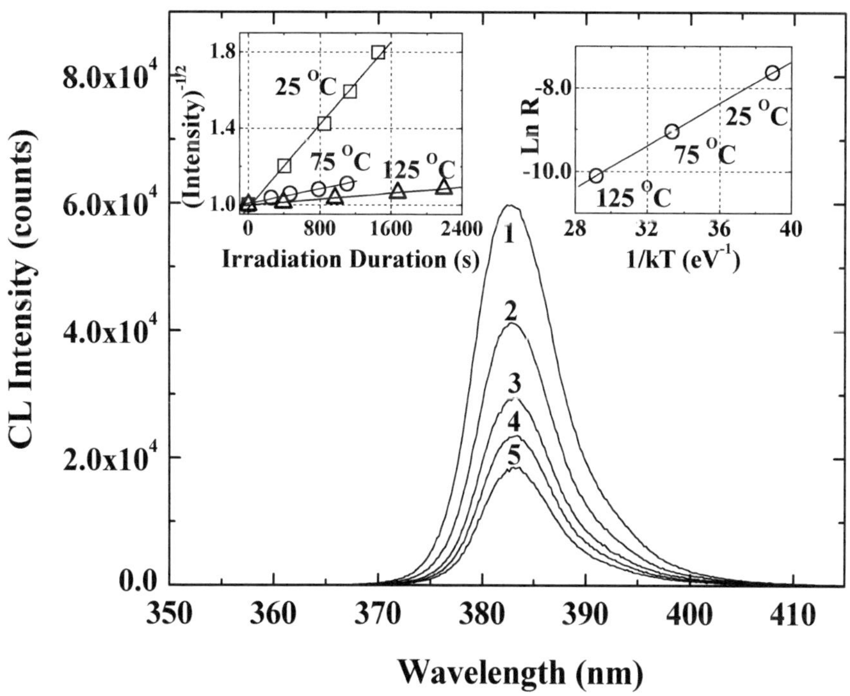

Fig. 14. Room temperature cathodoluminescence spectra of ZnO:Li taken under continuous excitation by the electron beam. **1** is the pre-irradiation spectrum and **5** is the spectrum after 1450 s of electron irradiation. **Left Inset:** Variable-temperature dependence of inverse square root of normalized intensity on duration of electron irradiation and the linear fit with the rate R. **Right Inset:** Arrhenius plot of R as a function of temperature yielding an activation energy $\Delta E_{A,I}$ of 283±9 meV. After Ref. [24].

CL measurements conducted at elevated temperatures confirmed the same trend for the irradiation-induced change of luminescence intensity. It can be seen from the left inset of Fig. 14 that the inverse square root of intensity increases linearly for all temperatures. The temperature dependence of rate, R, can be used to determine the activation energy of the irradiation-induced processes according to equation (15). On the other hand, our earlier studies of the *temperature*-induced CL intensity decay (Sec. 4) yielded the activation energy of about 60 meV [38], thus allowing us to separate the two components as followed from equation (14): the activation energy of electron irradiation effect $\Delta E_{A,I}$; and, the previously determined, activation energy of thermally induced intensity decay $\Delta E_{A,T}$. This treatment yielded the value for $\Delta E_{A,I}$ of 283 ± 9 meV. Incidentally, Ref. [25] also reports high concentration of ~0.3 eV electron traps found in the same material by Deep Level Transient Spectroscopy (DLTS). The significant difference between $\Delta E_{A,I}$ and $\Delta E_{A,T}$ observed in bulk ZnO suggests that, unlike in other materials, temperature- and irradiation-induced changes of the minority carrier transport characteristics are two distinctly different processes.

Although several theoretical works have predicted a very shallow Li_{Zn} level [56,57], these predictions have not been substantiated experimentally, as most studies find a rather deep Li-acceptor with activation energies of several hundred meVs [34,58]. In fact, recent first-principles calculations by Wardle *et al.*, also suggest that the Li_{Zn} acceptor state lies at about 0.2 eV above the valence band maximum [58], which is in reasonable agreement with the $\Delta E_{A,I}$ of 283 ± 9 meV obtained in this work.

It should be clarified that the weak n-type character of the sample is not necessarily in contradiction with the dominant behavior of acceptor states observed in electron trapping phenomena. As was mentioned, the n-type conductivity in nominally undoped ZnO is due to the *shallow* donor states, whereas in presence of deep electron traps the Fermi level may lie far below these states. Although shallow donors may capture non-equilibrium electrons under excitation, those are quickly released if the temperature is sufficiently high. Therefore, if the difference in the energetic position between the donor and trap states is large, the latter dominate the kinetics of electron trapping [59].

5.2.3 Optical studies of ZnO Doped with nitrogen

CL measurements performed on MBE-grown, nitrogen-doped p-ZnO revealed behavior similar to that of bulk ZnO [40] and PLD-grown ZnMgO [38]. Room temperature ZnO:N spectra are shown in Fig. 15 and feature a NBE luminescence band with a maximum at about 388 nm (~3.20 eV). This band includes the (e, A^0) transition as well as the donor-acceptor pair (DAP) recombination, with nitrogen identified as the acceptor in both processes [60]. Additionally, a violet band centered on 435 nm has been attributed to the radiative recombination of the electrons trapped at grain boundaries with the holes in the valence band [61].

As expected, irradiation with electron beam resulted in decay of the intensity of NBE luminescence, indicating increasing lifetime. One can observe from the inset of Fig. 15 that, in agreement with the diffusion length measurements (cf. Fig. 10 in Sec. 5.1), the inverse square root of the peak normalized intensity of the NBE transition changes linearly with irradiation time, yielding the rate R. Note that the intensity of the violet band is not affected by electron irradiation, which suggests that electron trapping at the grain boundaries does not play a significant role in the irradiation-induced increase of carrier lifetime.

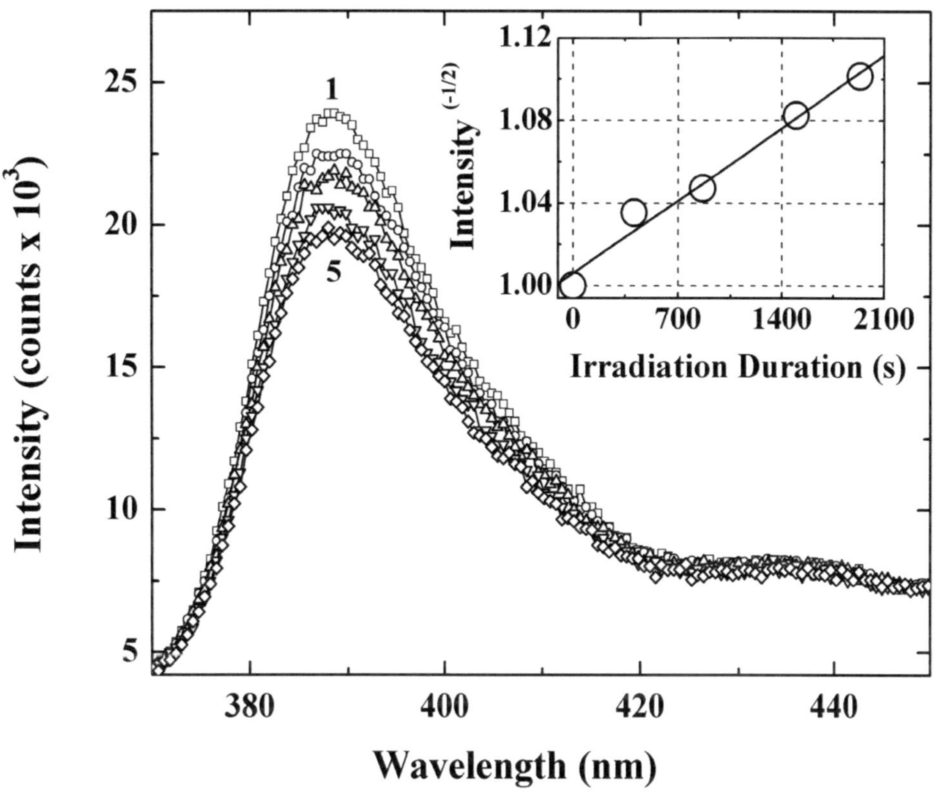

Fig. 15. Room temperature CL spectra of ZnO:N taken under continuous excitation. Trace 1 corresponds to the pre-irradiation spectrum and trace **5** to the spectrum after 1940 s of electron irradiation. **Inset**: Linear dependence of the inverse square root of normalized peak intensity at room temperature. After Ref. [49].

CL measurements at elevated temperatures (Fig. 16) confirmed that R decreases with temperature, indicating a thermally activated process that counteracts the effects of electron injection, similar to what occurs in bulk ZnO and ZnMnO:P. Note that while the values of R were obtained based on the intensity normalized with respect to its initial value, the data displayed in Fig. 16 are offset by shifting the normalized results along the y-axis to avoid the overlap of the data points. The activation energy ($\Delta E_{A,I}$) of about 134 ± 10 meV was determined from the Arrhenius plot shown in the inset of Fig. 16, based on equation (15) and using $\Delta E_{A,T} = 118$ meV obtained from the temperature-dependent CL measurements (Sec. 4). This value is in reasonable agreement with the ionization energy of the nitrogen acceptor in ZnO [34,35 ,60,62] and indicates that electron trapping by these levels plays an important role in the recombination dynamics of minority carriers.

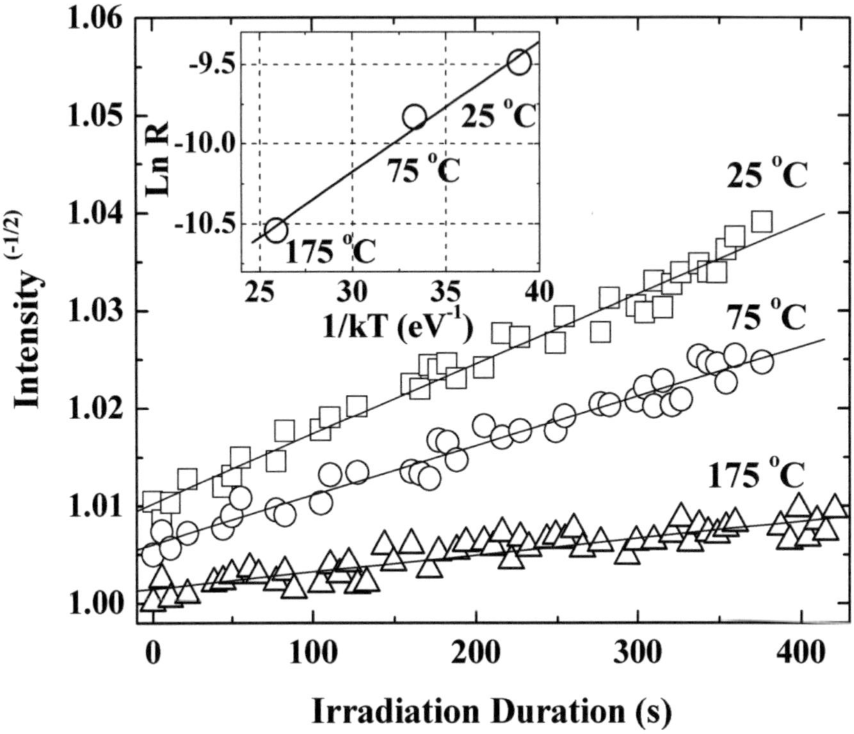

Fig. 16. Variable-temperature dependence of the inverse square root of normalized intensity in ZnO:N on duration of electron irradiation and the linear fit with a rate R. The data are offset for clarity. **Inset:** Arrhenius plot of R as a function of temperature with a fit yielding $\Delta E_{A,I}$=134 ± 10 meV. After Ref. [49].

6. Summary

Issues affecting minority carrier transport in ZnO have been discussed, with special attention given to the temperature dependence of minority carrier diffusion length and lifetime, as well as to the recombination dynamics of non-equilibrium minority carriers. The mechanisms governing temperature- and irradiation-induced effects have been presented.

7. References

[1] F.Oba, S.R.Nishitani, S.Isotani, H.Adachi, and I.Tanaka, *J. Appl. Phys.*, 90, 824 (2001)
[2] G.W.Tomlins, J.L.Routbort, and T.O.Mason, *J. Appl. Phys.*, 87, 117 (2000)
[3] C.G.Van de Walle, *Phys. Rev. Lett.*, 85,1012 (2000)
[4] T. Gruber, C. Kirchner, R. Kling, F. Reuss and A. Waag, *Appl. Phys. Lett.*, 84, 5359 (2004).
[5] K. Ip, Y W. Heo, D. P. Norton, S. J. Pearton, J. R. LaRoche and F. Ren, *Appl. Phys.Lett.*, 85, 1169 (2004).
[6] K.G. Chen, Z. Z. Ye, W. Z. Xu, B. H. Zhao, L. P. Zhu and J. G Lv, *J. Cryst. Growth*, 281,458 (2005).
[7] V. Vaithianathan, B. T. Lee and S. S. Kim, *J. Appl Phys.*, 98, 043519 (2005).

[8] E. J. Egerton, A. K. Sood, R. Singh, Y R. Puri, R. F. Davis, J. Pierce, D. C. Look, and T. Steiner, *J. Electron. Mater.*, 34, 949 (2005).
[9] V. Vaithianathan, B. T. Lee, and S. S. Kim, *Appl Phys. Lett.*, 86, 062101 (2005).
[10] F. X. Xiu, Z. Yang, L. J. Mandalapu, D. T. Zhao, J. L. Liu, and W. P. Beyermann, *Appl.Phys. Lett.*, 87, 152101 (2005).
[11] S. H. Lim, D. Shindo, H. B. Kang and K. Nakamura, *J. Vac. Sci. Technol B*, 19, 506 (2001).
[12] K. Miyamoto, M. Sano, H. Kato and T. Yao, *J. Cryst. Growth*, 265, 34 (2004).
[13] Y Wang, X. L. Du, Z. X. Mei, Z. Q. Zeng, Q. Y Xu, Q. K. Xue and Z. Zhang, *J. Cryst. Growth*, 273, 100 (2004).
[14] M. W. Cho, A. Setiawan, H. J. Ko, S. K. Hong and T. Yao, *Semicond. Sci. Technol*,20, S13 (2005).
[15] E. Monroy, F. Calle, J. L. Pau, F. J. Sanchez, E. Munoz, F. Omnes, B. Beaumont, and P. Gibart, *J. Appl. Phys.*, 88, 2081 (2000).
[16] J. F. Muth, R. M. Kolbas, A. K. Sharma, S. Oktyabrsky and J. Narayan, *J. Appl. Phys.*,85, 7884 (1999).
[17] H. Yoshikawa and S. Adachi, *Jpn. J. Appl. Phys. Part 1 - Regul. Pap. Short Notes Rev. Pap.*, 36, 6237 (1997).
[18] D. E. loannou and S. M. Davidson, *J. Phys. D-Appl. Phys.*, 12, 1339 (1979).
[19] D. E. loannou and C. A. Dimitriadis, *IEEE Trans. Electron. Devices*, 29, 445 (1982).
[20] J. Boersma, J. J. E.Iindenkleef, and H. K. Kuiken, *J. Eng. Math*, 18, 315 (1984).
[21] D. S. H. Chan, V. K. S. Ong, and J. C. H. Phang, *IEEE Trans. Electron. Devices*, 42, 963 (1995).
[22] L. Chernyak, A. Osinsky and A. Schulte, *Solid-State Electron.*, 45, 1687 (2001).
[23] R. K. Ahrenkiel, *Solid-State Electron.*, 35, 239 (1992).
[24] O. Lopatiuk, L. Chernyak, A. Osinsky, J. Q. Xie, and P. P. Chow, *Appl. Phys. Lett.*, 87, 214110 (2005)
[25] A. Y. Polyakov, N. B. Smirnov, A. V. Govorkov, E. A. Kozhukhova, S. J. Pearton, D. P. Norton, A. Osinsky, and A. Dabiran, *J. Electron. Mater*, 35, 663 (2006).
[26] M. Eckstein and H. U. Habermeier, *J. Phys. IV*, 1, 23 (1991).
[27] L. Chernyak, A. Osinsky, H. Temkin, J. W. Yang, Q. Chen and M. A. Khan, *Appl Phys. Lett.*, 69, 2531 (1996).
[28] K. Thonke, T. Gruber, N. Teofilov, R. Schonfelder, A. Waag, and R. Sauer, *Physica B*, 308, 945 (2001).
[29] D. S. Jiang, H. Jung and K. Ploog, *J. Appl. Phys.*, 64, 1371 (1988).
[30] A. Ortiz, C. Falcony, J. Hernandez, M. Garcia and J. C. Alonso, *Thin Solid Films*, 293, 103 (1997).
[31] J. I. Pankove, *Optical Processes in Semiconductors*, Prentice-Hall, Englewood Cliffs, New Jersey (1971).
[32] D. C. Look, D. C. Reynolds, J. R. Sizelove, R. L. Jones, C. W. Litton, G. Cantwell and W. C. Harsch, *Solid State Commun.*, 105, 399 (1998).
[33] O. Lopatiuk-Tirpak, "Influence of Electron Trapping on Minority Carrier Transport Properties of Wide Band Gap Semiconductors", Ph.D. Dissertation, University of Central Florida, Orlando, 2007.
[34] B. K. Meyer, H. Alves, D. M. Hofmann, W. Kriegseis, D. Forster, F. Bertram, J. Christen, A. Hoffmann, M. Strassburg, M. Dworzak, U. Haboeck, and A. V. Rodina, *Phys. Status Solidi B - Basic Res.*, 241, 231 (2004).

[35] S. Yamauchi, Y. Goto, and T. Hariu, *J. Cryst. Growth,* 260, 1 (2004).
[36] T. Aoki, Y. Shimizu, A. Miyake, A. Nakamura, Y. Nakanishi and Y. Hatanaka, *Phys. Status Solidi B-Basic Res.,* 229, 911 (2002).
[37] S. Limpijumnong, S. B. Zhang, S. H. Wei and C. H. Park, *Phys. Rev. Lett.,* 92, 155504 (2004).
[38] O. Lopatiuk, W. Burdett, L. Chernyak, K. P. Ip, Y. W. Heo, D. P. Norton, S. J. Pearton, B. Hertog, P. P. Chow and A. Osinsky, *Appl. Phys. Lett.,* 86, 012105 (2005).
[39] O.Lopatiuk-Tirpak, W.V.Schoenfeld, L.Chernyak, F.X.Xiu, J.L.Liu, S.Jang, F.Ren, S.J.Pearton, A.Osinsky, P.Chow, *Appl. Phys. Lett.,* 88, 202110 (2006).
[40] M. G. Cheong, K. S. Kim, C. S. Kim, R. J. Choi, H. S. Yoon, N. W. Namgung, E. K. Suh and H. J. Lee, *Appl. Phys. Lett.,* 80, 1001 (2002).
[41] P. Kozodoy, H. L. Xing, S. P. DenBaars, U. K. Mishra, A. Saxler, R. Perrin, S. Elhamri and W. C. Mitchel, *J. Appl. Phys.,* 87, 1832 (2000).
[42] W. Gotz, R. S. Kern, C. H. Chen, H. Liu, D. A. Steigerwald and R. M. Fletcher, *Mater. Sci. Eng. B-Solid State Mater. Adv. Technol.,* 59, 211 (1999).
[43] P. P. Debye and E. M. Conwell, *Phys. Rev.,* 93, 693 (1954).
[44] S. Dhar and S. Ghosh, *Appl. Phys. Lett.,* 80, 4519 (2002).
[45] V. Ursaki, I. M. Tiginyanu, P. C. Ricci, A. Anedda, S. Hubbard and D. Pavlidis, *J. Appl. Phys.,* 94, 3875 (2003).
[46] J. Ryan, D. P. Lowney, M. O. Henry, P. J. McNally, E. McGlynn, K. Jacobs and L. Considine, *Thin Solid Films,* 473, 308 (2005).
[47] L. Chernyak, G Nootz and A. Osinsky, *Electron. Lett.,* 37, 922 (2001).
[48] C. Look and B. Claftin, *Phys. Status Solidi B - Basic Res.,* 241, 624 (2004).
[49] O. Lopatiuk, A.Osinsky, L.Chernyak, in *Zinc Oxide Bulk, Thin Films and Nanostructures,* pp.241-265, Edited by C. Jagadish and S. Pearton, Elsevier Ltd. (2006)
[50] L. Chernyak, W. Burdett, M. Klimov and A. Osinsky, *Appl. Phys. Lett.,* 82, 3680 (2003).
[51] W. Burdett, O. Lopatiuk, L. Chernyak, M. Hermann, M. Stutzmann, and M. Eickhoff, *J. Appl. Phys.,* 96, 3556 (2004).
[52] L. Chernyak, A. Osinsky, V. Fuflyigin and E. F. Schubert, *Appl. Phys. Lett.,* 77, 875 (2000).
[53] O.Lopatiuk-Tirpak, L.Chernyak, F.X.Xiu, J.L.Liu, S.Jang, F.Ren, S.J.Pearton, K.Gartsman, Y.Feldman, A.Osinsky, P.Chow, *J.Appl. Phys.,*100, 086101 (2006).
[54] Y. W. Heo, K. Ip, S. J. Pearton and D. P. Norton, *Phys. Status Solidi A - Appl. Res.,* 201, 1500 (2004).
[55] S. J. Pearton, D. P. Norton, K. Ip, Y. W. Heo and T. Steiner, *Prog. Mater. Sci,* 50, 293 (2005).
[56] C. H. Park, S. B. Zhang and S. H. Wei, *Phys. Rev. B,* 66, 073202 (2002).
[57] E. C. Lee and K. J. Chang, *Phys. Rev. B,* 70, 115210 (2004).
[58] M. G. Wardle, J. P. Goss and P. R. Briddon, *Phys. Rev. B,* 71, 155205 (2005).
[59] M. Salis, A. Anedda, F. Quarati, A. J. Blue and W. Cunningham, *J. Appl. Phys.,* 97, 033709 (2005).
[60] F. Reuss, C. Kirchner, T. Gruber, R. Kling, S. Maschek, W. Limmer, A. Waag, and P. Ziemann, *J. Appl. Phys.,* 95, 3385 (2004).
[61] R. Ghosh, B. Mallik, S. Fujihara, and D. Basak, *Chem. Phys. Lett.,* 403, 415 (2005).
[62] G. Xiong, K. B. Ucer, R. T. Williams, J. Lee, D. Bhattacharyya, J. Metson, and P. Evans, *J. Appl. Phys.,* 97, 043528 (2005).

2

What is the Pulsed Cathodoluminescence?

Vladimir Solomonov and Alfiya Spirina
The Institute of Electrophysics of Ural Branch of the Russian Academy of Sciences
Russia

1. Introduction

A cathodoluminescence phenomenon was revealed in the 19th century. The first explanation of this phenomenon was given by Julius Plücker in 1858, and two decades later by Sir William Crookes (1879). A cathodoluminescence intensity is defined by the concentration of electrons (holes) – $n_{e(h)}$, generated by electrons injected into substance. Generally, this concentration is described by the kinetic equation

$$\frac{dn_{e(h)}}{dt} = G - An_{e(h)} - Bn_{e(h)}^2, \quad G = \frac{E_0 j_e}{ed_e \varepsilon_i}. \tag{1}$$

Here A and B are the coefficients of electrons (holes) recombination processes which obey to linear and quadratic laws. Coefficient A is a generalized characteristic of the processes which relate to formation and dissociation of electron and hole centres. These processes are accompanied with the capture or the release of free electron (hole). The luminescence intensity of these centres is proportional to concentration, $I_{e(h)} \sim A \cdot n_{e(h)}$. Coefficient B is defined by the annihilation of free electron-hole pairs which resulted in origination of interband, excitonic, and intracentre kinds of luminescence with intensity of $I \sim B \cdot n^2_{e(h)}$. Thus, the coefficients A and B are the characteristics of certain substance. Usually for pure crystals (undoped) the coefficient A has an order of 10^5-10^6 s^{-1} whereas it is of 10^8 s^{-1} in case of crystals doped with donor (acceptor) ions. The value of coefficient B amounts to 10^{-10} $cm^3 \cdot s^{-1}$ for the interband transition in large-band-gap semiconductors [Bogdankevich et al., 1975, Galkin, 1981]. Coefficient G is the generation rate of electron-hole pairs inside a sample that is irradiated by primary electrons. The energy and current density of these electrons are determined as follows

$$E_0 = eU_0, j_e = \frac{i_e}{S}. \tag{2}$$

In Eqs. (1) and (2) U_0, e, i_e, S, d_e are an accelerating potential, electron charge, a current of accelerated electrons, an irradiated area of sample and an electron penetration depth, respectively. An average ionization energy ε_i in Eq. (1) can be approximately estimated as $\varepsilon_i \approx 3 \cdot E_g$, where E_g is a band-gap energy.

The current of accelerated electrons (i_e), being injected into the sample, can be determined with the help of equivalent circuit that is shown in Fig. 1.

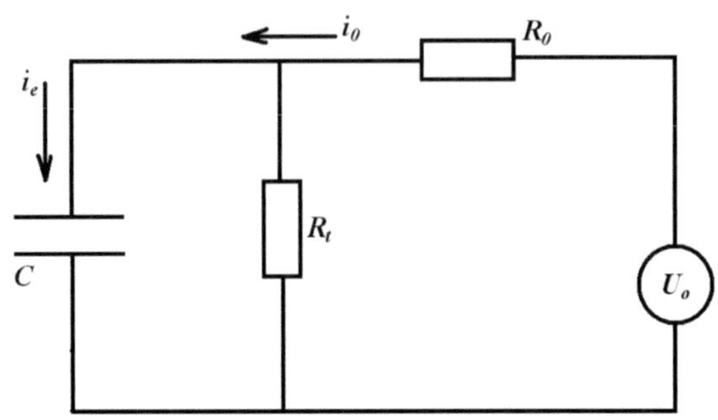

Fig. 1. The equivalent circuit of electron current

Here electron accelerator is represented as a source of accelerating potential U_0 with internal resistance $R_0=U_0/i_0$=const. The sample with irradiated surface $S=\pi r_0^2$ forms a capacitor with $C=2\pi r_0 \varepsilon \varepsilon_0$. Here ε_0 is the dielectric constant and ε is the permittivity of environment (e.g. air). Resistor R_t is introduced as a shunt for capacitor C and it provides the discharge of the sample surface. For this circuit at R_t=const and initial condition $i_e(0)=i_o$, the electron current is described by the equation

$$i_e(t)=\frac{U_0}{R_0+R_t}\left[1-\exp\left(-\frac{t}{\tau_c}\right)\right]+i_{e0}\exp\left(-\frac{t}{\tau_c}\right), \tag{3}$$

where $\tau_c=CR_0R_t/(R_0+R_t)$ is a typical charge time of capacity C.

For example, in the cathodoluminescent microscope [Ramseyer et. al., 1989, Petrov, 1996] the shunting of capacity C is provided by the emission of secondary electrons over the irradiated sample surface. This emission results in setting the finite value of shunting resistance R_t. Dynamic balance between the primary electrons which are injected into the sample, and the secondary electrons which leave the sample is equilibrated at $t>>\tau_c$. After that time the current of injected electrons i_e tends to achieve the value defined as $i_e=U_0/(R_0+R_t)$ and the constant generation rate of electron-hole pairs G is equilibrated. These conditions are realized for the narrow energy range of the primary electrons $1<E_e<12$ keV resulted in a small depth of electron penetration d_e=0.1-1.5 μm. Thus, the lower energy of the primary electrons is limited by the work function of the secondary electrons through the sample surface. The upper one is limited by the energy loss of secondary electrons, which appeared on the large depth inside the sample, under diffusion to the sample surface.

The solution of Eq. (1) using G=const reveals that under irradiation by electron beam the concentration of electron-hole pairs inside the sample volume which is determined by the beam cross-section and penetration depth of electrons is saturated rapidly with time according to the following equation:

$$n_{e(h)}=2G\tau_i\frac{\exp(t/\tau_i)-1}{\delta\exp(t/\tau_i)+\gamma}\xrightarrow{t\to\infty}2G\frac{\tau_i}{1+A\tau_i},\tau_i=\frac{1}{\sqrt{4BG+A^2}}. \tag{4}$$

Here τ_i is an ionization time of substance, $\delta=1+A\tau_i$, $\gamma=1-A\tau_i$, where $A\tau_i<1$. From the Eq. (4) it can be seen that the concentration of electron-hole pairs increases with increasing the coefficient G. Therefore in order to achieve a high brightness of luminescence, the electron beam is focused on the sample surface in the spot with a diameter of 1-50 μm and current density j_e in the range from 10^{-2} to 10 A/cm^2. The coefficient G amounts to 10^{23}-10^{26} cm^{-3}·s^{-1} and the range of typical time τ_i has an order of 10^{-6}-10^{-9} s.

Since the 1980-s the electron beams with the energy increased to 20-70 keV and density of electron current of j_e=0.1-10 A/cm^2 are applied. These electrons are able to penetrate into the sample on a depth of 3-30 μm [Chukichev et. al., 1990, Yang et. al., 1992]. The secondary electrons originated on such a depth inside the sample dissipate their own energy while moved to the outer surface and can't emit outside. In this case the dynamic balance is provided by the flow of the surplus charge via thin metal film previously deposited on the sample surface being irradiated and served as a ground. Now the shunt R_t is determined by the contact electroconductivity of the irradiated area with the metal film. At these parameters of electron beam the coefficient G, the ionization time τ_i and the charging time of capacity τ_c have values of the same order those were mentioned above. However, at increased energy of electrons the luminescence intensity increases due to the deeper penetration of electrons and as a result the larger excited volume of substance.

To decrease the thermal load on the irradiated surface at the electron energy of 20-70 keV the pulsed electron beam with the pulse duration of 1-10 ms [Chukichev et. al., 1990] or modulated electron beam with the modulation frequency of 100-300 Hz [Yang et. al., 1992] are applied. The luminescence, excited by such electron beams, is usually called the pulsed cathodoluminescence (PCL) [Chukichev et. al., 1990]. However, this PCL is the steady-state one since the pulse duration of injected electrons is much greater than τ_i and τ_c yet.

PCL [Solomonov et. al., 2003], to be talked about in the present chapter, is excited at the conditions when the dynamic balance between the injected and left electrons is absent, i.e. at $R_t \to \infty$. The Eq. (3) shows in this case that the current of injected electrons damps exponentially with characteristic time constant τ_v, as

$$i_e(t) = i_{eo}\exp(-t/\tau_v), \quad \tau_v = R_0 C = \frac{2\varepsilon\varepsilon_0 U_0}{j_{e0} r_0}. \tag{5}$$

This time, τ_v, increases with energy of injected electrons. Its value is about 3.5 ns at E_0=200 keV, j_{e0}=100 A/cm^2, r_0=1 mm, and ε=1 according to Eq. (5). Therefore the PCL excitation should be carried out by the electron beam with duration t_e of the same order as τ_v. The electrons with energy of 100-200 keV penetrate into dielectric solids on the depth of 100-150 μm. Due to the large penetration depth the coefficient G reaches the value of 10^{26}-10^{27} cm^{-3}·s^{-1}, which is similar to that realized at maximum excitation conditions of the steady-state cathodoluminescence. When $t_e \le \tau_v$ the concentration of electron-hole pairs comes to $n_{e(h)} \ge 0.5 n_{e(h)max}$ according to Eq. (4). Here $n_{e(h)max}$ is a maximal value of the concentration at $t_e \to \infty$. It means that PCL brightness is higher than that of steady-state cathodoluminescence excited at the maximal conditions.

PCL spectrum gives the information about the composition and crystal structure of the sample bulk rather than interface layer. Interface layers are usually characterized with presence of many absorbed molecules and defect of crystal structure and their properties are not inherent to the bulk of materials. In PCL the interface layers with a thickness up to 20 μm don't have a significant influence on the PCL spectrum quality [Ramseyer et. al., 1990].

It should be emphasized that despite of short time of electron beam impact PCL persistence occurs and its kinetics is ascribed by complicated laws. This is associated with that the primary source of luminescence excitation is the electron-hole pairs. Their concentration according to Eq. (1) over the time of electron beam impact (G=0) is given by

$$n_{e(h)} = \frac{A}{B} \cdot \frac{\alpha \cdot \exp(-At)}{1-\alpha \cdot \exp(-At)}, \alpha = \frac{B \cdot n_{eh0}}{A + B \cdot n_{eh0}}. \quad (6)$$

Here n_{eh0} is the concentration of electron-hole pairs introduced by the electron beam. The luminescence intensity of the electron and hole centres changes the same law. The intensity of interband luminescence falls proportionally to n^2_{eh}. These kinds of luminescence reach their maxima at the time moment when the excitation is over. The intensity of intracentre luminescence changes more difficulty. The first maximum is also reached at the same time moment but further behaviour depends on the life time of radiative level (τ_r). In the paper [Solomonov et. al., 1996, 2003] it has been shown that there is the second maximum of the intensity in the long persistence at $\tau_r > (\sqrt{2} - 1)/A$. After this maximum the intensity falls according to the exponential law with the characteristic time constant τ_r. Moreover the second maximum can be more intensive than the first one. If the $\tau_r < (\sqrt{2} - 1)/A$ the second maximum doesn't appear and an exponential decay of luminescence occurs but with characteristic time constant that is proportional to 1/(2A). It is worthy to note that in case of using nanosecond exciting electron beams the integral intensity of persistent luminescence is usually similar or even higher than that during excitation.

2. Apparatus for the PCL registration

The generation of high-current nanosecond electron beam with the energy higher than 100 keV became possible after creation of electron accelerators by G.A. Mesyats in the 1970-s. These accelerators are founded on the explosive electron emission [Mesyats, 1974]. The electrons having this energy extend at great distance (more than 10 cm) in air. The samples excited in air can be used in the form of pieces, powders, and solution. The irradiation in air furthers also to the partial compensation of injected charge into sample by the stream of positive air ions, created by the electron beam. The large penetration depth of these electrons into sample simplifies the sample preparation for analysis considerably, namely there is no need to undergo the sample to grinding and polishing procedures. Moreover irradiated surface doesn't require metallization. This is very important for the analysis of the finished product, in particular, jewels. It should be also noted that the problem of sample warming, which is typical for the steady-state cathodoluminescence, is solved due to the introduction of the small energy density (≤ 3 J/cm^3).

The investigation of PCL in the different mediums has shown that the portable nanosecond accelerators of RADAN [Mesyats et. al., 1992] are most applicable for its excitation. These accelerators include the sealed vacuum electron tube. The biological shielding from X-ray emission is provided by the design of analytic chamber, which is connected with the output of the accelerator. The analysed samples are placed into the chamber. The pulsed type of the luminescence allows using the different methods of its registration.

As a first, the traditional method of the registration with the help of optical monochromator, photoelectronic multiplier, and oscillograph is applied [Vaysburd et.al., 1982, Solomonov et. al., 1996]. The intensity kinetics of separated luminescence band is measured by this method. This is necessary for the identification of its nature. The application of the scanning monochromator allows registering the intensity distribution by wavelengths I (λ,t). However, two PCL features have to be kept in mind. The first feature is that the PCL is characterised by the certain degree of instability of the registered parameters because of the pulsed regime. Therefore the spectrum measurement has to be performed in the averaging mode. The second feature is caused by the different kinetics of PCL bands with the various nature and spectrum registered by such an approach strongly depends on time.

As a second, the time-integral intensity of the luminescent bands can be measured

$$I(\lambda) = \int_{t_1}^{t_2} I(\lambda,t)dt \,. \tag{7}$$

Here I (λ,t) is the current intensity, t_1 is the beginning of registration and t_2 is the ending of registration. This intensity is registered with the help of multichannel semiconductor photodetectors based on diode matrix and charge-coupled device [Solomonov et. al., 2003]. In this case optical spectrograph is applied instead of the scanning monochromator and the wide spectral range for one frame is measured. This method can be used for the PCL research when intracentre luminescence is dominant. Also the kinetic information about all registered spectrum can be obtained by means of changing of the integration limits t_1 and t_2.

Fig. 2 demonstrates the scheme of experimental setup for the receiving of PCL spectra. The setup consists of the luminescence excitation block (1), multichannel photodetector (2) and computer (3).

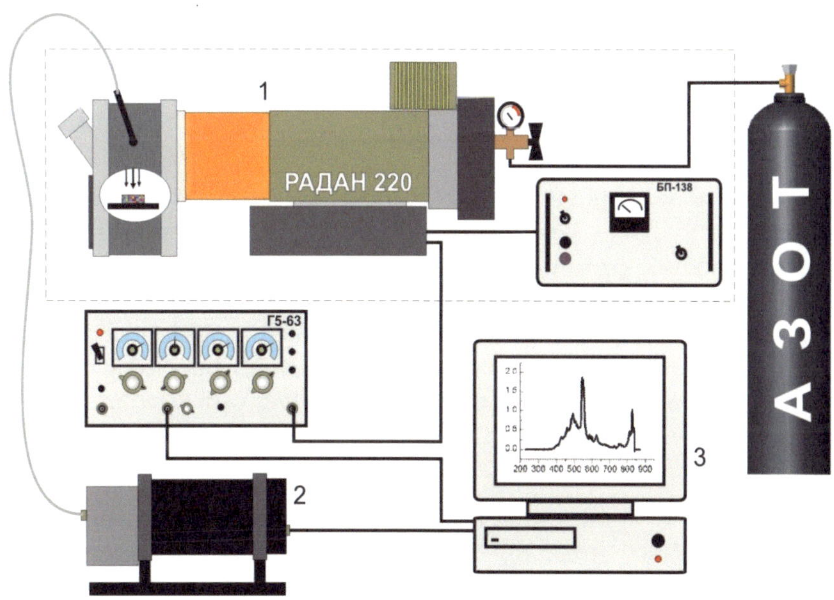

Fig. 2. Scheme of experimental setup

The excitation block (1) represents a combination of RADAN-220 pulsed electron accelerator and analytical chamber. The operating principle of the accelerator is based on the explosive emission of electrons from the cold cathode of accelerating tube. The RADAN-220 generates electron beam with the duration of 2 ns. The voltage that can be applied to the accelerating tube ranges from 150 to 220 keV. The commercially available IMA3-150E tube is placed in the analytic chamber. The generated electron beam is extracted to air through the beryllium foil and directed vertically downwards. The luminescence stream is transferred to the multichannel photodetector (2) by means of the silica multifiber. Computer (3) is the control system of the experimental setup. "Specad" software makes possible to realize various modes of the photodetector. It provides the calibration, registration, reviewing, processing and archiving of obtained spectra. The commercially available pulsed cathodoluminescent spectrograph "CLAVI" [Michailov et. al., 2001] was created on basis of this experimental setup.

3. The application of the pulsed cathodoluminescence for the luminescent analysis of $Nd^{3+}:Y_3Al_5O_{12}$ and $Nd^{3+}:Y_2O_3$

In the last year the intensive investigations in the field of the optical ceramics creation based on the metal refractory oxide doped with rare-earth ions, particularly $Nd^{3+}:Y_3Al_5O_{12}$ и $Nd^{3+}:Y_2O_3$ are carried out [Ikesue et. al., 1995, Lu et. al., 2001, Bagaev et. al., 2009]. The advantages of the laser ceramics against single crystals include the possibility of creating multilayer elements with sizes greater than those of single crystals, larger concentration of active ions, and lower manufacturing cost. The fitness of crystals or ceramics for active laser elements is determined usually by the photoluminescent methods in infrared region by means of lifetime measurement of upper laser Nd ion level $^4F_{3/2}$ [Hoskins et. al., 1963, Lupei et. al., 1995]. For this aim the method is effective, however it doesn't display couses of the lifetime decrease of the laser level. This is necessary to know to correct the conditions the conditions of crystal and ceramics synthesis of synthesis of crystalls and ceramics. Below the investigation of the PCL spectra is given. The possibility of realization of qualitative and quantitative luminescent analyses of $Nd^{3+}:Y_3Al_5O_{12}$, $Nd^{3+}:Y_2O_3$ laser materials is developed.

3.1 The luminescence of $Nd^{3+}:Y_3Al_5O_{12}$

The emission lines of neodymium ions in $Nd^{3+}:Y_3Al_5O_{12}$ in visible range correspond to the transitions from $4f^25d^1$ $^2F2_{5/2}$ level, which has three Stark components ν_0=37775 cm^{-1}, ν_1=37864 cm^{-1}, ν_2=38153 cm^{-1}, to the levels of $4f^3$ configuration of neodymium ion [Kolomiycev et. al., 1984]. The wavelengths of observed luminescent lines and their identification for $Nd^{3+}:Y_3Al_5O_{12}$ single crystal are presented in Table 1 in the first and the second columns, respectively. The numbers of Stark sublevels in according to nomenclature [Koningstein et. al., 1964] at increasing their energy, starting with zero, are pointed next to symbol of electron level in brackets; then emission band wavelengths, presented in [Kolomiycev et. al., 1982], are shown in brackets.

A conspicuous difference appears in PCL spectra of neodymium ions in yttrium aluminates in case of different crystal structure. The spectrum of orthorhombic $Nd^{3+}:YAlO_3$ single crystal together with the spectrum of cubic $Nd^{3+}:Y_3Al_5O_{12}$ are presented in Fig. 3 as an illustration of this difference.

$Nd^{3+}:Y_3Al_5O_{12}$		$Nd^{3+}:YAlO_3$	
λ, nm	Identification of optical trasfer	λ, nm	Identification of optical trasfer
		389,9	$^2F2_{5/2}(2)\rightarrow{}^2H_{9/2}(0)$, (390,0)
		394,6	$^2F2_{5/2}(0)\rightarrow{}^4F_{5/2}(2)$, (394,6)
399,2	$^2F2_{5/2}(0)\rightarrow{}^2H_{9/2}(2)$, (397,5)	398,1	$^2F2_{5/2}(0)\rightarrow{}^2H_{9/2}(3)$, (398,3)
401,6	$^2F2_{5/2}(0)\rightarrow{}^2H_{9/2}(4)$, (401,4)		
		422,5	$^2D_{5/2}(2)\rightarrow{}^4I_{9/2}(1)$, (423,0)
		426,9	$^2F2_{5/2}(2)\rightarrow{}^4F_{9/2}(3)$, (427,3)
429,9	$^2F2_{5/2}(1)\rightarrow{}^4F_{9/2}(0)$, (430,2)	429,9	$^2F2_{5/2}(1)\rightarrow{}^4F_{9/2}(0)$, (429,9)
435,6	$^2F2_{5/2}(0)\rightarrow{}^4F_{9/2}(2)$, (435,1)	435,2	$^2F2_{5/2}(1)\rightarrow{}^4F_{9/2}(1)$, (434,8)
		439,0	$^2D_{5/2}(0)\rightarrow{}^4I_{9/2}(4)$, (438,8)
		440,7	$^2P_{1/2}(0)\rightarrow{}^4I_{9/2}(3)$, (441,2)
450,4	$^2F2_{5/2}(2)\rightarrow{}^2H_{11/2}(3)$, (450,5)	450,4	$^2F2_{5/2}(2)\rightarrow{}^2H_{11/2}(3)$, (450,4)
455,4	$^2F2_{5/2}(0)\rightarrow{}^2H_{11/2}(1)$, (455,9)	456,0	$^2F2_{5/2}(1)\rightarrow{}^2H_{11/2}(3)$, (456,0)
458,8	$^2F2_{5/2}(0)\rightarrow{}^2H_{11/2}(3)$, (458,3)		
461,0	$^2F2_{5/2}(0)\rightarrow{}^2H_{11/2}(4)$, (461,4)		
479,3	$^2F2_{5/2}(1)\rightarrow{}^4G_{5/2}(1)$, (479,0)	480,7	$^2F2_{5/2}(1)\rightarrow{}^4G_{5/2}(2)$, (480,5)
487,5	$^2F2_{5/2}(0)\rightarrow{}^2G_{7/2}(0)$, (487,1)	487,8	$^2F2_{5/2}(0)\rightarrow{}^2G_{7/2}(2)$ (488,1)
494,4	$^4G_{11/2}(2)\rightarrow{}^4I_{9/2}(4)$ (494,2)		
525,2	$^2F2_{5/2}(0)\rightarrow{}^4G_{7/2}(0)$, (524,9)	525,4	$^2F2_{5/2}(1)\rightarrow{}^4G_{7/2}(1)$, (525,4)
		527,8	$^2F2_{5/2}(1)\rightarrow{}^4G_{7/2}(2)$, (527,8)
		538,0	$^2F2_{5/2}(1)\rightarrow{}^4G_{9/2}(2)$, (538,3)
540,6	$^2F2_{5/2}(0)\rightarrow{}^2K_{13/2}+{}^2G_{9/2}(2)$, (541,0)	539,5	$^2F2_{5/2}(0)\rightarrow{}^4G_{9/2}(0)$, (539,3)
		541,6	$^2F2_{5/2}(2)\rightarrow{}^2K_{13/2}(2)$, (542,0)
		545,7	$^2F2_{5/2}(2)\rightarrow{}^2K_{13/2}(4)$, (545,7)
		547,5	$^2F2_{5/2}(2)\rightarrow{}^2K_{13/2}(5)$, (547,5)
549,1	$^2F2_{5/2}(0)\rightarrow{}^2K_{13/2}+{}^2G_{9/2}(7)$, (549,4)	549,2	$^2F2_{5/2}(0)\rightarrow{}^2K_{13/2}(0)$, (549,0)
		554,5	$^2F2_{5/2}(0)\rightarrow{}^2K_{13/2}(3)$, (554,2)
		556,3	$^2F2_{5/2}(0)\rightarrow{}^2K_{13/2}(4)$, (556,2)
557,4	$^2F2_{5/2}(0)\rightarrow{}^2K_{13/2}+{}^2G_{9/2}(10)$,(557,0)		
562,9	$^2K_{15/2}(3)\rightarrow{}^4I_{13/2}(0)$, (562,6)	563,6	$^2K_{13/2}(4)\rightarrow{}^4I_{11/2}(1)$, (563,5)
576,4	Superposition $^2G_{7/2}(2,3)\rightarrow{}^4I_{9/2}(0,2)$, (576,3)		
		585,4	$^2F2_{5/2}(2)\rightarrow{}^4G_{9/2}(4)$, (585,2)
587,4	$^2F2_{5/2}(0)\rightarrow{}^4G_{9/2}(0)$, (586,8)		
596,2	Superposition $^2F2_{5/2}(0)\rightarrow{}^4G_{11/2}(0,1)$, (596,2)		
600,6	$^2F2_{5/2}(0)\rightarrow{}^4G_{11/2}(3)$, (600,1)		
		602,1	$^2F2_{5/2}(2)\rightarrow{}^4G_{11/2}+{}^2K_{15/2}+{}^2D_{3/2}(6)$,(601,8)
		610,4	$^2F2_{5/2}(0)\rightarrow{}^4G_{11/2}+{}^2K_{15/2}+{}^2D_{3/2}(4)$,(610,2)
		612,1	$^2F2_{5/2}(1)\rightarrow{}^4G_{11/2}+{}^2K_{15/2}+{}^2D_{3/2}(6)$,(611,8)
		615,4	$^2F2_{5/2}(2)\rightarrow{}^4G_{11/2}+{}^2K_{15/2}+{}^2D_{3/2}(13)$,(614,7)
620,1	$^2F2_{5/2}(0)\rightarrow{}^2K_{15/2}(2)$, (620,8)	620,1	$^2F2_{5/2}(0)\rightarrow{}^4G_{11/2}+{}^2K_{15/2}+{}^2D_{3/2}(9)$,(619,8)
		622,1	$^2F2_{5/2}(1)\rightarrow{}^4G_{11/2}+{}^2K_{15/2}+{}^2D_{3/2}(11)$,(621,4)
		625,4	$^2F2_{5/2}(1)\rightarrow{}^4G_{11/2}+{}^2K_{15/2}+{}^2D_{3/2}(13)$,(625,2)
		638,3	$^2H_{11/2}(1)\rightarrow{}^4I_{9/2}(2)$ (637,7)
		639,5	$^2H_{11/2}(0)\rightarrow{}^4I_{9/2}(2)$ (639,1)
		646,2	$^2H_{11/2}(3)\rightarrow{}^4I_{9/2}(3)$ (645,4)
		652,4	$^2H_{11/2}(3)\rightarrow{}^4I_{9/2}(4)$ (652,6)
		660,9	$^2G_{7/2}(3)\rightarrow{}^4I_{11/2}(4)$ (660,8)
		665,3	$^2G_{7/2}(0)\rightarrow{}^4I_{11/2}(3)$ (665,3)
		668,6	$^2G_{7/2}(0)\rightarrow{}^4I_{11/2}(4)$ (667,9)

Table 1. PCL lines and their identification for $Nd^{3+}:Y_3Al_5O_{12}$ and $Nd^{3+}:YAlO_3$ single crystals

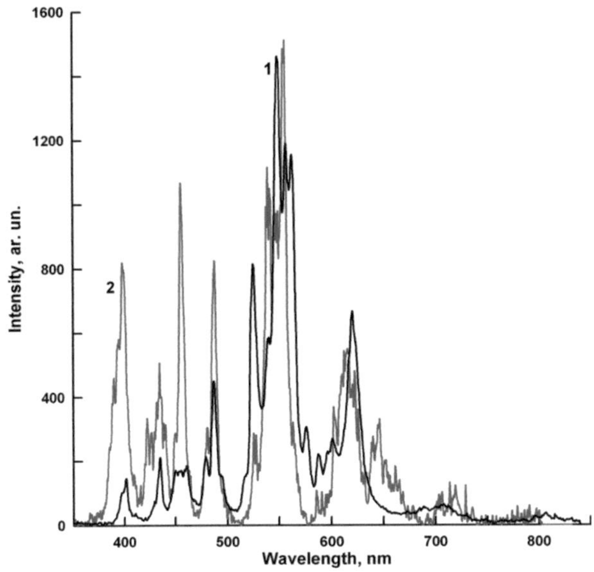

Fig. 3. The PCL spectra of $Nd^{3+}:Y_3Al_5O_{12}$ (1) and $Nd^{3+}:YAlO_3$ (2) single crystals

The wavelengths of fundamental neodymium luminescent lines and their identification for $Nd^{3+}:YAlO_3$ single crystal are presented in Table 1 in the third and the fourth columns, respectively [Osipov et. al., 2011].

The Fig. 3 and Table 1 show that the principal change is manifested in the considerable increase of the luminescent band numbers in the yttrium monoaluminate spectrum. This takes place due to activation of d-f transitions between the different Stark sublevels and the appearance of f-f transitions. These changes arise from the distortion of crystalline field symmetry in positions of individual neodymium ions that leads to the modifications of the oscillator strength and optical transition probability. Thus, the distortion of crystalline field symmetry appears in the spectrum as the change in intensity and numbers of emission bands.

The differences in the spectra can be used for the determination of the second phase content in $Nd^{3+}:Y_3Al_5O_{12}$ [Osipov et. al., 2011]. The luminescence lightsum in the spectrum region from λ_1 to λ_2 ($S = \int_{\lambda_1}^{\lambda_2} I(\lambda)d\lambda$) can be presented by the additive function depended from the dominant phase content C_g (cubic $Nd^{3+}:Y_3Al_5O_{12}$) and the second phase C_{im}=1-C_g

$$S = \alpha \cdot C_g + \beta \cdot (1 - C_g), \tag{8}$$

where α and β are the coefficients of proportionality. They are determined by the integration range and excitation conditions. To eliminate the influence of the intensity instability it is necessary to use the ratio of lightsums (S_1/S_2) as the analytical parameter calculated for two ranges of the spectrum. The lightsums in the ranges 350-500 nm (S_1) and 501-650 nm (S_2) to obtain the functional relation between the C_g and the luminescence intensity of neodymium ions have been chosen. In that case in accordance with Eq. (8) the content of cubic phase into $Nd^{3+}:Y_3Al_5O_{12}$ is defined the following equation

$$C_g = \frac{\beta_1 - \beta_2 \, {}^{S_1}\!/\!_{S_2}}{(\alpha_2 - \beta_2)\, {}^{S_1}\!/\!_{S_2} - (\alpha_1 - \beta_2)} \tag{9}$$

In Fig. 4 the correlation between the C_g and S_1/S_2, calculated for the samples with known content of cubic phase is shown.

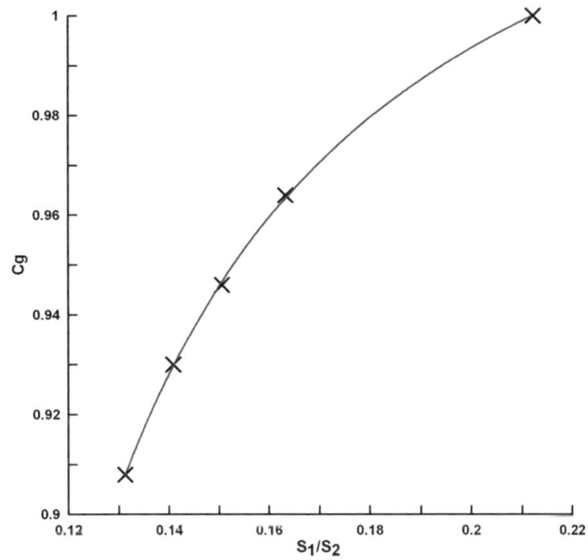

Fig. 4. The correlation between the C_g and S_1/S_2.

This dependence (Fig. 4) is approximated by the following equation with the $r^2 > 0.99$

$$C_g = \frac{1.071 \cdot {}^{S_1}\!/\!_{S_2} - 0.084}{{}^{S_1}\!/\!_{S_2} - 0.069}. \tag{10}$$

Moreover the obtained data validity was checked out by the analysis of samples with electron and optical microscopes.

3.2 The luminescence of Y_2O_3, Nd^{3+}:Y_2O_3

The wide band of intrinsic radiation in visible range is a visiting luminescent card of pure yttria. Earlier the other authors observed this band at different excitation type [Conor, 1964, Kuznetsov et. al., 1978, Bordun et. al., 1995]. Even at cryogenic temperature of the samples the unresolved band was registered.

We investigated commercially available yttria powders with a particle sizes of 1-3 μm and 5-10 μm. All the powders have a cubic lattice of the α-Y_2O_3. From these commercial powders, nanopowders with the average particle size of 10-12 nm were prepared by the laser evaporation method. Particles were crystallized in the metastable monoclinic phase γ-Y_2O_3. After annealing they transformed to the α-Y_2O_3.

The spectrum of the powder with particle size of 1-3 μm has a broad asymetric band peaked at 437 nm and long - wavelength wing shows local maxima (Fig. 5, curve 1).

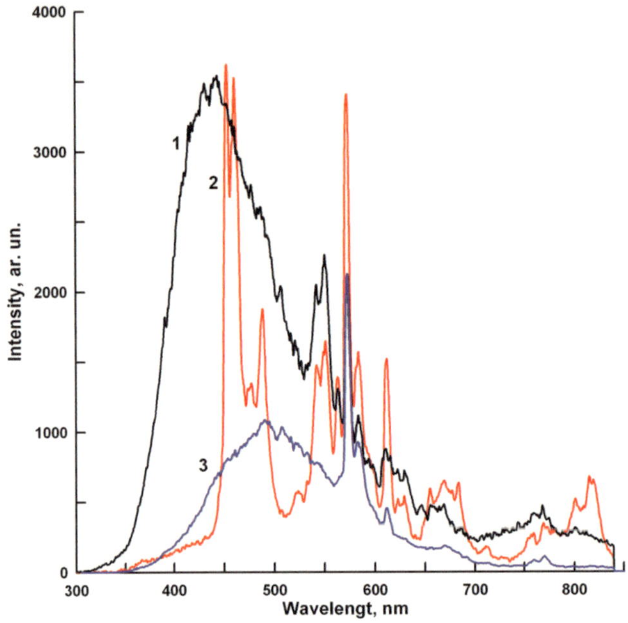

Fig. 5. PCL spectra of commercial yttria powders with particle size 1-3 μm (1), 5-10 μm (2) and nanopowders with the average particle size of 10-12 nm (3).

In the spectrum of the powder with the particle size of 5-10 μm almost all local maxima are transformed into narrow bands (Fig. 5, curve 2). They are grouped into four series 435 - 510 nm (the blue series), 515 - 640 nm (the orange series), 645 - 700 nm (the red series) and 785 - 840 nm (the infrared series). The PCL spectra of nanopowders, irrespective of the crystal phase (either the γ-Y_2O_3 or the α-Y_2O_3 phase) and of the initial coarse powder, have a similar structures (Fig. 3, curve 3). The broad band with the maximum at 485 nm dominates in these spectra. The peak range of this band exhibits local maxima of the blue series. Also the lines of orange series at 573, 583, 612 nm become apparent. The red series is weak, while the infrared series is hardly seen.

The range of the band series observed in the spectra of pulsed cathodoluminescence corresponds to the range of intrinsic radiation of yttria, which is identified as the radiation of associated donor-acceptor pairs Y^{3+} - O^{2-} [Bordun, 2002].

Since the luminescence wavelengths of narrow bands of commercial powders, nanopowders coincide, we can assume that these materials contain intrinsic luminescence centers of the same type.

The series of PCL bands of yttria resemble the radiation of free YO radicals, which is observed, for example, in laser plume of yttria-containing target [Osipov et. al., 2005]. This radical has been fairly well studied [Pearse et. al., 1949]. The Table 2 shows the wavelengths of the bands observed in PCL spectra and their identification. In the second column of this table the wavelengths of the strongest bands are in boldface.

Intrinsic luminescence center	
V′→V″	**λ, nm**
Blue band series, the electronic transition $B^2\Sigma \rightarrow X^2\Sigma$	
0→0	**453.8**
2→2	458.6
3→3	**461.1**
0→1	470.6
2→3	475.2
0→2	**488.7**
Orange band series, the electronic transition $A^2\Pi \rightarrow X^2\Sigma$	
2→0+(T_g+A_g)=380	**542.8**
1→0	**551.6**
1→0+(T_g+A_g)=380	**563.5**
0→0	**572.9**
3→3	**583.6**
0→1	600.0
3→4	**612.2**
0→2	629.3
Red band series, the electronic transition $A^2\Pi \rightarrow X^2\Sigma$	
0→3-(T_g+A_g)=162	655.3
0→3	662.4
0→3+(T_g+A_g)=162	669.6
0→3+T_g=469	683.8
Infrared band series, the electronic transition $A^2\Pi \rightarrow X^2\Sigma$	
0→5+(T_g+A_g)=380	760.8
0→6	785.0
	801.1
	818.4

Table 2. Parameters of the PCL lines in the yttria spectrum

Based on these data, we constructed the energy scheme of the intrinsic luminescence center (Fig. 6). Qualitatively this scheme coincides with that of the YO free radicals. In this scheme the configuration curves were calculated in the harmonic oscilator approximation as

$$E_i = E_{0i} + \frac{2\pi^2 c v_i m_0}{h} \cdot 10^{-16} (r - \rho_i)^2 , \tag{11}$$

where i =X, A and B denotes the electronic states $X^2\Sigma$, $A^2\Pi$, and $B^2\Sigma$, E_{0i}, ρ_I, and ν_i are the minimal energy, the equilibrium distance, and the wavenumber of the vibration mode of ith electronic state, respectively; m_0 is the mass of the oxygen atom; c and h are the light speed and Planck's constant. The energy E and wavenumbers ν_i are expressed in (11) in reciprocal centimetres, the amplitudes of the (r-ρ_i) vibrations are given in nanometres.

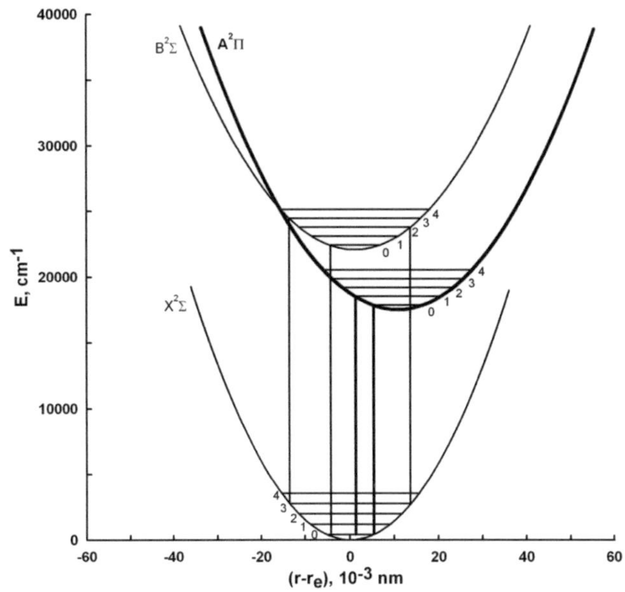

Fig. 6. Energy scheme of the intrinsic luminescence center

For the configuration curves of the $X^2\Sigma$, $A^2\Pi$, and $B^2\Sigma$ states (Fig. 6) E_{0i}=0, 17510, and 22090 cm^{-1} and ν_i=786, 675, and 675 cm^{-1}, respectively. With these parameters, up to the measurement error, the wavelengths of pulsed cathodoluminescence coincide with the wavelengths of the optical transitions shown in Table 2. The Franck - Condon principle for molecular transitions is most precisely implemented at $\rho_A=\rho_X+10.8709\cdot10^{-3}$ nm and $\rho_B=\rho_X+1.2351\cdot10^{-3}$ nm, and ρ_X can be estimated as half of elementary cube edge, ρ_X=0.1385 nm. For electronic - vibration transitions for which one of the vibration quantum numbers V is large, this principle is implemented only if these transitions involve the most strong phonons [Schaak et. al., 1970], Table 2. Under these parameters, the configuration curves of the electronic states $A^2\Pi$ and $B^2\Sigma$ intersect at the point with E=25256 cm^{-1}.

The qualitative coincidence of the emission bands and the energy structure of intrinsic luminescence centres observed by us and YO free radicals allow us to conclude that intrinsic luminescence centres in yttria contain bound YO radicals [Osipov et. al., 2008]. Consider the possibility of formation of such intrinsic luminescence centre. It is known [Schaak et. al., 1970] that the cubic yttria has unit cell composed of 16 formula units Y_2O_3. Twenty four cations occupy positions with C_2 symmetry and eight cations occupy the positions with C_{3i} symmetry (Fig. 7). Every cation is surrounded by six oxygen ions which are positioned on the corners of deformed cube with the edge size of 0.2702 nm, at that two corners is unoccupied. Thus in one-third of the cubes (YO_6) two oxygen vacancies are located at the cube corners along the face diagonal, while, in the remaining cubes, they are located along the body diagonal (Fig. 7).

For such packing a structure, presented in Fig. 8. ,can be formed at the outer cube face that contains two oxygen vacancies and that is located at the crystal boundary. In essence, this structure is the YO radical bound to the crystal lattice by the yttriun ion. On such surfaces the fraction of faces with two oxygen vacancies is 1/3×1/6=1/18 and the average distance between them is about 5 nm. All of this leads to the dependence of the luminescence spectrum of such bound radicals on the particle size of yttria mainly via their shape.

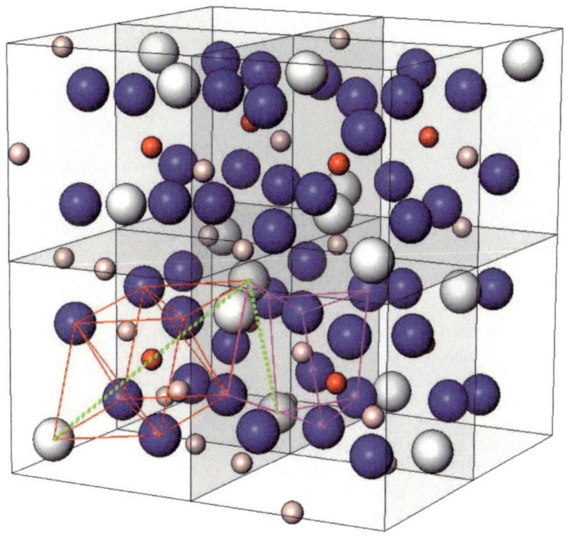

Fig. 7. The unit cell of yttria. The yttrium positions with C_{3i} and C_2 symmetry designated by red and pink balls respectively. The oxygen and the vacant positions designated by blue and grey balls. The vacant positions are associated by green dot line.

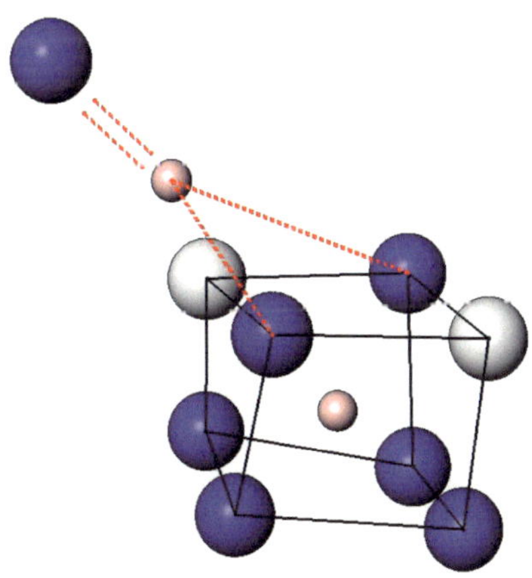

Fig. 8. The structure of intrinsic luminescence centre.

The considered above intrinsic luminescence centre also presents in Nd^{3+}:Y_2O_3. However the presence of neodymium results in decrease of the intrinsic band intensity and distortion of its profile. We studied the monoclinic and cubic Nd^{3+}:Y_2O_3 nanopowders. The nanopowders were prepared using a mixture of micropowder α-Y_2O_3 phase and 1 mol.% Nd_2O_3 powder. After evaporation of this mixture by CO_2 laser Nd^{3+}:Y_2O_3 nanoparticles were crystallized into monoclinic phase. To transfer nanopowders into cubic phase annealing in air was carried out above 900°C [Kotov et. al., 2002].

The PCL spectra of all $Nd^{3+}:Y_2O_3$ samples contain emission lines of neodymium ions, which have not been previously observed in photoluminescence. Namely, neodymium doped nanopowders exhibit a strong band peaking at 825 nm (Fig. 9).

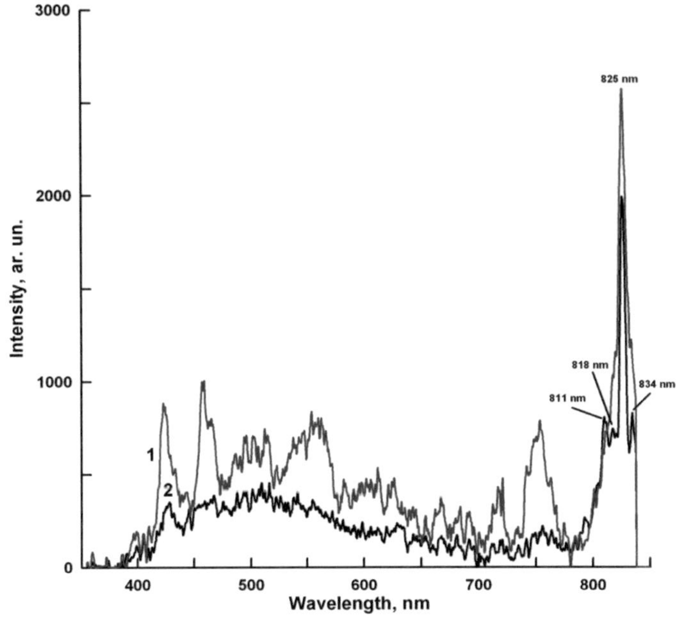

Fig. 9. The PCL spectra of monoclinic (1) and cubic (2) $Nd^{3+}:Y_2O_3$ nanopowders

Its components correspond well to the $^4F_{5/2}\rightarrow{}^4I_{9/2}$ (825, 811, 834 nm) and $^2H_{9/2}\rightarrow{}^4I_{9/2}$ (818 nm) optical transitions between Nd^{3+} Stark's sublevels with the energies in Y_2O_3 cubic lattice [Chang, 1966]. These components are resolved only in cubic samples (Fig. 9). Therefore the splitting of Stark's components allows us to conclude about the presence of the dominant phase into $Nd^{3+}:Y_2O_3$.

To check this assumption the additional investigations were made. In Fig 10. the PCL spectra of pressed nanopowders (compacts) are presented. The compacts were annealed at 530, 750, 950, 1100, and 1300°C. The X-ray analysis for this samples showed that the unannealed compact and annealed compact at 530°C have monoclinic phase, all remaining compacts are cubic samples. It is shown that the splitting of neodymium band at region of 800-840 nm only takes place in cubic samples and one component appears at 825 nm in monoclinic samples.

In addition to the band in the region of 800-840 nm two emission bands of neodymium ions arise in the $Nd^{3+}:Y_2O_3$. These are a weak band at 720 nm due to $^4F_{9/2}\rightarrow{}^4I_{9/2}$ transition and stronger band at 750 nm with the components due to the transitions between the Stark sublevels: $^4F_{7/2}\rightarrow{}^4I_{9/2}$ and $^4S_{3/2}\rightarrow{}^4I_{9/2}$.

The intensity weaking of intrinsic band into $Nd^{3+}:Y_2O_3$ is associated with the quantitative decrease of this centres, since the part of yttrium ions are replaced by the neodymium ions. The distortion of the intrinsic band is determined by the neodymium absorption of it. The most absorption is observed in region at 560-613 nm [Osipov et. al., 2009].

In addition to the present bands into $Nd^{3+}:Y_2O_3$ the appearance of four well-resolved components in the range of 610-660 nm can be seen. The specta of this band for compacts annealed at 950 and 1300°C are presented in Fig. 11. The band contains the following four narrow lines at 620.6, 630.6, 645.3, 655.6 nm which we identify using Raman spectra, as luminescence of oxygen molecular ion O_2^- [Solomonov et. al., 2011].

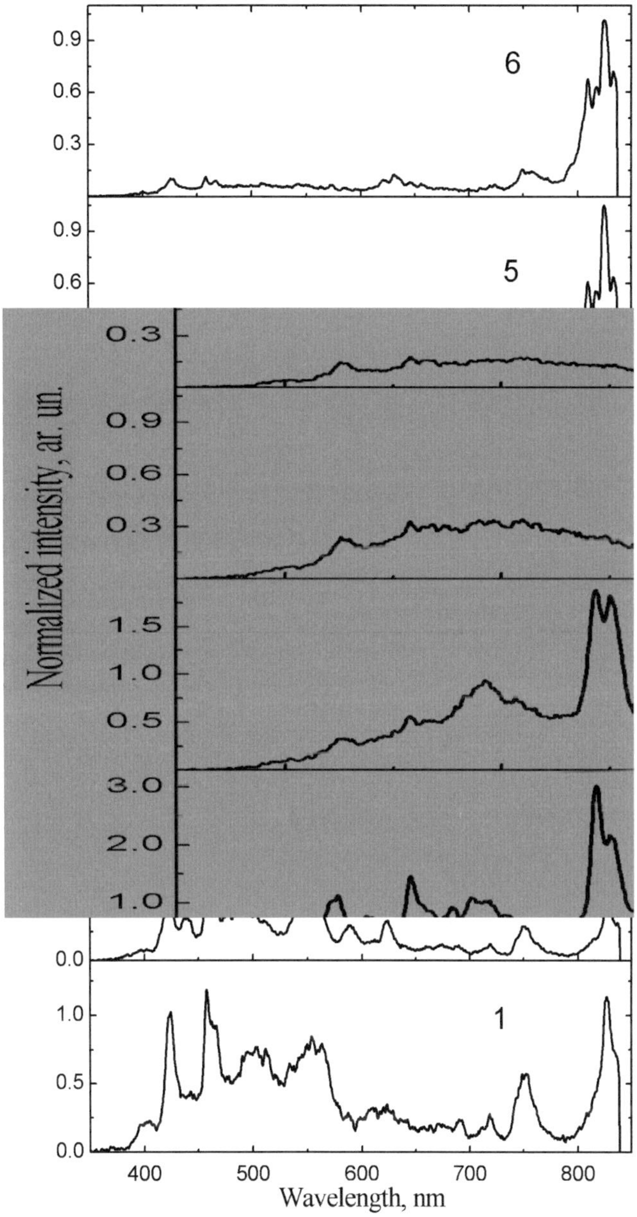

Fig. 10. The PCL spectra of unannealed compact (1) and compacts annealed at 530°C (2), 750°C (3), 950°C (4), 1100°C (5), 1300°C (6).

The bands with the frequencies of 1615 and 1702 cm^{-1} correspond to vibrations of the molecular ion in the ground state, while the frequencies at 966 and 993 cm^{-1} correspond to the excited state for the two sites of the oxygen molecular ion in the yttria lattice.

Fig. 12 demonstrates two sites of defects O_2^- taking into account the occurrence of two types of natural vacancies about we talked earlier.

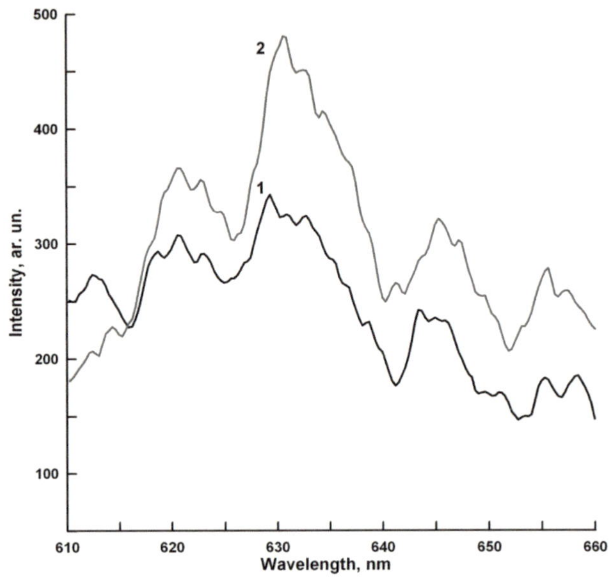

Fig. 11. The PCL of oxygen molecular ions.

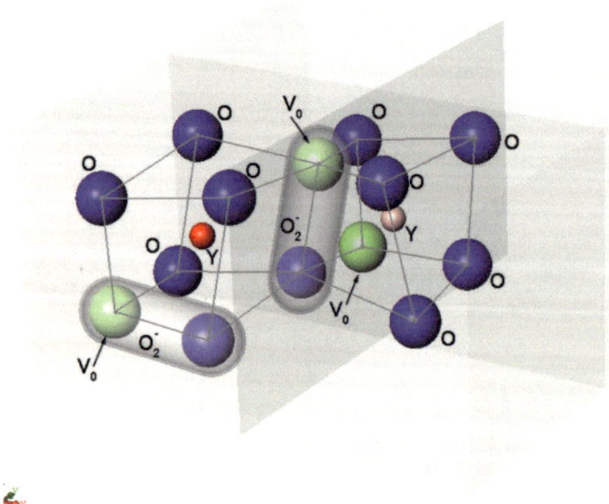

Fig. 12. Two sites of oxygen molecular ions defects in lattice of cubic Y_2O_3

Fig. 13 presents two potentional curves and formation of luminescence bands. The lowest excited state $A^2\Pi_u$ of the oxygen molecular ion can be stabilized.

The observed luminescence band results of the transition from the vibration level V′=0 of the excited electronic state $A^2\Pi_u$ to one vibration level V′′=3 of the ground state $X^2\Pi_g$. The bands at 630.6 and 655.6 nm are formed due to the transition to the vibration level of the groun state with the participation of lattice phonons. For the first curve the phonon energy is 255 cm, while, for the second curve it is 244 cm^{-1}. These phonons are observed in the Raman spectrum. The excited electronic state of the molecular ion is spaced from the ground state by 20660-21580 cm^{-1} [Solomonov et. al., 2011].

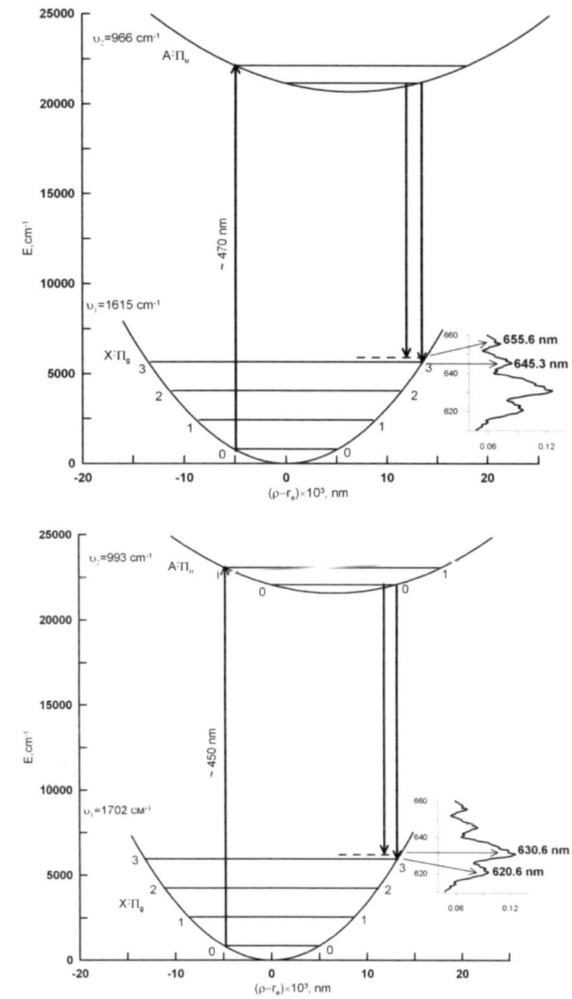

Fig. 13. Configuration curves for two sites of oxygen molecular ions

On the basis of the qualitative luminescent analysis of $Nd^{3+}:Y_2O_3$ the determination of neodymium concentration by means of the calibration curve construction is possible because of intensity of neodymium lines is proportional to it concentration $I_{Nd}(\lambda_i)=a_i\cdot C_{Nd}$ (for example 750 or 825 nm). However we can't use this equation because PCL spectrum is characterized by the instability. Therefore we also chose lightsums ratio as analytic parameter in regions of 730-840 nm and 350-840 nm. The first region of $Nd^{3+}:Y_2O_3$ spectrum involves only neodymium bands, but the second one includes in addition the intrinsic band which is distorted by the neodymium absorption and can be ascribed as follows

$$I_{YO}(\lambda_i) = I_{0i}\frac{1-\exp(-k_i \cdot C_{Nd} \cdot l)}{k_i \cdot C_{Nd}} \quad (12)$$

Here I_{0i}, l, k_i are the intensity of i - intrinsic band without absorption, the thickness of samples, coefficient of absorption of i - band, respectively. Hence relation of lightsums ratio (S(350-840)/S(730-840)) with C_{Nd} has to include the equation (12). Really this relation is decribed by the following equation

$$\frac{S(350-840)}{S(730-840)} = A \cdot \frac{1-\exp(-B \cdot C_{Nd})}{C_{Nd}^2} + D. \quad (13)$$

Fig. 14 demonstrates the calibration curve ($r^2 > 0.99$) for the determination of neodymium concentration in region of 0.11 - 1.07 at. %.

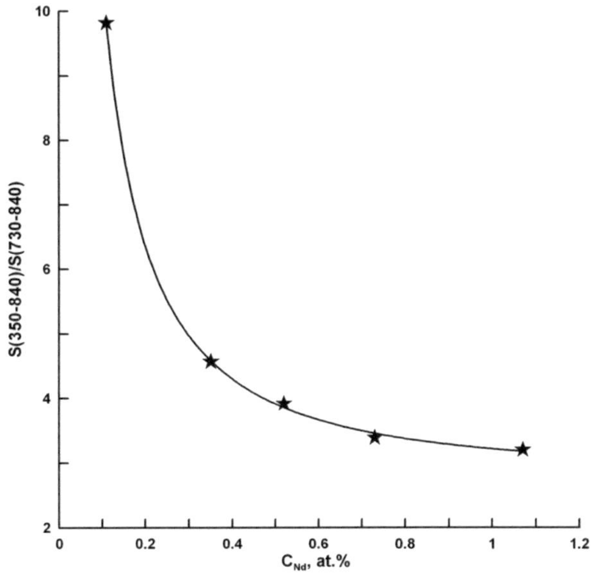

Fig. 14. The calibration curve for the determination of neodymium concentration in region of 0.11 - 1.07 at. %.

4. Conclusion

Thus, the possibility of realization of rapid, nondestructive, qualitative and quantitative luminescent analyses of laser materials, in particular Nd^{3+}:$Y_3Al_5O_{12}$, Nd^{3+}:Y_2O_3, with the help of pulsed cathodoluminescence was shown.

5. References

Bagaev, S.; Osipov, V.; Ivanov, M.; Solomonov, V.; Platonov, V.; Orlov, A.; Rasuleva, A. & Vatnik, S. (2009).Fabrication and characteristics of neodymium - activated yttrium oxide optical ceramics. *Optical Materials*, Vol.31, pp. 740-743, ISSN 0925-3467

Bogdankevich, O.; Darznik, S. & Eliseev, P. (1975). *Semiconductor lasers,* Nauka, Moscow, Russia

Bordun, O.; Bordun, I. & Novosad, S. (1995). Luminescence centres into Y_2O_3. *Journal of applied spectroscopy,* Vol.62, No.6, pp. 91-95, ISSN 0021-9037

Bordun, O. (2002). Influence of oxygen vacancies on the luminescence spectra of thin Y_2O_3 films. *Journal of applied spectroscopy,* Vol.69, No.3, pp. 371-374, ISSN 0021-9037

Chang, N. (1966). Energy levels and crystal - field splittings of Nd^{3+} in yttrium oxide. *J. Chem. Phys.,* Vol.44, pp. 4044-4050, ISSN 0021-9606

Chukichev, M.; Sabri, D.; Sokolov, V. & Surcova T. (1990). Cathodoluminescence of solid solution $Zn_{1-x}Mn_xSe$. *Optics and Spectroscopy,* Vol.68, pp. 2000-2021, ISSN 0030-4034

Connor, O. (1964). A theory of thermoluminescence of fluorite (CaF_2:Y). Radiative recombination from highly associated electron - hole pairs. *Appl. Phys. Lett.,* Vol.4, No.1, pp. 126-129, ISSN 0003-6951

Galkin, G. (1981). The interband processes of the recombination into semiconductors at high excitation level. *Proceedings of the P.N. Lebedev Physical Institute (Trudi FIAN),* Vol.128, pp. 3-64

Hoskins, H. & Soffer, B. (1963). Stimulated emission from $Y_2O_3:Nd^{3+}$. *Appl. Phys. Lett.,* Vol.4, pp. 22-23, ISSN 0003-6951

Ikesue, A.; Kinoshita, T.; Kamata, K. & Yoshida, K. (1995). Fabrication and optical properties of high-performance polycrystalline Nd:YAG ceramics for solid-state lasers. *J. Amer. Ceram. Soc.,* Vol.78, pp. 1033-1040, ISSN 1551-2916

Kolomiycev, A.; Meylman, M.; Volodina, I.; Chukichev, M.; Smagin, A. & Bagdasarov H. (1982). Luminescence of neodymium - activated yttrium - aluminum garnet crystals in ultra-violet and visible ranges at high energy excitation. *Deposition in VINITI,* No. 5995-82, pp. 1-15.

Kolomiycev, A.; Meylman, M.; Volodina, I.; Chukichev, M.; Smagin, A. & Bagdasarov H. (1984). Luminescence of neodymium - activated yttrium - aluminum garnet crystals in ultra-violet and visible ranges at high energy excitation. *Optics and Spectroscopy,* Vol.56, pp. 365-367, ISSN 0030-4034

Koningstein, J. & Geusic, J. (1964).Energy levels and crystal - field calculations of neodymium in yttrium aluminum garnet. *Phys. Rev.,* Vol.136, No. 3A pp. 711-716, ISSN 1893 1912

Kotov, Yu.; Osipov, V.; Ivanov, M.; Samatov, O.; Platonov, V., Azarkevich, E.; Murzakayev, A. & Medvedev. A. (2002). Properties of oxide nanopowders prepared by target evaporation with a pulse - periodic CO_2 laser. *Technical Physics,* Vol.47, No.11 pp. 1420-1426, ISSN 1063-7842

Kuznetsov, A.; Abramov, V.; Rooze, N. & Savikhina, T. (1978). Autolocalized excitons into Y_2O_3. *JETP Letters,* Vol.28, No.10, pp. 602-605, ISSN 0021-3640

Lu, J.; Murai, T.; Takaichi, K.; Ueda, K.; Yagi, H.; Yanagitani, T. & Kaminskii, A. (2001) $Nd^{3+}:Y_2O_3$ ceramic laser. *Jpn. J. Appl. Phys.,* Vol.40, pp. 1277-1279, ISSN 0021-4922

Lupei, V.; Lupei, A.; Tiseanu, C.; Georgescu, S.; Stoicescu, C. & Nanau, P. (1995) High - resolution optical spectroscopy of YAG: Nd: A test for structural and distribution models. *Phys. Rev. B,* Vol.51, No.1 pp. 8-17, ISSN 0003-6951

Mesyats, G. (1974). *Generating of highpower nanosecond pulses,* The Soviet radio, Moscow, Russia

Mesyats, G.; Shpak, V.; Yalandin, M. & Shunailov S. (1992). Compact RADAN electron accelerators for testing new radiation technologies and sterilization. *Radiat. Phys. Chem.*, Vol.46, pp. 489-492, ISSN 0969-806X

Michailov, S.; Osipov, V. & Solomonov V. (2001). Pulsed cathodoluminescent analyzer of materials. *Pribory i tekhnika eksperimenta,* No.3, pp. 164-165, ISSN 0032-8162

Osipov, V.; Solomonov, V.; Platonov, V.; Snigireva, O.; Ivanov, V. & Lisenkov, V. (2005). Laser plume spectroscopy. 2. Graphite yttrium - stabilized and zirconium oxide targets. *Quantum Electron.*, Vol.35, No.7 pp. 633-637, ISSN 0368-7147

Osipov, V.; Rasuleva, A. & Solomonov, V. (2008). Luminescence of pure yttria. *Optics and Spectroscopy,* Vol.105, No.4 pp. 524-530, ISSN 0030-400X

Osipov, V.; Solomonov, V.; Spirina, A.; Ivanov, M. & Orlov, A. (2009). Luminescence of yttrium oxide doped with neodymium. *Optics and Spectroscopy,* Vol.106, No.1 pp. 78-83, ISSN 0030-400X

Osipov, V.; Solomonov, V. & Spirina, A. (2011). Luminescent investigation of neodymium doped yttrium aluminates. *J. of Optical Technology,* Vol.78, No.6 pp. 81-87, ISSN 0030-4042

Pearse, R. & Gaydon A. (1949). *The identification of molecular spectra,* Inostrannaya Literatura, Moscow, Russia

Petrov, V. (1996). Cathodoluminescent microscopy. *Physics - Uspekhi,* Vol.39, pp. 807-818, ISSN 0042-1294

Ramseyer, K.; Fisher, J.; Matter, A.; Eberhardt, P. & Geiss J. (1989). A cathodoluminescence microscope for low intensity luminescence. *Journal of Sedimentary Research,* Vol.59, pp. 619-622, ISSN 0022-4472

Ramseyer, K. & Millis J. (1990). Factors influencing short-lived blue cathodoluminescence in α - quartz. *Journal of Sedimentary Research,* Vol.59, pp. 619-622, ISSN 0003-004X

Schaak, G. & Koningstein, J. (1970). Phonon and electronic raman spectra of cubic rare-earth oxides and isomorphous ittrium oxide. *J. Opt. Soc. Amer.* Vol.60, No.8, pp. 1110-1115, ISSN 1551-2916

Solomonov, V. & Michailov, S. (2003). *Pulsed cathodoluminescence and its application for the analysis of condensed substances,* UB RAS, ISBN 5-7691-1357-X, Yekaterinburg, Russia

Solomonv, V. (2003). Kinetics of pulsed cathodoluminescence. *Optics and Spectroscopy,* Vol.95, pp. 248-254, ISSN 0030-4034

Solomonv, V.; Michailov, S; & Deykun, A. (1996). About the mecanism of excitation and the structure of the pulsed cathodoluminescence bands of Cr^{3+}, Mn^{2+} impurity ions into minerals. *Optics and Spectroscopy,* Vol.80, pp. 447-458, ISSN 0030-4034

Vaysburd, D.; Semin, B.; Tavanov, E.; Matlis, S.; Balichev, I. & Gering G. (1982) *Highenergy electronics of solid,* Nauka, Novosibirsk, Russia

Yang, B.; Luff, B. & Townsend F. (1992). Chatodoluminescence of natural zircons. *J. Phys.: Condens. Matter.,* Vol.4, pp. 5617-5624, ISSN 0953-8984

3

Pulsed Cathodoluminescence of Natural and Synthetic Diamonds Excited by Nanosecond and Subnanosecond Electron Beams

E.I. Lipatov[1], V.M. Lisitsyn[2], V.I. Oleshko[2],
E.F. Polisadova[2], V.F. Tarasenko[1] and E.H. Baksht[1]
[1]*High Current Electronics Institute, Tomsk,*
[2]*Tomsk Polytechnic University, Tomsk,*
Russian Federation

1. Introduction

Research on the luminescence of crystals due to ionizing radiation is an important field in spectral analysis of solids that allows determination of the chemical composition of material, its structure (the presence of defects), state (internal stress), etc. [Marfunin; Gritsenko *et al.*; Kravchenko *et al.*; Solomonov&Mikhailov]. The luminescence spectra and kinetics of various crystals are most commonly studied using pulsed accelerators with vacuum diodes [Gritsenko *et al*; Korepanov *et al.*, 2000; Korepanov *et al.*, 2005; Kravchenko *et al.*; Lipatov *et al.*, 2007 (1); Lisitsyna *et al.*, 2002; Lisitsyna *et al.*, 2008; Lisitsyna *et al.*, 2011; Solomonov&Mikhailov]. These accelerators produce electron beams of varying energy, current density, and pulse duration: the beam current pulse duration is varied by varying the voltage pulse parameters; the electron energy in a vacuum diode is determined by the voltage across the interelectrode gap. However with short beam current pulses (~ 100 ps), the design of accelerators with vacuum diodes is added much complexity [Zheltov]. In this case, electron accelerators with gas diodes are more preferable to use [Tarasenko *et al.*, 2008; Tarasenko *et al.*, 2009; Tarasenko *et al.*, 2010]; they are simpler in design and make it possible to produce an electron beam of current density ~ 100 A/cm^2 and to control the pulse duration in the range from 100 to 500 ps [Baksht]. In this work, experiments were performed on two electron accelerators with vacuum and gas diodes. Compact electron accelerators based on vacuum and gas-filled diodes ensure high-intensity pulsed cathodoluminescence (PCL) typical of high-current electron beams [Solomonov&Mikhailov].

In our previous works, we demonstrated intense PCL of spodumene [Lipatov *et al.*, 2007 (1)], ruby [Lipatov *et al.*, 2005 (1)], calcite [Baksht *et at*], fluorite [Baksht *et al*], and diamond [Baksht *et al.*; Lipatov *et al.*, 2005 (2)] excited by a subnanosecond avalanche electron beam and measured the integrated PCL spectra [Baksht *et al.*; Lipatov *et al.*, 2005 (1); Lipatov *et al.*, 2005 (2)] and the luminescence decay kinetics [Baksht *et al.*; Lipatov *et al.*, 2007 (1)].

The objective of the present paper was to study the luminescence of two specimens of diamond (natural and synthetic) excited by nanosecond and subnanosecond electron beams with a current density of 10–200 A/cm^2 in the spectral range from 200 to 850 nm. The use of compact nanosecond and subnanosecond high-current electron accelerators as excitation sources allowed pulsed luminescence spectrometry with high temporal resolution, high light signal intensity, and hence high sensitivity and accuracy of measuring the decay characteristics over a wide time interval ($10^{-10} \div 10^{-1}$ s).

The characteristics under study were the PCL spectra and decay kinetics of the diamond specimens, including the time evolution of the spectra. The both specimens revealed a PCL band due to intrinsic structural defects (Band-A). In the PCL spectra of the synthetic diamond, because of the lower impurity content, a radiative recombination band of free excitons was found, whereas in the PCL spectra of the natural diamond this band escaped detection because of the high N impurity content. In the spectrum of the natural diamond, the N3 system due to nitrogen-containing centers was observed.

2. Experimental setups, methods, and specimens

The integrated PCL spectra of the specimens were measured as shown in Fig. 1*a*. The electron beam produced by accelerator *1* (the characteristics of the accelerators are presented in Table 1) excited pulsed cathodoluminescence in specimen *2*. The luminescence was transferred through optical fiber *3* to spectrometer *5* (EPP-2000C Stellar-Net Inc.) and converted to an electrical signal which was transmitted to computer *9* and represented as a spectrum. In PCL spectrum processing, the optical absorption spectra of the specimen *2* and fiber *3* and the spectral sensitivity of the spectrometer *5* were taken into account.

Accelerator	E_m, keV	$t_{0.5}$, ns	j_e, A/cm^2	E, mJ/cm^2
SLEP-150	100	0.1–0.65	10–100	1.5–8.0
RADAN-EXPERT - IMA3-150E	150	1.5	100	20
RADAN-220 - IMA3-150E	220	2	200	60
NORA	240	4	200	120
Vacuum diode with GIN-600	360	10	30	100

Table 1. Characteristics of the accelerators and electron beams that excited PCL in the specimens.

The optical absorption spectra were also measured as shown in Fig. 1*a*, except that a KrCl-lamp [Lipatov *et al.*, 2010] was used instead of the accelerator *1*. The optical radiation was delivered to the spectrometer *5* through the optical waveguide *3*. Once the reference spectrum was measured, the entrance aperture of the waveguide *3* was covered with the specimen *2*. The plane-parallel specimen *2* was arranged on a special cooled holder with forevacuum pumping such that the output facet was parallel to the entrance aperture of the waveguide *3*. Because the radiation intensity of the KrCl-excilamp at 200–260 nm was much higher than that at 260–800 nm, the foregoing procedure was followed for each range separately. Two parts of the optical absorption spectrum was joined through multiplying the short-wave part by an appropriate factor.

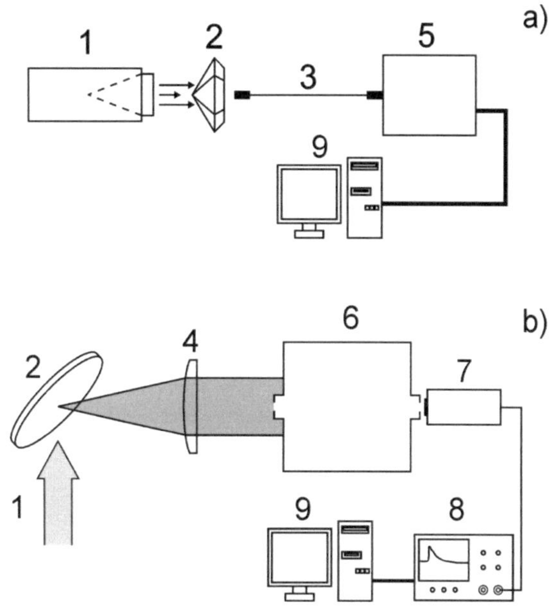

Fig. 1. Measurement scheme a) the integral spectra of OA and PCL samples and b) PCL spectra with a time resolution. 1) light source or an electron accelerator, 2) the test specimen of diamond, 3) fiber, 4) the collecting lens, 5) spectrometer, 6) monochromator, 7) photomultiplier or photodiode, 8) digital oscilloscope, 9) PC.

The PCL spectra and decay kinetics of the specimens were studied using a MDR-23 monochromator, a FEU-100 (FEU-97) photomultiplier tube, and a FEK-22 photodiode. The measuring procedure is shown in Fig. *1b*. The electron beam produced by the accelerator *1* excited luminescence in the specimen 2, which was fixed on a special cooled holder with forevacuum pumping. The irradiated part of the specimen surface was in the focus of collecting lens *4*. The collimated light beam was delivered to the entrance slit of monochromator *6*. The entrance and exit slits were normally 10 μm wide. Spectrum scanning was provided by rotation of a diffraction grating of 600 and 1200 groove/mm with a stepping motor. The spectral resolution was about 1 nm. The optical signal was converted to an electrical signal by photomultiplier tube (or photodiode) *7*, was recorded as a pulse by TDS-3032 and TDS-6604 Tektronix digital oscilloscopes *8*, and was transmitted to the computer *9*. The luminescence pulse was recorded in a separate file for each wavelength. The luminescence spectrum at an arbitrary point in time was reconstructed by software processing. The resulting spectrum was normalized to the spectral sensitivity of the photomultiplier tube or photodiode. The time interval of measurement of the luminescence decay kinetics (nano- or milliseconds) was determined by choosing the load on the oscilloscope – 50 Ohm or 1 MOhm.

The luminescence relaxation kinetics was determined from approximation of the light pulse fall time by a simple exponential function for each wavelength. The thus obtained values were used to reconstruct the luminescence relaxation time spectrum of the specimens.

We had ten diamond specimens at our disposal. The main experiments were performed with one natural diamond specimen and one synthetic diamond specimen whose description and characteristics are given in Table 2.

No	Description	Dimensions, mm	Type*	Synthesis method
1	Disk	Ø5 × 0.25	2a	natural
2	Square plate	10 × 10 × 0.1	2a	CVD**

* standard physical classification [Zaitsev];
** chemical vapor deposition.

Table 2. Number, description, dimensions, type, and method of synthesis of the diamond specimens.

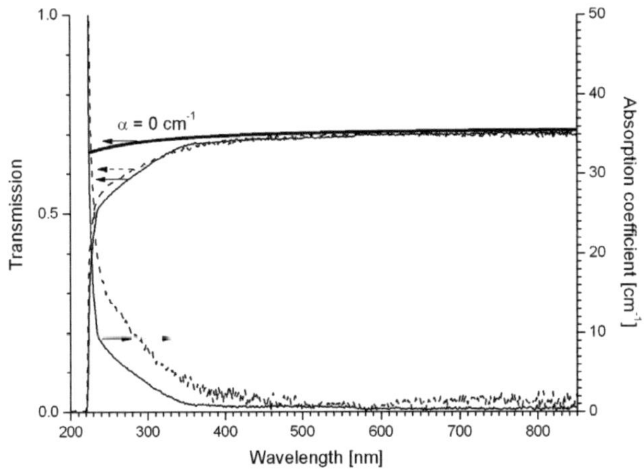

Fig. 2. The left axis: the transmission spectra of natural (continuous curve) and synthetic (broken curve) samples of diamond, measured at room temperature, and the transmission spectrum, calculated according to expression (1), for an ideal (non-absorbing) plane-parallel specimen of the diamond. The right axis: absorption spectra, calculated according to expression (1) and (2), of natural (continuous curve) and synthetic (broken curve) samples of the diamond.

Figure 2 shows the optical absorption spectra of the specimens measured by the foregoing procedure. The absorption spectra of the specimens was calculated by the dichotomy technique with resort to the available data [Zaitsev] on dispersion of the refractive index of diamond and to the expressions:

$$T(\lambda)=\frac{\left(1-r(\lambda)\right)^{2}\cdot e^{-\alpha(\lambda)\cdot d}}{\left(1-r(\lambda)^{2}\cdot e^{-2\cdot\alpha(\lambda)\cdot d}\right)}, \quad (1)$$

$$r(\lambda)=\frac{\left(1-n(\lambda)\right)^{2}}{\left(1+n(\lambda)\right)^{2}}, \quad (2)$$

where $n(\lambda)$ is the refraction index, $r(\lambda)$ is the reflection coefficient, $\alpha(\lambda)$ is the absorption coefficient [cm^{-1}], and d is the optical thickness of a specimen. The calculated transmission spectrum of a perfect plane-parallel diamond shown in Fig. 2. Noteworthy is the absence of narrow-band peculiarities in the absorption spectra of the diamond specimens, the presence of a clearly defined fundamental absorption edge at 225 nm, and high imperfection of the synthetic diamond. According to the standard physical classification of diamonds [Zaitsev], the specimens are referred to type 2a.

3. Results and discussion

Figure 3 shows integrated PCL spectra of the natural (continuous spectrum) and synthetic (dotted spectrum) diamonds cooled with liquid nitrogen.

The integrated PCL spectrum of the natural diamond is dominated by the N3 vibronic system with a zero-phonon line (ZPL) at 415.2 nm and its phonon replicas at 420–450 nm due to the presence of N_3V defects [Zaitsev]. The N3 system is thus superimposed on a structureless luminescence band-A in the spectral range 300–700 nm. According to [Takeuchi], the luminescence band-A owes to intrinsic defects of the diamond lattice - sp^2-hybridized carbon bonds.

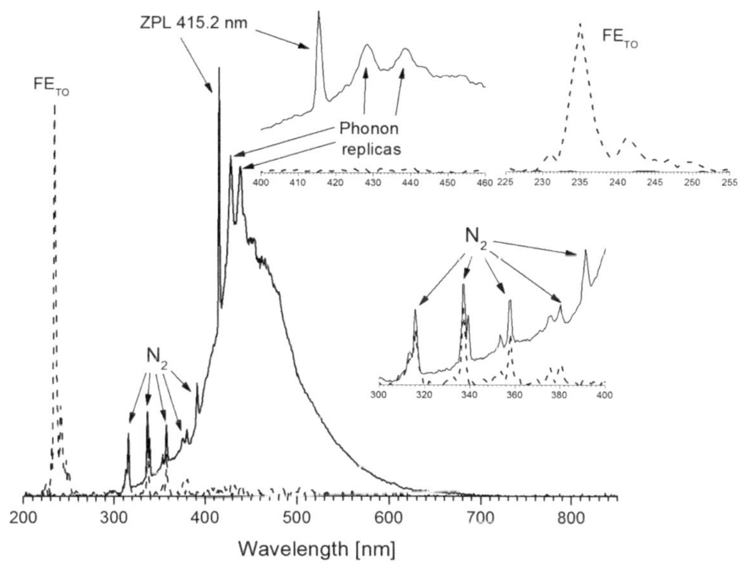

Fig. 3. Integrated PCL spectra of natural (continuous line) and synthetic (broken curve) of diamond samples, measured at liquid nitrogen temperatures. Inset: the band of the radiative recombination of free excitons (FE_{TO} at 235 nm), the zero-phonon line of N_3V-defects (ZPL 415.2 nm), the emission bands of the second positive system of molecular nitrogen in the air.

On the contrary, the integrated PCL spectrum of the synthetic diamond revealed a very weak band-A and was dominated by a radiative recombination band of free excitons FE_{TO} at 235 nm. Because diamond is an indirect semiconductor, radiative recombination of free excitons occurs during the generation of phonons or a cascade of phonons [Fujii]. In the radiative recombination band of free excitons in the diamond, local maxima are observed according to the following expression:

$$E_{h\nu} = E_g - E_b - E_p \, , \tag{3}$$

where $E_{h\nu}$ is the maximum energy in the band, E_g is the width of the band gap of diamond (5.49 eV), E_b is the binding energy of a free exciton (80 meV), and E_p is the energy of the corresponding phonon. The radiative recombination spectrum of free excitons in pure diamond normally reveals local maxima corresponding to the following phonons of the diamond lattice [Fujii]: transverse acoustic phonon (TA, 87 meV) at 232.8 nm, transverse optical phonon (TO, 141 meV) at 235.2 nm, longitudinal optical phonon (LO, 163 meV) at 236.2 nm, and their cascades TO+LO at 242.7 nm and TO+LO+LO at 250.7 nm. The dominant component is thus the transverse acoustic phonon FE_{TO} at 235.2 nm against the background of which the longitudinal optical phonon at 236.2 nm is not distinguished.

In the PCL spectrum of the synthetic diamond in Fig. 3, the classical pattern of radiative recombination of free excitons is observed.

In the PCL spectrums of the both specimens in Fig. 3, luminescence of the second positive system of molecular nitrogen is detected in the form of a series of bands at 305–395 nm. The luminescence of the second positive system of molecular nitrogen was observed earlier in studies of the luminescence properties of natural spodumene ($LiAlSi_2O_6$ with Mn^{2+} impurity) in [Lipatov et al., 2007 (1)].

The considerable difference in the spectra of the natural and synthetic diamonds owes to the difference in their impurity contents. The ratio between radiative recombination bands of free excitons and luminescence band-A in a diamond allows one to judge the specimen perfection [Kawarada], i.e., as the defect or impurity concentration increases, the exciton luminescence intensity decreases and the band-A intensity increases due to amplification of radiationless recombination of free excitons and amount of sp^2- hybridized carbon bonds. According to [Kawarada], the ratio between the exciton and band-A intensities (FE_{TO}/A) is close to unity at a defect concentration of ~ 10^{17} cm^{-3}.

The natural diamond contains nitrogen in sufficient concentration (~ 10^{18} cm^{-3}) to demonstrate the vibronic luminescence band of N_3V defects. It was assumed that the excitation from N_3V defects is transferred to the band-A centers thus increasing the band-A contribution to the integrated luminescence spectrum [Lipatov et al., 2007 (2)].

In the natural diamond, exciton luminescence bands were observed, but at $FE_{TO}/A < 0.01$ [Dean], i.e., the exciton luminescence under our conditions is impossible to distinguish against the background of the noise component.

3.1 Integrated PCL spectra of natural diamond

Pulsed cathodoluminescence of the natural diamond was excited by electron beams of duration 0.1–10 ns. Figure 4 shows integrated PCL spectra of the natural diamond excited by electron beams of duration 0.1, 2, and 4 ns.

The general PCL characteristics depend almost not at all on the duration of electron beam pulses. The band-A was found to dominate at 300–650 nm and was superimposed by the vibronic N3 system with a ZPL at 415.2 nm and phonon replicas at 420–480 nm. The maximum of the integrated spectra was between 430 and 440 nm.

It is found that the longer the beam duration and the higher the beam current density, the larger the contribution of the N3 system to the integrated PCL spectrum. Presumably, increasing the beam current density increased the duration of luminescence of the N3 system.

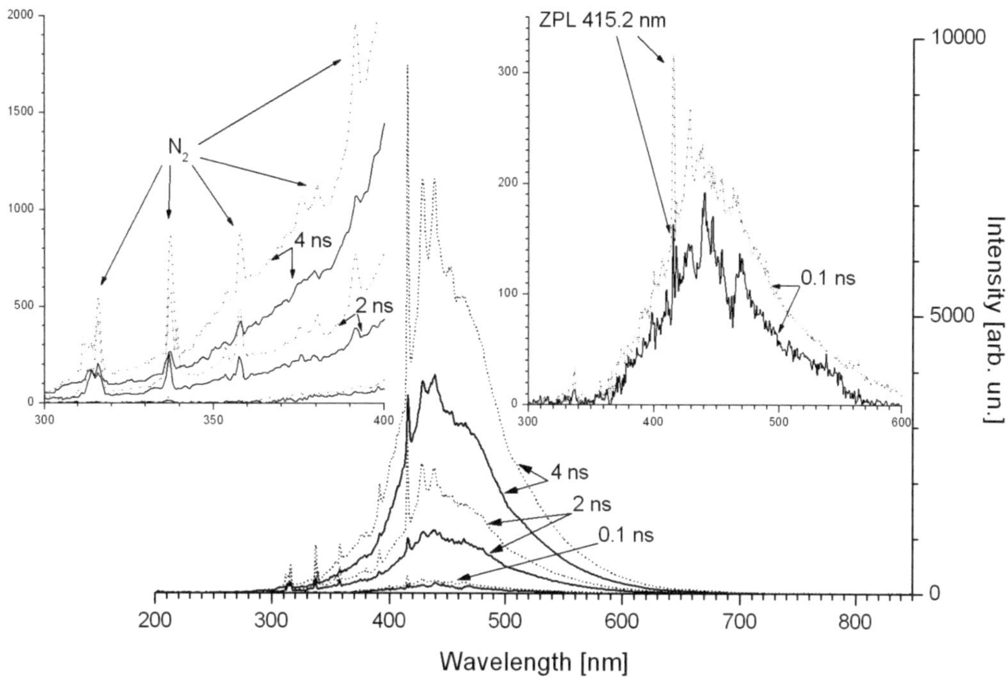

Fig. 4. Integrated luminescence spectra of natural diamond sample at various durations of the electron beam (0.1 ns, 2 ns, 4 ns) at room temperature (continuous lines) and at cooling with liquid nitrogen (broken lines). Zero-phonon line of N_3V-defects (ZPL 415.2 nm) is marked. Inset: the spectral range of luminescence bands of the second positive system of molecular nitrogen in the air, and integral luminescence spectra of natural diamond sample under excitation by the electron beam with a duration 0.1 ns on a larger scale.

Cooling of the natural specimen with liquid nitrogen increased the ZPL intensity at 415.2 nm by a factor of ~ 2.5, ~ 8.8, and ~ 6.8 with a beam of duration 0.1, 2, and 4 ns, respectively. The intensity of phonon replicas of the ZPL of the N3 system increased on cooling with liquid nitrogen by a factor of ~ 2.1, 6.2, and 6.3, respectively. The band-A intensity was nearly the same and increased ~ 1.8 times at all three beam durations.

Thus, the temperature quenching of luminescence was weaker for recombination radiation (the band-A) than for intracenter transitions (the N3 system) [Solomonov].

On cooling of the natural specimen, the intensity of lines of the second positive system of molecular nitrogen also increased. The causes for this phenomenon were discussed for mineral spodumene earlier in [Lipatov et al., 2007 (1)] and are beyond the scope of the present study.

3.2 Time-resolved PCL spectra of natural diamond

The PCL spectra of the natural diamond were measured at room temperature by the procedure illustrated in Fig. 1*b*. Figure 5 shows PCL spectra of the natural diamond excited by an electron beam of duration 0.1 and 10 ns; the spectra were reconstructed at the maxima of light pulses and within 2 ms after the beginning of the pulse.

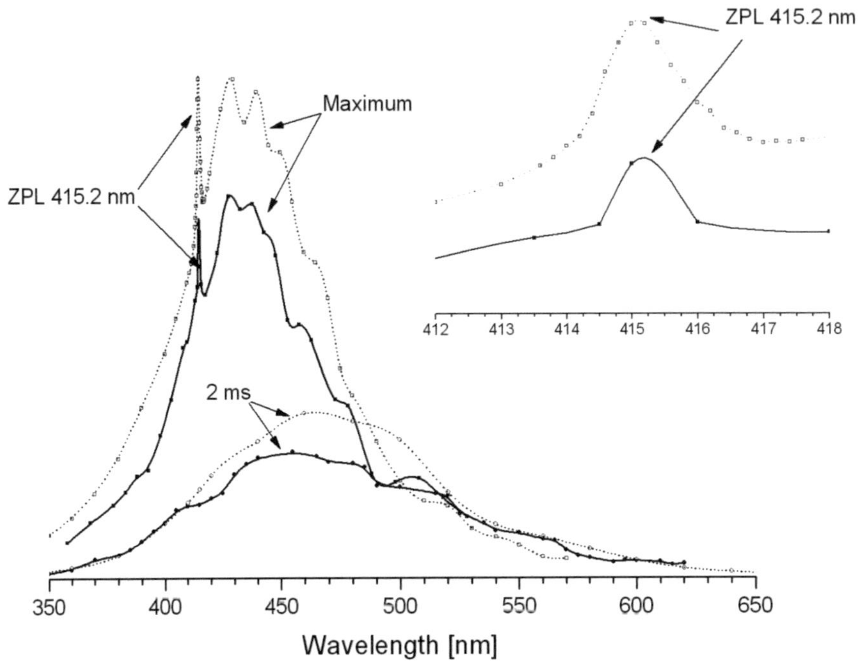

Fig. 5. PCL spectra of natural diamond sample recovered from the amplitudes of the light signal at the maximums, and 2 ms after the onset of a light signal when excited by an electron beam of duration 0.1 ns (continuous lines) and 10 ns (broken line). For ease of comparison, the spectra measured at 2 ms after the start of the pulse, are enlarged in 1000 times. Inset: the spectral range of zero-phonon line 415.2 nm in an enlarged scale. The measurements were performed at room temperature.

With time, the PCL spectrum of the natural diamond underwent considerable transformations. At the maxima of the light pulse (within 10–15 ns after the beginning of the pulse), the PCL spectrum contained an intense band of the N3 system with a ZPL at 415.2 nm and phonon replicas at 420–480 nm.

Within 2 ms after the beginning of the excitation pulse, the PCL band intensity of the natural diamond decreased three orders of magnitude (for convenient comparison, the spectra are magnified by the corresponding factor). In this case, the N3 system escaped detection, whereas the band-A was detected in its undistorted form with a maximum at 450–470 nm.

For two excitation pulse (electron beam) durations of 0.1 and 10 ns at comparable current densities, the reconstructed spectra were almost the same. With a duration of 10 ns, the phonon replicas of the ZPL at 415.2 nm in the spectra of maximum light signals were much more pronounced. With a duration of 10 ns, the ZPL intensity at 415.2 nm was thus 1.6 times higher than that found with a duration of 0.1 ns. These differences were likely to be due to transient processes.

Within 2 ms after the beginning of the excitation pulse, the band-A was the same for both excitation pulse durations.

3.3 Integrated PCL spectra of synthetic diamond

With an excitation pulse duration of 2 and 1.5 ns, the integrated PCL spectra of the synthetic diamond contained a weak band-A, an intense recombination band of free excitons, and lines of the second positive system of atmospheric nitrogen (see Fig. 6).

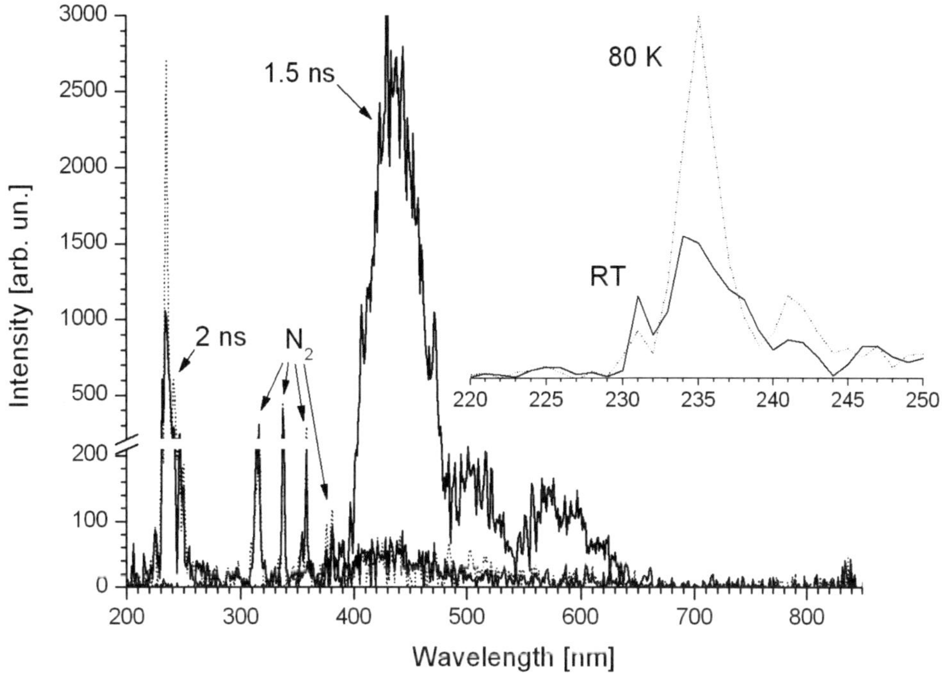

Fig. 6. Integrated luminescence spectra of synthetic diamond at room temperature (RT, continuous lines) and measured at temperature of liquid nitrogen (80 K, broken lines) under excitation by electron beam with a duration 2 ns and 1.5 ns. Spectra of PCL for the duration of the electron beam of 1.5 ns is shown in an enlarged scale. The second positive band system of molecular nitrogen of air is marked. Inset: the spectral range of the band radiative recombination of free excitons (FE_{TO} at 235 nm).

Cooling of the specimen with liquid nitrogen increased the exciton luminescence intensity ~ 2.5 times with a modest increase in band-A intensity.

For observation of the band-A of the synthetic diamond, a setup incorporated an image converter tube with a spectral range of 350–750 nm and excitation pulse duration of 1.5 ns was used [see Lipatov et al., 2007 (1) and references herein]. Note that the thus obtained band-A had a spectral maximum at 430–450 nm, i.e., was shifted by 20 nm toward the short-wave region with respect to the band-A of the natural diamond (see Fig. 5). The short-wave wing of the band-A of the synthetic diamond terminated steeply at 400 nm, whereas that of the natural diamond decreased gradually down to 350 nm.

The long-wave wing of the band-A was observed to 650 nm for both specimens; however, the intensity for the synthetic diamond decreased rapidly from 450 to 480 nm, whereupon it decreased gradually to 650 nm.

It is conceivable that the described distinctions of the band-A for the synthetic diamond owe to the peculiar spectral sensitivity of the multichannel setup with an image converter tube.

The recombination band of free excitons at room temperature (RT) and on cooling with liquid nitrogen (80 K) was similar. All phonon components inherent in the luminescence of free excitons in diamond were observed (FE_{TA} at 232.8 nm, FE_{TO} at 235.2 nm, FE_{LO} at 236.2 nm, FE_{TO+LO} at 242.7 nm, and $FE_{TO+LO+LO}$ at 250.7 nm) [Fujii]. Cooling with liquid nitrogen increased the intensity (2.5 times for the FE_{TO} component) with a decrease in the contribution of the FE_{TA} component.

At 80 K, the intensity of lines of the second positive nitrogen system also increased, as was the case for the natural diamond (see Fig. 4). However at an excitation pulse duration of 0.1 ns, the measured spectrum was free of lines of the second positive nitrogen system (see Fig. 7).

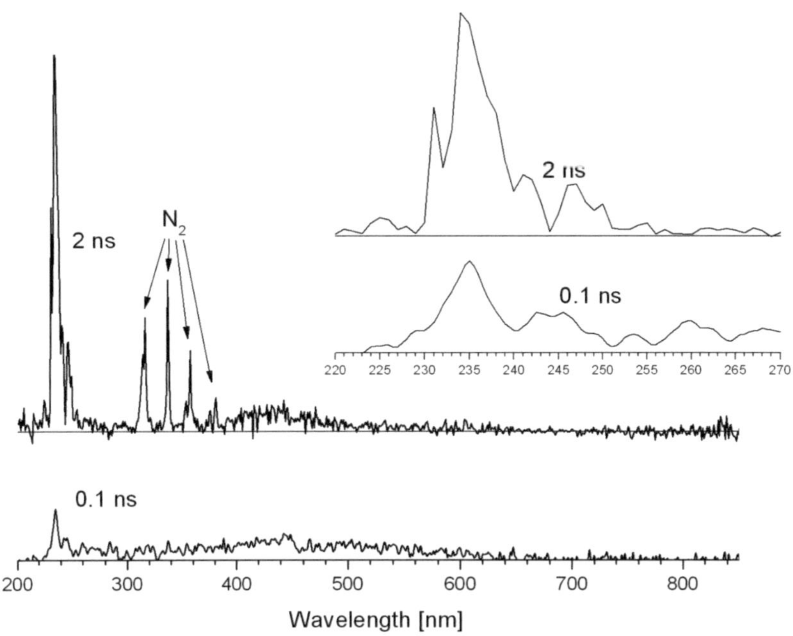

Fig. 7. Integrated luminescence spectra of synthetic diamond sample, measured at room temperature, under excitation by electron beam with durations 0.1 ns and 2 ns. The second positive band system of molecular nitrogen of the air is marked. Inset: the spectral range of the band radiative recombination of free excitons (FE_{TO} at 235 nm).

The recombination luminescence bands of free excitons with excitation pulse durations of 0.1 and 2 ns were distinct (Fig. 7). So in both cases, the FE_{TO} component of exciton luminescence of the diamond at 235.2 nm dominates; however, with an excitation pulse duration of 0.1 ns, the FE_{TA} component at 232.8 nm is not distinguished and mixing of multiphonon components with a common maximum at 242–246 nm in the long-wave region of the band is found. The difference of the radiative recombination band of free excitons in the diamond from the classical form at an excitation pulse duration of 0.1 ns presumably owes to transient processes.

At both excitation pulse durations, the band-A was weak and structureless.

The intensity of the FE_{TO} component of exciton luminescence of the synthetic specimen increased nonlinearly with increasing the beam current density, as evidenced in Fig. 8. Approximation of the dependence by a power function gave an exponent of 1.4 (from 1.2 to 1.6 within confidence intervals). The nonlinear dependence of the output PCL on the electron beam current density made the radiative recombination band of free excitons undetectable against the background of the noise component in the low-impurity diamond at low excitation intensities. Apparently, this effect impeded the observation of exciton luminescence in the synthetic diamond excited by spontaneous UV sources (excilamps), whereas the excitation of photoluminescence by a pulsed KrCl-laser allowed observation of an intense FE_{TO} component of exciton luminescence [Lipatov et al., 2010].

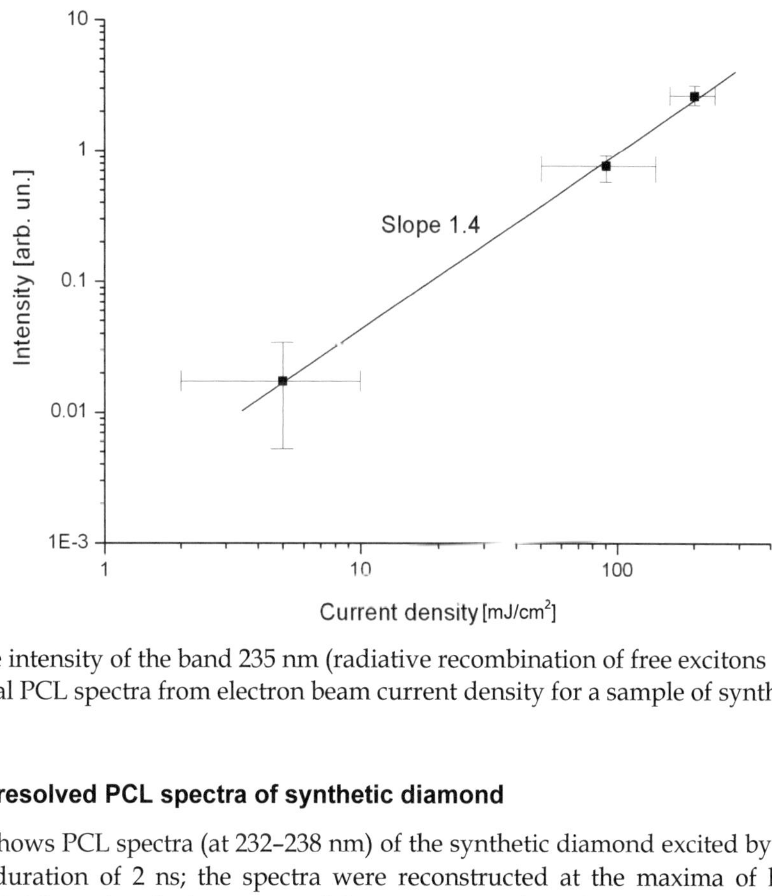

Fig. 8. The intensity of the band 235 nm (radiative recombination of free excitons FE_{TO}) in the integral PCL spectra from electron beam current density for a sample of synthetic diamond.

3.4 Time-resolved PCL spectra of synthetic diamond

Figure 9 shows PCL spectra (at 232–238 nm) of the synthetic diamond excited by an electron beam of duration of 2 ns; the spectra were reconstructed at the maxima of light pulses (within 0.6–0.9 ns after the beginning of the excitation pulse); the measurements were taken at room temperature and at 80 K.

The spectra reveal only the FE_{TO} component at 235.2 nm and the other phonon components present in the integrated luminescence spectra of this diamond (see Figs. 6 and 7) escape detection at an excitation pulse duration of 2 ns. The absence of phonon components of exciton luminescence, except for the FE_{TO} component, in the time-resolved PCL spectrum of the synthetic diamond confirms the effect of transient processes on the luminescence of phonon components the assumption of which was made in analyzing the integrated PCL spectrum of the diamond excited by an electron beam of duration 0.1 ns (Fig. 7). So the integrated PCL spectrum was free of the FE_{TA} component at 232.8 nm and mixing of multiphonon components with a common maximum at 242–246 nm was observed in the spectrum.

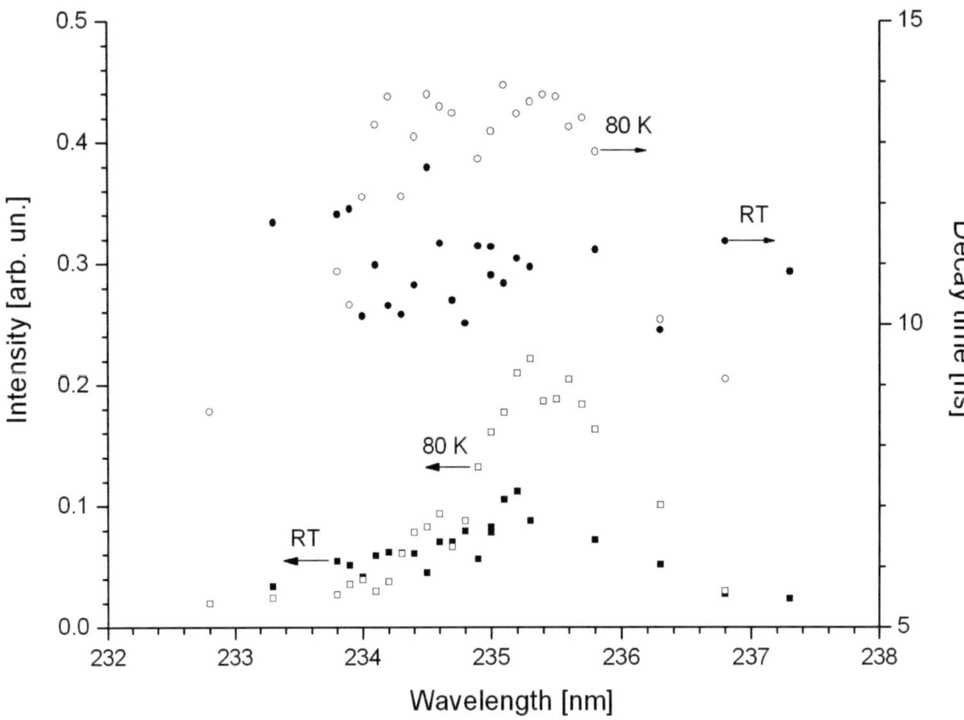

Fig. 9. Luminescence spectra and the decay time PCL of synthetic diamond sample in the spectral range of the band radiative recombination of free excitons (FE_{TO} at 235 nm). The intensity of luminescence is the maximum amplitude of the light pulse, which corresponds to the time 1 ns after the beginning of the excitation pulse. Duration of the electron beam was 2 ns. The decay time was calculated from the approximation of relaxation curve by an exponential function. The measurements were performed at room temperature (RT, full symbols) and at cooling with liquid nitrogen (80 K, empty symbols).

Cooling of the diamond to 80 K increased the intensity of the FE_{TO} components two times with a decrease in the contribution of the short-wave wing of the exciton band, as is the case in the integrated spectrum in Fig. 6.

The PCL spectra of synthetic diamond sample excited by the electron beam of duration of 10 ns are shown in Figure 10. The spectra were reconstructed from the amplitudes of the light pulse, measured at 0.5, 1 and 2 ms after the start of the excitation pulse. As we can see, the band-A with maximum about 450 nm in this time range is dominant in the spectrum of synthetic diamond; in addition, is registered a weak peak at 415 nm.

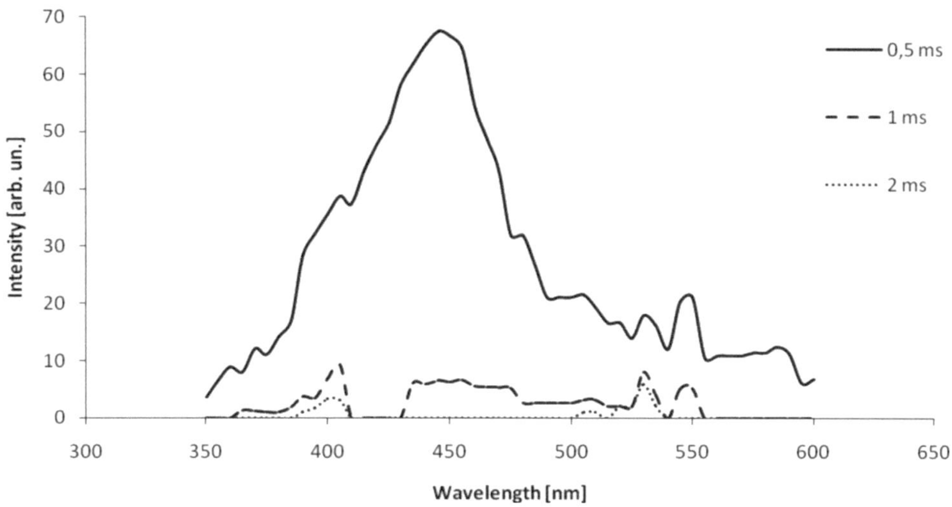

Fig. 10. PCL spectra of synthetic diamond sample under excitation by electron beam with a duration 10 ns, shown at 0.5, 1, 2 ms after the beginning of the excitation pulse.

3.5 PCL decay kinetics in natural and synthetic diamonds

The PCL decay kinetics in the diamonds was studied at an electron beam duration of 0.1–10 ns by the procedure shown in Fig. 1b. The electron accelerator was operated in two modes: a single-pulse mode at a FWHM of 0.1–10 ns and a two-pulse mode at a FWHM of 0.65 ns with a 3-ns interval. Figure 11 shows oscillograms of the electron beam current at a beam duration of 0.2 and 0.65 ns (two-spike mode). The same figure presents oscillograms of PCL at 300–650 nm (the region of the band-A and N3 system) with no spectral resolution (for which we removed the monochromator 6 from the circuit shown in Fig. 1b and replaced the photomultiplier tube 7 by the photodiode for the natural (atop) and synthetic (at the middle) diamonds.

For the natural diamond, the luminescence duration at 300–650 nm was an order of magnitude longer than that for the synthetic diamond. In the two-pulse excitation mode, the luminescence duration was also much longer than that in the single-pulse mode (Fig.11).

For the synthetic diamond, the light pulse decay time is nearly the same in the single- and two-pulse excitation modes, i.e., in PCL excitation by an electron beam of duration 0.25–0.65 ns, the light pulse duration was constant and was ~ 2 ns (Fig.11, middle).

Figure 9 shows PCL decay spectra of free excitons in the synthetic diamond excited by a pulse of duration 2 ns at room temperature and at 80 K; the spectra were obtained through approximation of the light pulse decay time in the single-pulse excitation mode by a first-order exponential function. At room temperature, the decay time was nearly constant at 233–237.5 nm and was 10–13 ns.

Cooling to 80 K caused an increase of the PCL decay time of the excitons to 13-14 ns in spectral range 234-236 nm. Outside this range the PCL decay time decreased to 8-11 ns.

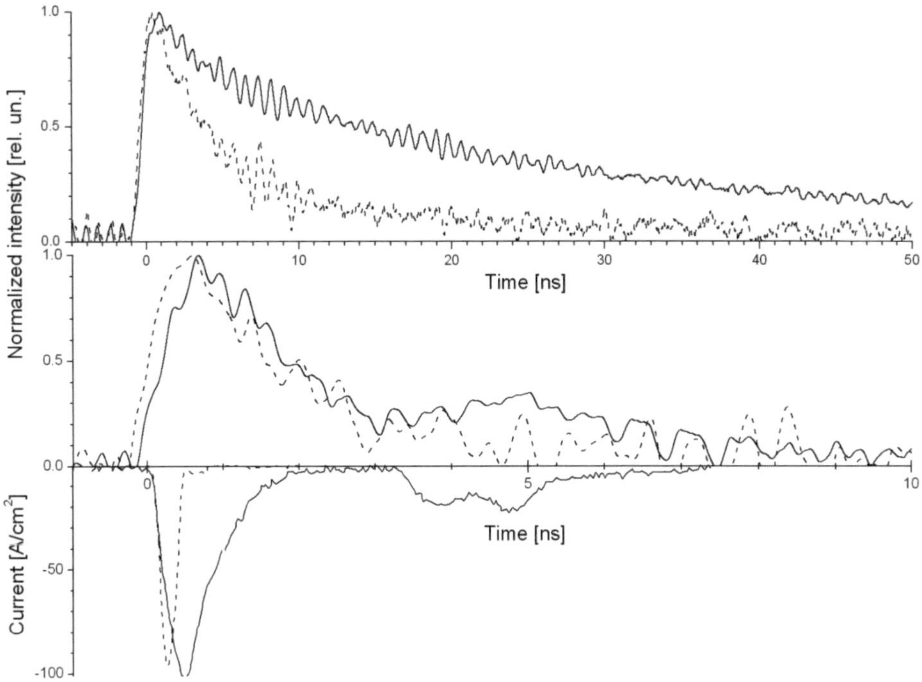

Fig. 11. Oscillograms of the luminescence of natural (atop) and synthetic (middle) diamond samples under electron beam excitation at single-pulse operation (duration of pulse 0.25 ns, broken curves) and two-pulse operation (duration of pulse 0.65 ns, continuous curves). Oscillograms of the beam current duration 0.25 ns (broken curve) and 0.65 ns (continuous curves) are shown down.

The PCL decay in the natural diamond was studied in the single-pulse mode at three excitation pulse durations: 0.1, 2, and 10 ns. The PCL decay spectra of the natural diamond are shown in Fig. 12.

In the nanosecond range, the decay time was 15–60 ns. With all excitation pulse durations, an increase in PCL decay time to 40–45 ns was observed in the vicinity of the ZPL at 415.2 nm. In the region of phonon replicas of the ZPL at 415.2 nm (420-480 nm), the PCL decay time increased to 35–45 ns with an excitation pulse duration of 0.1 and 10 ns and to 45–60 ns with an excitation pulse duration of 2 ns.

Thus, in the nanosecond range the N3 system dominates, and this is evidenced by the PCL spectrum (Fig. 5) and the spectra of decay time (Fig. 12).

The electron beam current densities at a beam duration of 0.1 and 10 ns were comparable, as can be seen from Table 1. However with an excitation pulse duration of 2 ns, the current density increased several-fold and the PCL decay time in the nanosecond range also increased compared to the decay times with excitation pulse durations of 0.1 and 10 ns.

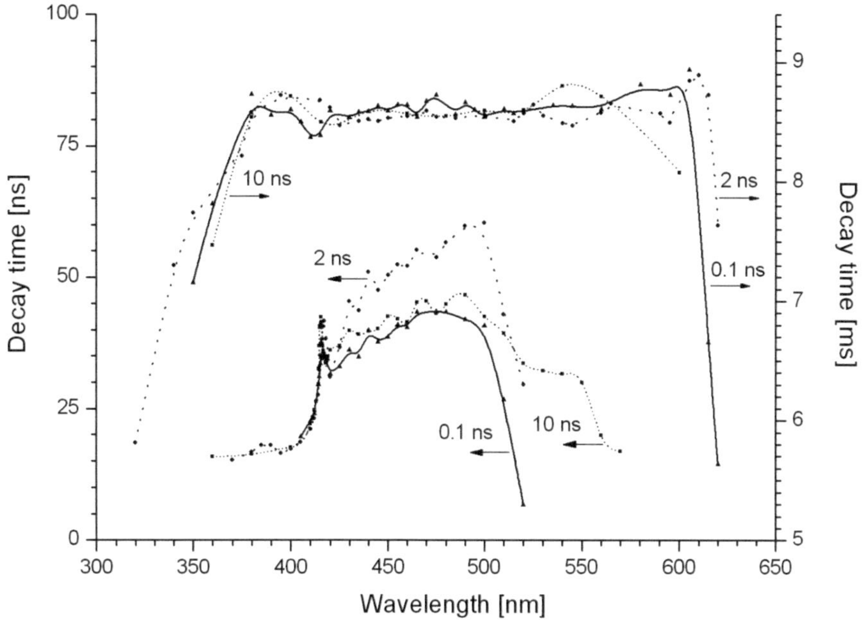

Fig. 12. The spectra of decay time of PCL of natural diamond sample under excitation by electron beams of duration 0.1, 2 and 10 ns at room temperature in nano- and millisecond ranges.

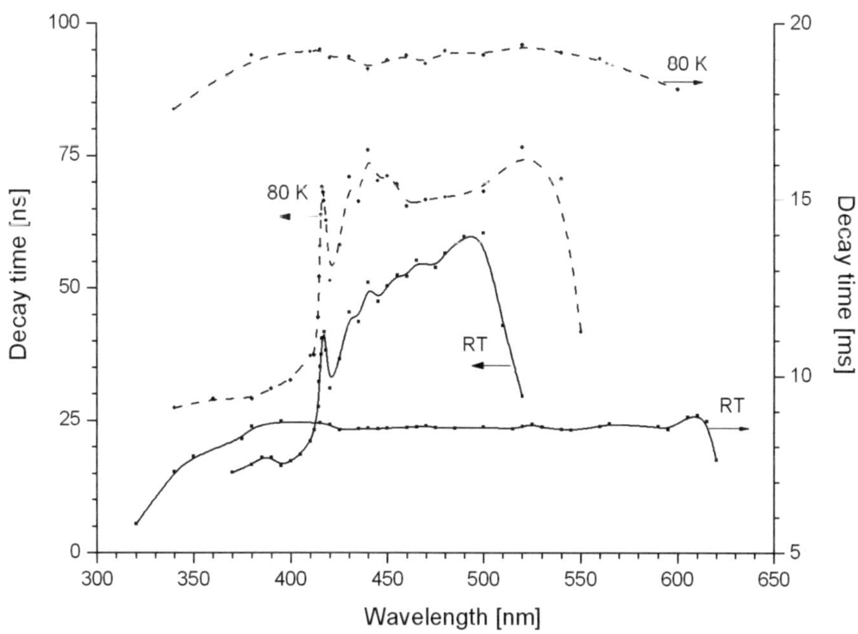

Fig. 13. The spectra of decay time of PCL of natural diamond sample under excitation by electron beams of duration 2 ns at room temperature (continuous curves) and at cooling with liquid nitrogen (80 K, broken curves) in nano- and millisecond ranges.

Note that increasing the excitation intensity (the beam current) increases the luminescence duration of the N3 system, and this is confirmed by the increase in PCL duration in the two-pulse mode (Fig. 11) compared to that in the single-pulse mode.

In the millisecond range (Fig. 13), the PCL decay time in the natural diamond was 8.5–9 ms in the spectral luminescence region of the band-A and decreased steeply beyond this region. The PCL decay time thus did not depend on the excitation pulse duration. Actually in the millisecond range, only the luminescence of the band-A was observed in the natural diamond and that of the N3 system was entirely absent (see Figs. 5 and 12 see).

Cooling of the natural diamond to 80 K increased the decay time 1.8-2 times both for the N3 systems (nanosecond range) and for the band-A (millisecond range).

Figure 14 shows the decay curves of PCL at a wavelength of 235 nm for samples of natural and synthetic diamond under excitation by electron-beam duration of 10 ns. Rise time of light pulse is the same for both samples. Decay times are slightly different: about 25 ns for the synthetic sample and 45 ns for the natural sample.

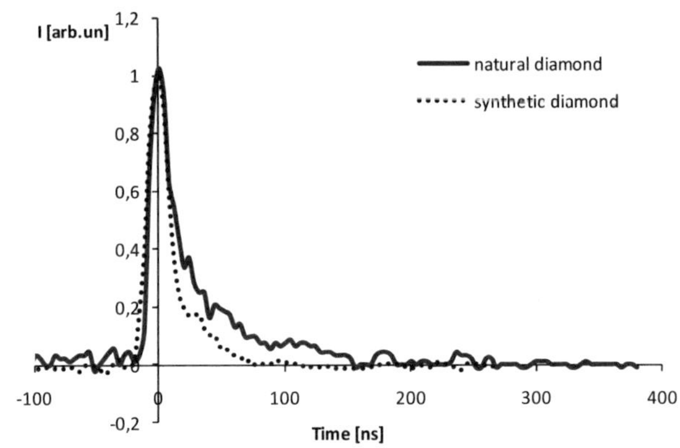

Fig. 14. Decay curve of PCL of natural and synthetic diamond sample at 235 nm under excitation by electron beams of duration 10 ns at room temperature.

Studies of the decay kinetics of 415 nm (Fig. 15) under excitation of samples by electron pulse with a duration of 10 ns showed that in this spectral range the decay time of PCL for synthetic sample is also smaller than for natural sample (16 and 50 ns, respectively) under identical conditions of excitation. That difference of the decay time may be due to the presence the channel of nonradiative energy transfer from the excited radiative levels in crystals of synthetic diamond.

Luminescence decay times of band-A in natural and synthetic diamond samples are also different (Fig. 16) under excitation by electron pulse of duration 10 ns. PCL decay in the natural sample has been slower, τ is about 50 ns, whereas in the synthetic one is about 10 ns.

Decay kinetics of PCL for emission 235 nm, 415 nm, 450 nm when changing the excitation density from 6 to 300 mJ/cm^2, were investigated. It is established that the decay kinetic of luminescence at 235 and 415 nm has changes weak with increasing density of excitation for both natural and synthetic diamond samples. The decay time is independent of excitation density.

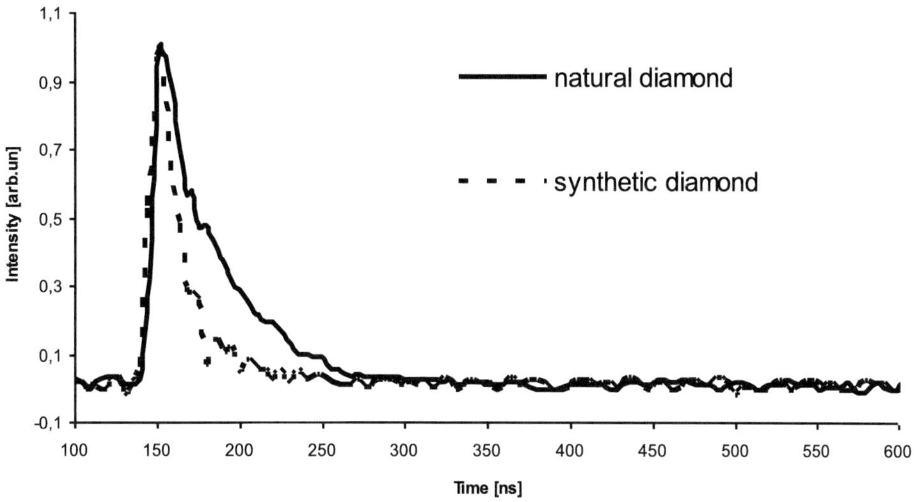

Fig. 15. Decay curve of PCL of natural and synthetic diamond sample at 415 nm under excitation by electron beams of duration 10 ns at room temperature.

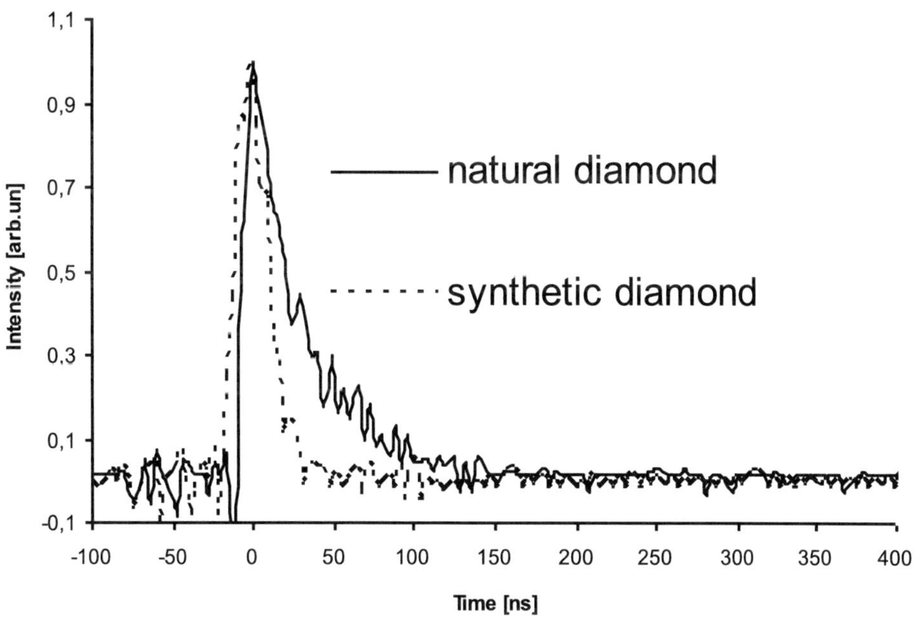

Fig. 16. Decay curve of PCL of natural and synthetic diamond sample at 450 nm under excitation by electron beams of duration 10 ns at room temperature.

Investigation of the decay of luminescence at band 450 nm showed that the decay time of PCL in nanosecond range for synthetic diamond does not depend on the excitation density (Fig. 17). It was found that the decay time of PCL for natural sample increases from 12 to 50 ns with increasing excitation density. Dependence is shown in Figure 17. This feature is characteristic for the recombination luminescence.

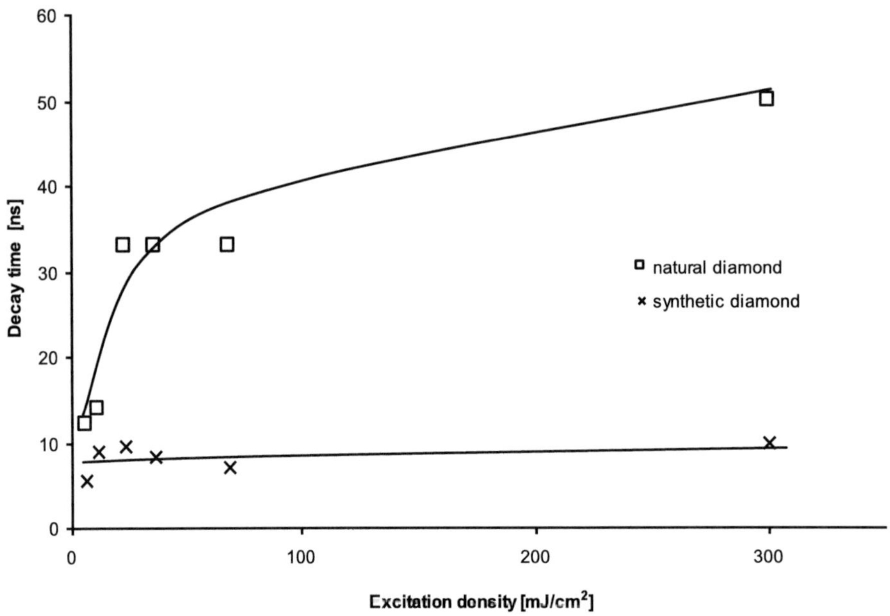

Fig. 17. The dependence of the decay time of PCL in the band 450 nm from excitation density for a sample of natural and synthetic diamond. The sample was excited by the electron beam of duration 10 ns.

4. Conclusion

In the work, we studied the integrated and time-resolved spectral-kinetic characteristics of pulsed cathodoluminescence (PCL) of natural and synthetic diamonds. The PCL was excited by electron beams of duration from 0.1 to 10 ns and current density from 10 to 200 A/cm^2. According to the physical classification of diamonds, the diamonds under study were of type 2a, i.e., their optical transmission spectra revealed a clearly defined fundamental absorption edge at 225 nm, transparency from the UV to IR range, and absence of resolvable features.

The PCL of the natural diamond displayed two luminescence bands differing in nature, spectral composition, and time response. In the nanosecond range, the vibronic N3 system due to N_3V defects dominated. The N3 system displayed a ZPL at 415.2 nm and phonon replicas at 420–480 nm. The characteristics decay time for the N3 system was 15–65 ns. In the vicinity of the ZPL at 415.2 nm, an increase in decay time to 45 ns was observed. In the region of phonon replicas of the ZPL, an increase in decay time to 40–65 ns was also found. Increasing the electron beam current density increased the intensity of the ZPL at 415.2 nm and its phonon replicas. The N3 system was reliably detected in the integrated and time-resolved PCL spectra, though it was not resolved in the absorption spectra of this diamond.

In the millisecond range, the N3 system escaped detection and the PCL spectra of the natural diamond revealed a broad structureless luminescence band-A at 350–650 nm with a decay time of 8.5–9 ms being dependent neither on the excitation pulse duration, nor on the electron beam current density.

Cooling of the diamond increased the intensity and the decay time two-fold at all excitation pulse durations and beam current densities. The N3 system thus revealed a double increase in decay time. The intensity of the ZPL at 415.2 nm and its phonon replicas also increased on cooling. The higher the current densities, the more considerable the increase in intensity was.

The synthetic diamond also revealed the luminescence band-A in the PCL spectra. However, its intensity was too weak to study its time characteristics. The band-A was structureless in its spectral composition and was in the same spectral range as the band-A in the natural diamond.

The spectra of the synthetic diamond were dominated by a radiative recombination band of free excitons with several single-phonon and multiphonon components in which the dominating component was the FE_{TO} component at 235.2 nm. At excitation pulse durations of 0.1–2 ns, this component was the main one, and the FE_{TA}, FE_{LO}, FE_{TO+LO}, and $FE_{TO+LO+LO}$ components (at 232.8, 236.2, 242.7, and 250.7 nm, respectively) were affected by transient processes.

The intensity of the exciton FE_{TO} component depended on the beam current density according to the power law with an exponent of ~ 1.4.

Cooling of the synthetic diamond increased the PCL intensity and decay time by a factor of 1.3–1.6 with a modest increase in band-A intensity. Cooling of the diamond also froze out the phonon FE_{TA} component of exciton luminescence.

It is shown that the decay time of PCL at band-A depends on the excitation density and has changes from 12 ns (at 6 mJ/cm^2) to 50 ns (at 300 mJ/cm^2) for a sample of natural diamond.

5. References

Baksht E. K.; Burachenko A.G. and Tarasenko V.F. (2010) *Tech. Phys. Lett.*, Vol.36, pp. 1020-1023, ISSN 0320-0116.

Dean P.J., Jones I.H. Recombination radiation from diamond (1964) *Physical review*, Vol.133, №6A, pp.A1698-A1705.

Fujii A., Takiyama K., Maki R., Fujita T. (2001) Lifetime and quantum efficiency of luminescence due to indirect excitons in a diamond. *Journal of luminescence*, 94-95, pp.355-357. ISSN 0953-4075

Gritsenko B.P.; Lisitsyn V.M., Stepanchuk V.N. (1981) *Phys. of the Solid State*, Vol.23, pp. 393-396. ISSN 0367-3294

Kawarada H.; Tsutsumi T.; Hirayama H.; et al. (1994) *Appl. Phys. Lett.*, Vol.64, pp. 451-453. ISSN 0003-6951.

Korepanov V. I.; Lisitsyn V. M. and. Oleshko V. I. (2000). *Russian Physics Journal*, Vol.43, pp. 185-192. ISSN 0021-3411.

Korepanov V.I.; Vil'chinskaya S.S.; Lisitsyn V.M.; Kuznetsov M.F. (2005) *Optics and Spectr.*, Vol.98, pp. 401-404. ISSN 0030-4034.

Kravchenko V.D.; Lisitsyn V.M.; Yakovlev V.Yu. (1985) *Phys. of the Solid State.*, Vol.27, pp. 2181-2183. ISSN 0367-3294

Marfunin A.S.(1975) *Spectroscopy, Luminescence and Radiation Centers in Minerals*; Nedra: Moscow, 327 p. (in Russian).

Lipatov E.I.; Tarasenko V.F.; Orlovskii V.M.; Alekseev S.B. and Rybka D.V. (2005) *Tech. Phys. Lett.* Vol.31, pp. 231–232. ISSN 0320-0116.

Lipatov E.I., Tarasenko V.F., Orlovskii V.M., Alekseev S.B. (2005) Luminescence of crystals excited by a KrCl laser and a subnanosecond electron beam. *Quantum electronics*, Vol.35, №8, pp. 745-748. ISSN 0368-7147

Lipatov E.I., Orlovskii V.M., Tarasenko V.F., Solomonov V.I. (2007) Comparison of luminescence spectra of natural spodumene under KrCl laser and e-beam excitation. *Journal of luminescence*, Vol.126, №2, pp.817–821. ISSN 0953-4075.

Lipatov E.I.; Lisitsyn V.M.; Oleshko V.I.; and Tarasenko V. F. (2007) *Russian Phys. J.*, Vol.50, pp. 52-57. ISSN 0021-3411.

Lipatov E.I., Avdeev S.M., Tarasenko V.F. (2010) Photoluminescence and optical transmission of diamond and its imitators. *Journal of luminescence*, 130, pp.2106-2112. ISSN 0953-4075

Lisitsyna L.A.; Korepanov V.I.; Lisitsyn V.M. (2002). *Phys. of the Solid State.* Vol. 44, pp. 2235-2239. ISSN 0367-3294

Lisitsyna L.A.; Oleshko V.I.; Putintseva S.N.; Lisitsyn V.M. (2008) *Optics and Spectr.* Vol. 105, pp. 531-537. ISSN 0030-4034

Lisitsyna L.A.; Korepanov V.I.; Lisitsyn V.M. Eliseev A.E. at al. (2011) *Optics and Spectr.* Vol.110, pp. 568-573. ISSN 0030-4034

Solomonov V. I. and Mikhailov S. G. (2003) *Pulsed Cathodoluminescence and Its Application to Analysis of Condensed Substances*; Publisher: UrO RAN, Yekaterinburg, p. 1-181.

Takeuchi D.; Watanabe H.; Yamanaka S.; et al. (2001) *Physical review B.* Vol. 63, 245328. ISSN 0163-1829

Tarasenko V.F.; Baksht E.K.; Burachenko A.G.; Kostyrya I.D.; Lomaev M.I. and. Rybka D.V. (2008) *Plasma Devises and Operation.* Vol.16, pp.267-298. ISSN 1051-9998

Tarasenko V.F.; Baksht E.K.; Burachenko A.G.; Kostyrya I.D.; Lomaev M.I. and Rybka D.V. (2009) *IEEE Trans. of Plasma Science.* Vol.37, pp.832-838. ISSN 0093-3813.

Tarasenko V. F.; Baksht E. K.; Burachenko A.G.; Kostyrya I. D., Lomaev M. I. and Rybka D. V. (2010) *IEEE Transactions on Plasma Science.* Vol.38, 741-750. ISSN 0093-3813.

Zaitsev A.M. (2001) *Optical properties of diamond*; Springer:Berlin; 502 p. ISBN 354066582X

Zheltov K.A. (1991) *Pikosekundnye sil'notochnye elektronnye uskoriteli* (Picosecond High-Current Electron Accelerators). Energoatomizdat. Moscow, Russia, 1991. 114 p. ISBN: 5-283-03978-1

4

Study of Defects by Cathodoluminescence Measurements

A. Djemel[1] and R-J. Tarento[2]
[1]*LPCS, Université. Mentouri Constantine,*
[2]*LPS, Université. Paris-Sud, Orsay,*
[1]*Algeria*
[2]*France*

1. Introduction

Scanning electron microscopy (SEM) can be used to obtain images of a large variety of materials resulting from secondary electron. The SEM can also be used to detect different signals that provide composition information , surface morphology and characteristics of the local electronic structure. The signals provide different processes in the electron-matter interaction. The combination of different signals produces an detailled physical image and qualitative and quantitative analysis of the electronic properties of studied samples.

The cathodoluminescence (CL) is one of different signals that has frequently been used within SEM to study the semiconductor materials. This technique avoids destruction of the sample ,has a high resolution and an image depth field which are determined by the beam current , the beam energy and the beam size.

This method has been used to investigate and to identify the particular features of the crystal defects(dislocation,precipitate,boundaries (Djemel , Castaing et al ,1990); Djemel , Castaing et al ,1992).

The CL method allows the determination of quantitative information on local electronic and optical parameters of materials such as the diffusion length (L) ,absorption coefficient (α) , the dopping levels and the defects parameters such as the recombination velocity (V_s) ,the defects density (N_t) ,the capture cross section (σ) and the energy level (E_t) associated to defects.

The cathodoluminescence is based on the study of the interaction of the electron beam with the semiconductor. This interaction gives rise to electron-hole (e-h) pairs generation within the sample . The density of e-h pairs generated is limited by the scattering process of the electron beam within the sample. The distribution of the e-h pairs created depends on the diffusion length and the recombination behaviour at the surface and in the bulk. The recombination can be either radiative in which case a photon is emitted or non-radiative generating phonons , Auger electron ....At the surface ,the recombination process is generally non-radiative. The generated photon submits to an absoption in the escape from the sample and those that finally emerge can be collected and subsequently detected to provide the cathodoluminescence signal . The radiative recombination results from band to band transition and band to energy level transition. The energy level is associated with impurities or with crystal defects.

2. Quantitative study

The quatitative determination of CL requires an accurate simulation of the CL signal as a function of the electron beam paramaters , the beam current (I_p) and energy beam (E_0). However, this depends on various stages :

- a better description of the electron-semiconductor interaction
- the diffusion and the distribution of the generated (e-h) pairs within the sample
- the boundary conditions and the recombination at the surface and in the bulk.

2.1 Generation function

Many models can describe the electron-semiconductor interaction : the polynomial form (Everhoff&Hoff,1971) , the point or spherical models , the modified Gaussian approximation (Wu & Wittry,1978 ; Hergert , Reck et al,1987) and the Monte Carlo method (Phang, Pey et al ,1992).

For GaAs, the electron-semiconductor interaction is approximeted by the modified Gaussian function which gives the local generation rate G(z)

$$G(z) = (\rho/R)\,\varphi(u) \tag{1}$$

$$\varphi(u) = A\exp[-(u-u_0)^2/\Delta u] - B\exp-(bu/u_0) \tag{2}$$

where $u = \rho z/R$, ρ is the density of semiconductor (g/cm^3) ,z is the depth (cm) and R is the maximum electron range (g/cm^2).R has been deduced for GaAs as e function of the beam energy E_0 (keV):

$$R = 2.56.10^{-3}\,(E_0/30)^{1.7} \tag{3}$$

For GaAs , $u_0 = 0.125$, $\Delta u = 0.350$, $b = 4.0$, $B/A = 0.4$.The constant A and B are determined using the normalization condition.(Wu & Wittry,1978).

2.2 Diffusion and distribution

Consider a n-type semiconductor doped with concentration N_d .

The transport of (e-h) pairs generated has been controlled by the continuity equation which has the following classic form :

$$\text{div}J = G(z) - r(z) \tag{4}$$

where J is the flux of the carrier excess, r(z) is the recombination rate in the bulk and at the semiconductor surface.

The recombination rate in the bulk (neutral region) is expressed by:

$$r(z) = \Delta p(z)/\tau_p \tag{5}$$

where $\Delta p(z)$ is the excess hole carrier and τ_p is the hole lifetime .

In the litterature there are two definitions of the semiconductor surface . In the first definition, the surface is defined as a dead-layer (non-radiative region) with a thickness Z_T (Hergert , Reck et al 1987 ; Wittry & Kyser 1967).

The distribution of minority excess carriers is determined by the continuity equation in the bulk or neutral region:

$$D_{p)} d^2\Delta p(z)/dz^2 = G(z) - \Delta p(z)/\tau_p \quad (6)$$

In the second definition , the surface is described by a defects density N_t and an energy level E_t in the band gap (Djemel , Tarento et al , 1998).The consequence of electronic surface states associated to surface defects consists in the existence of a depletion region that is linked to a potential barrier E_b between the free semiconductor surface and the bulk.This second definition of surface is used in cases of dislocation (Tarento & Marfaing , 1992) and grain boundary (Oualid , Singal et al , 1984).

The distribution of excess carrier is governed by :

- the continuity equations of both excess majority and minority carrier in the depletion region (the recombination is neglected in the region r(z)=0) (Djemel , Tarento et al ,1998) :

$$-d^2\Delta n(z)/dz^2 + 2\alpha(z-Z_d)\, d\Delta n(z)/dz + 2\alpha\Delta n(z) = G(z)/D_n \quad (7)$$

$$-d^2\Delta p(z)/dz^2 - 2\alpha(z-Z_d)\, d\Delta p(z)/dz - 2\alpha\Delta p(z) = G(z)/D_p \quad (8)$$

Using the Einstein relation ($D/\mu = kT/e$), α becomes equal to $N_d\, e^2/2kT\varepsilon$, D and μ are the diffusion coefficient and mobility of carriers and ε is the electric permitivity.

- and by the continuity equation of excess minority carrier in the neutral region (in the bulk)

$$D_p\, d^2\Delta p(z)/dz^2 = G(z) - \Delta p(z)/\tau_p \quad (9)$$

2.3 Surface recombination

The recombination rate at the semiconductor surface is treated by two methods.The free surface of semiconductor is defined as a dead-layer (non-radiative region) with a thickness Z_T and a surface recombination velocity V_S (Hergert , Reck et al , 1987 ; Wittry & Kyser , 1967). In this case V_S is found from the condition:

$$D_P\, [d\Delta p(z)/dz]_{z=0} = V_S\, \Delta p(0) \quad (10)$$

In the second method ,the free surface of semiconductor is described by defects density N_t and an energy level E_t in the band gap with an occupation probability f given by the Shockley-Read-Hall theory.

$$f = [\Delta n(0) + n_0 + n_i \exp((E_i - E_t)/ kT)]/[\Delta n(0) + n_0 + \Delta p(0) + p_0 + 2n_i \cosh((E_t - E_i)/kT)] \quad (11)$$

n_0 , n_i , p_0 , E_i are defined in (Djemel , Nouiri et al ,2002).

$\Delta n(0)$, $\Delta p(0)$ are the excess carriers at the surface (z=0) of the electron and hole respectively.

The recombination rate U(0) at the surface (z=0) is given by:

$$U(0) = C\,N_t\,(\Delta n(0)\,\Delta p(0) + p_0\,\Delta n(0) + n_0\,\Delta p(0)) \,/\, (n_t + n_0 + \Delta n(0) + p_t + p_0 + \Delta p(0)) \quad (12)$$

Where $C = \sigma V_{th}$ is the capture coefficient of electron and hole (σ is the capture cross section which is generally linked to both defect and the environment (Bourgoin & Corbett ,1972), and V_{th} is the thermal velocity) .

n_t , p_t , are given in Ref (Djemel , Nouiri et al , 2004).

The relation between V_S (first definition of surface) and (N_t , E_t) (second definition of surface) can be established (Djemel , Nouiri et al , 2002).

2.4 Signal of Cathodoluminescence (CL)

The CL signal is calculated from the excess minority carrier in the neutral region (in the bulk).The photons are generated within the sample when the carriers recombine radiatively. It is assumed that the CL signal is proportional to the integral of the carriers excess over the generation volume. The main optical loss mechanism taken into account is the absorption within the sample. The attenuation of photons propagation towards the surface is given by an exponential law.

In the case where the suface is considered as a dead-layer with a thickness Z_T , the CL signal is given by: (Hergert , Hildebrandt et al , 1987 ; Hildebrandt , Schreiber et al ,1988) .

$$I_{CL}\ \alpha \int (1/\tau_r)\ \sin\theta_c\ F(\alpha_c)d\theta\ ;\ \alpha_c = \alpha/\cos\theta \quad (13)$$

$$F(x) = \tau\, G_0 \exp(-\alpha Z_T)\ [\ \varphi(x,Z_T) - (Lx+S)\ \varphi(1/L\,,Z_T)/(1+S)]/(\,1 - x^2L^2) \quad (14)$$

$$\varphi(x\,,Z_T) = \int \exp(\,-\alpha z\,)\ G(z + Z_T)dz \quad (15)$$

where θ_c denote the critical angle of the total reflexion at the surface , G_0 is the total generation rate, τ and τ_r are the total and radiative lifetimes , respectively , L is the diffusion length , α is the optical absorption coefficient , S is the reduced surface recombination velocity.

In the second definition where the surface is described by defects density N_t and an energy level E_t in the band gap ,the CL signal is written as follow (Djemel , Kouissa et al , 2008)

$$I_{CL}\ \alpha \int [\ \Delta p(z)/\tau_p\,]\ \exp(\,-\alpha_b\,z\,)\ \eta\ dz \quad (16)$$

τ_p is the lifetime of excess minority carriers , α_b is the optical absorption coefficient that depends on the wavelength , η is the efficiency of radiative recombination and written as (Yacobi & Holt , 1990) :

$$\eta = \tau_{nr} \,/\, (\,\tau_r + \tau_{nr}\,) \quad (17)$$

τ_r and τ_{nr} are the lifetimes of radiative and non-radiative recombination, respectively .

$\Delta p(z)$ is the minority carriers excess solution of equations (7) , (8) and (9).

$$\Delta p(z) = B_p \exp\left[-(z - Z_d)/L_p\right] + \\ + L_p / 2D_p \int G(z) \left\{ \exp\left(-\left|z - Z_d\right| / L_p\right) - \exp\left[-\left(z + z' - 2Z_d\right) / L_p\right]\right\} dz \quad (18)$$

$B_p = \Delta p(z = Z_d)$ is a constant and represents the excess carrier at the limit between neutral region and depletion region .

The influence of parameters of electron beam (beam intensity I_p and energy beam E_0) , of parameters of semiconductor (N_d , L_p , α_b) and of parameters of surface (V_s , Z_T) or (N_t , Et, σ) on the CL signal has been intensively studied (Wittry & Kyser , 1967 ; Djemel , Nouiri et al ; 2002;Ben Nacer,Matoussi et al ,2009).

On the other hand,there are correlations between the theoretical CL signal and CL measurement to determine the parameters of surface (V_s , Z_T) (Hergert , Reck et al , 1987) or (N_t ,Et ,σ) (Djemel , Nouiri et al , 2000) .

To illustrate the CL calculation where the surface is defined by N_t ,E_t and σ , one sample of p-type GaAs is investigated experimentally. The p-type conduction is obtained by thermal treatment of semi-insulating GaAs (heated at 1019 °C, 24 h), the acceptors concentration $Na=10^{16}$ cm^{-3} has been measured by Hall effect (Djemel, Castaing et al , 1992).

Two surfaces are compared: one is non-treated and the other is chemically polished with H_2SO_4:H_2O_2:H_2O (9:1:1) . CL measurements have been performed at room temperature. The primary electron beam intensity is 15 nA, with an electron energy ranging from 5 to 40 KeV.

To understand the influence of E_t on the CL intensity, we have reported in figure 1, the variation of the occupation probability f (equation 11) as a function of E_t for different excess carrier at the surface Δn (0). This figure shows that f increases when the energy level E_t moves away from the conduction band E_c ; it means that the number of the occupied centres is increased, consequently, the surface recombination velocity increases. Therefore, the change in energy level E_t explains the surface recombination. On the other hand, the variation of f is important when the excess carrier at the surface $\Delta n(0)$ increases.

Figure 2 shows the CL curves (equation 16) for different E_t. The influence of E_t on the CL curves is important for the low electron beam energy. The CL intensity increases and the maximum moves to low acceleration energy when E_t is near to conduction band (weak surface recombination). This behaviour is similar to that of the surface recombination velocity (Hergert, Reck et al, 1987) .

In figure 3, the CL intensity is measured for two surface states. After chemical treatment of the surface (open circles), the CL intensity increases in the low electron energy region .In the high electron energy region, the difference of CL signal between both surface states (treated and non-treated chemically) is reduced...

The adjustment between the numerical results and experimental data has been obtained by a change in energy level E_t. Thus, the two surface states can be characterised by a same defect density N_t = 2.10^9 cm^{-2}, and an energy level E_t=1.30 eV (Fig.3, 1), and by an energy level E_t=1.33 eV (Fig.3, 2). The similar surface states are obtained in p-GaAs deformed and non-deformed (Djemel, Nouiri et al, 2000).

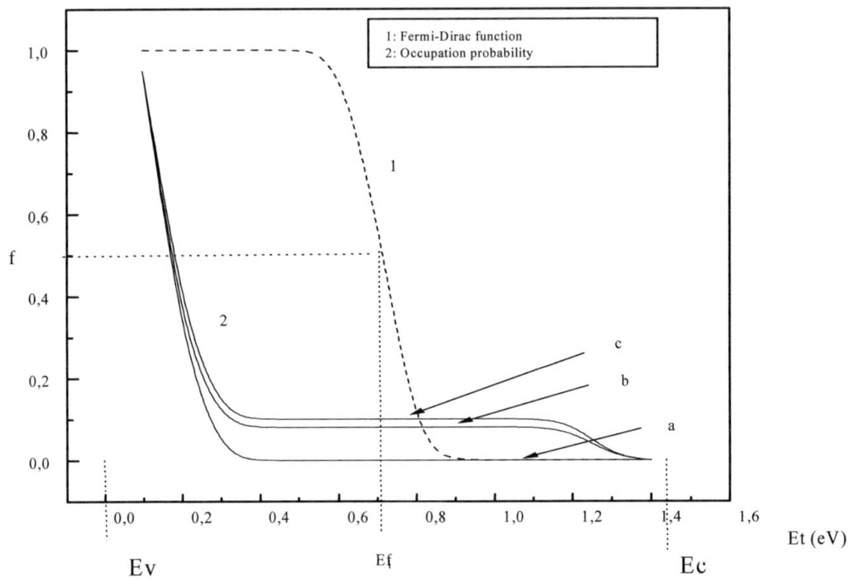

Fig. 1. The occupation probability f as a function of E_t for different excess carrier at the surface ($\Delta n(0)=10^{12}$ cm^{-3} (a), 10^{13} cm^{-3} (b), 10^{14} cm^{-3} (c))

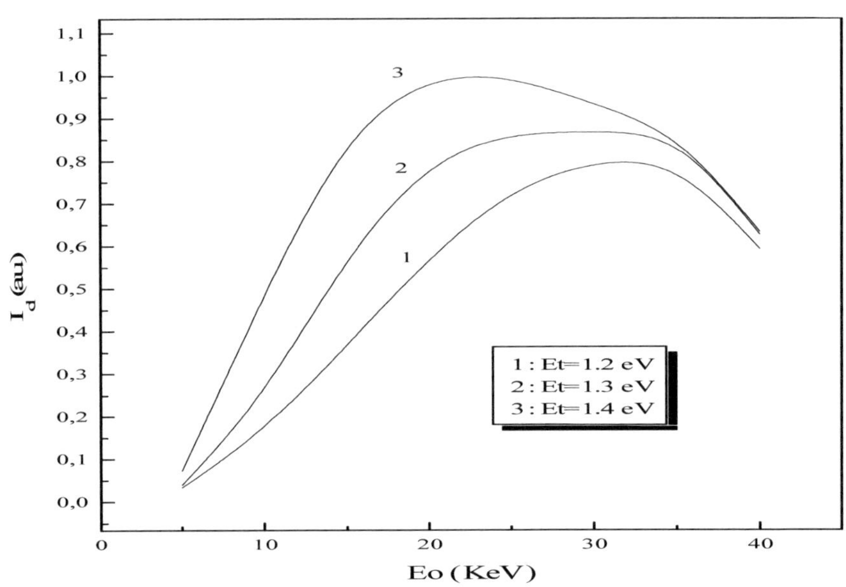

Fig. 2. CL intensity as a function of electron beam energy for different E_t.

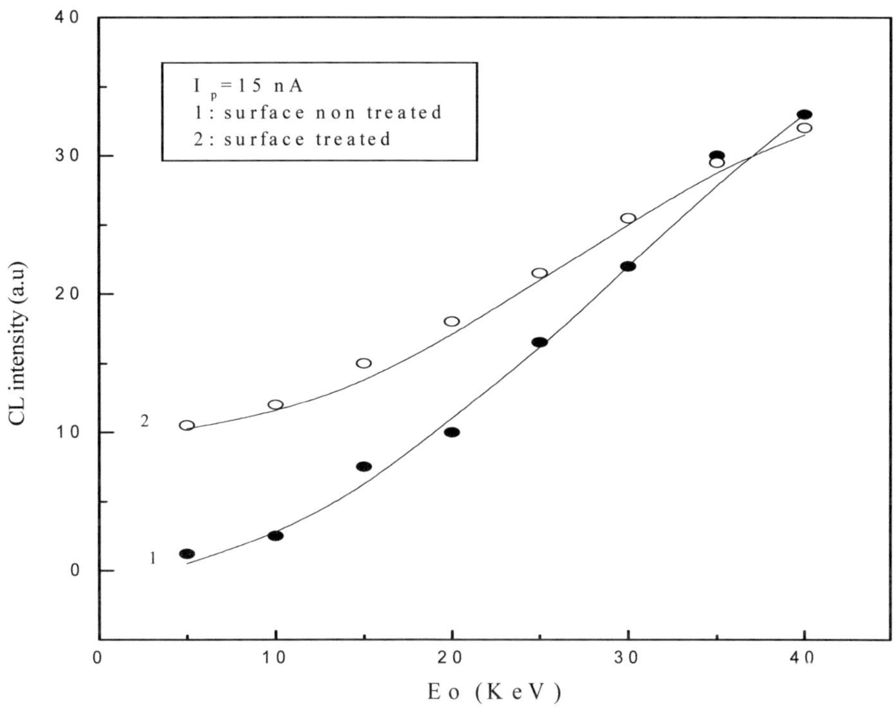

Fig. 3. CL intensity as a function of electron beam energy E_0 before and after surface treatment for p-GaAs. Full lines are the results of proposed model (Djemel, Tarento et al, 1998).

3. Electrical activity of defects

The electrical activity of defects can be imaged by cathodoluminescence .It reveals the regions where the recombination is radiative or non radiative. This electrical activity can be observed by the measurement of the contrast (C_{NR}) which is defined by the relative difference between the measured CL intensity over the defect and the measured CL intensity in the region without defect. The contrast is changed by the modification of the electronic properties of defects which are induced by the external treatment (hydrogenation, thermal treatment.. (Djemel , Castaing et al ,1990 ;Djemel , Castaing et al ,1992).

Figure 4 illustreates the modifications of the thermal treatment on the electrical activity of defects in GaAs.

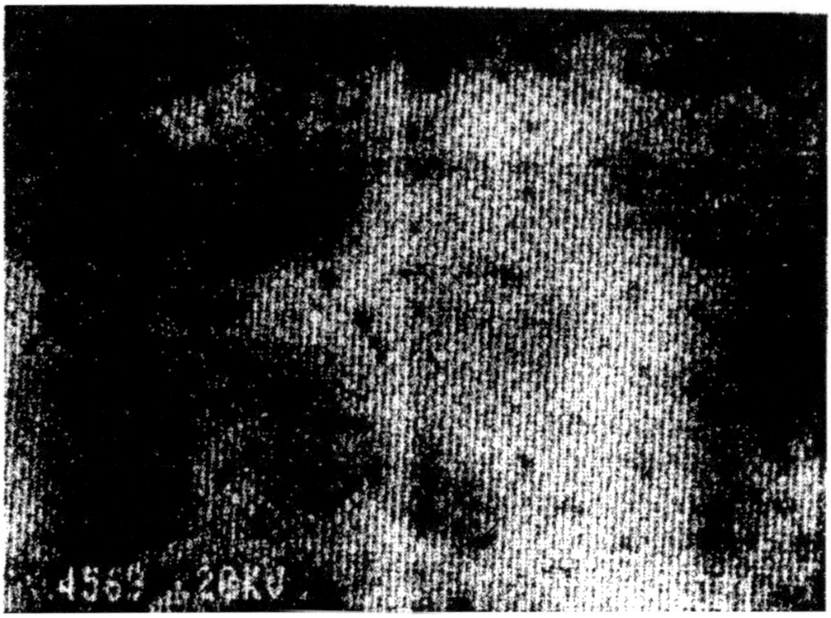

Fig. 4. CL observation after a thermal treatment of 15 hours at 850°C followed by quenching, with Si_3N_4 encapsulation (Djemel , Castaing et al ,1990)(x 1000)

Two kinds of contrast are essentially observed : dark contrast or halo contrast such as illustrated in fig.5 and white contrast. The two kinds of contrast are explained (i) by the existence of non radiative recombination centres or energy elimination by any other process (phonons , Auger ...) (or ones emitting out of the spectral band of detector) for the dark contrast and (ii) by the existence of radiative recombination centres or by an increase of carrier concentration for the white contrast .

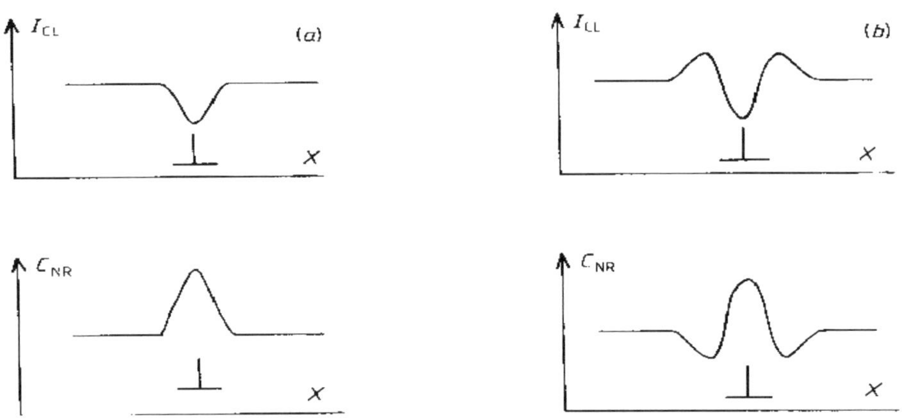

Fig. 5. Typical variation of CL Intensity (I_{CL}) for a line scan through a defect and variation of contrast (C_{NR}) : a) dark contrast , b) halo contrast. (Djemel , Castaing et al ,1990)

Figure 6 illustreates the modifications of the hydrogenation on the electrical activity of subgrain boundaries in GaAs.

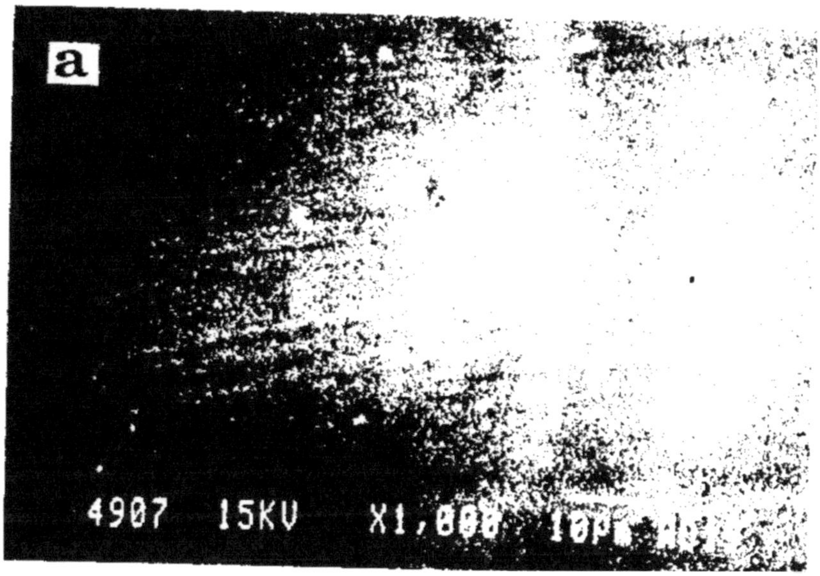
a
4907 15KV X1,000 10Pm

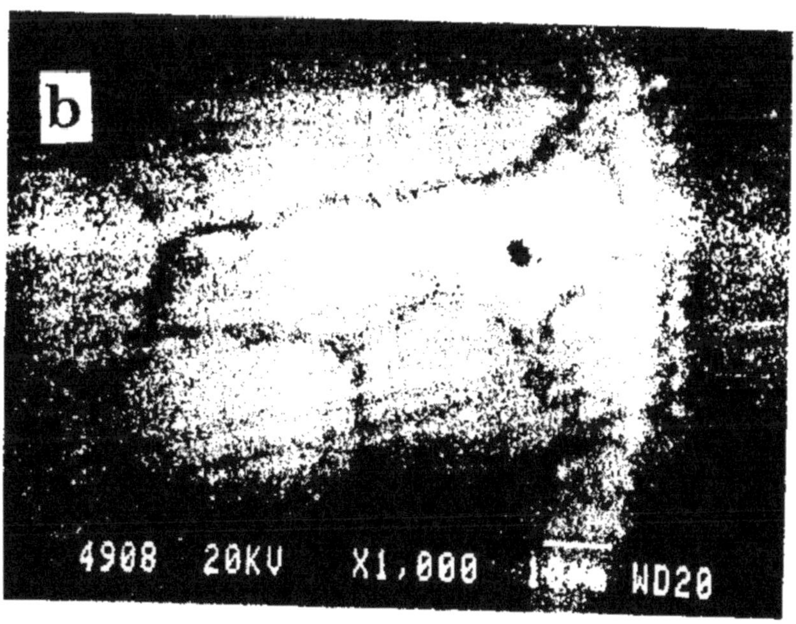
b
4908 20KV X1,000 WD20

Fig. 6. CL migrographs of deformed GaAs covered with a layer of GaAlAs and exposed to a hydrogen plasma. Observations for various beam accelerating voltages: a) 15 KeV , b) 20 KeV , c) 25 KeV.(Djemel , Castaing et al ,1992)

Deuterium is introduced in deformed GaAs by RF plasma and the Deuterium profil is determined by secondary ion mass analysis (SIMS).We have checked that the CL emission originates from the subsurface region containing Deuterium, by changing the range penetration of electrons. The values are 1.6 µm , 2.6 µm, 3.7 µm for 15 KeV , 20 KeV , and 25 KeV beam voltages respectively .The subgrain boudaries are invisible at the low beam voltage (Fig.6c) when electron-hole pairs are generated in the Hydrogenated region of the specimen (1-2 µm obtained by the Deuteurium profil SIMS). For high voltages , the contrast is strong (Fig.6c) because the hydrogen penetration is small compared to the electrons one.

4. Spectroscopic study

Defects and Complex defects in semiconductors are still attractive topics.They influence the electronic and optical properties of the materials. Complex defects are formed between the defects impurities(doping,already present in the starting material) and the created intrinsic defects (vacancy,interstitial and substitution position ,anti-site).These defects include nearest- and second-neighbor point-defect pairs.These defects and complex defects introduce generally the energy levels in the band gap. They can be radiative or nonradiative. In the case of radiative defects , the cathodoluminescence (Yacobi,Holt,1986;Lei,Leipner et al ,2004) and the photoluminescence (Sauncy,Palsule et al,1996 ;William,1968) allow to obtain a quantitative spectroscopy of the energy levels associated with defects.

Figure 7 shows the liquid-He spectra in semi-insulating GaAs. (Yacobi, Holt,1986)

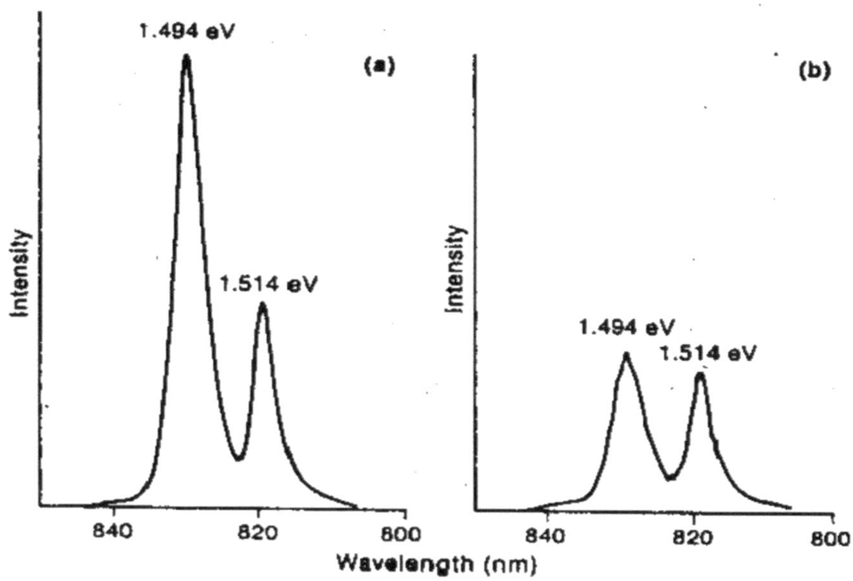

Fig. 7. Liquid-He CL spectra from (a) bright area and (b) an adjacent dark area (Fig.8) in semi-insulating GaAs (Yacobi, Holt, 1986)

Using the monochromatic CL micrograph, we can determine the spatial distribution of the defect corresponding to the emission .Fig.8.

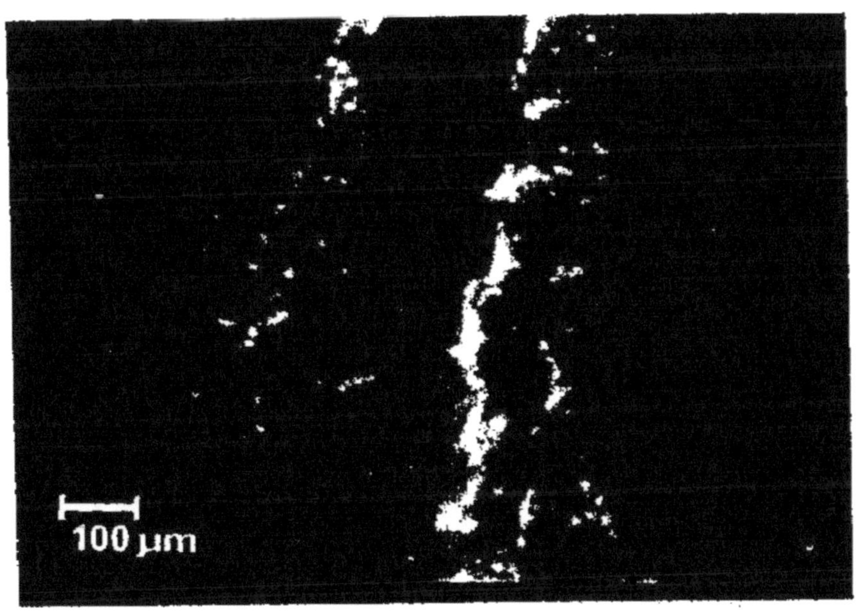

Fig. 8. Liquid-He monochromatic CL micrograph of a cell boundary in semi-insulating GaAs using the 1.494 eV emission due to residual Carbon (Yacobi, Holt, 1986)

5. Impact of electron beam irradiation

The cathodoluminescence is an effective technique for the analysis of the electronic properties of semiconductors. However, some materials (GaAs , ZnO , GaN ..) are sensitive to the irradiation electron beam (Djemel, 1988 ; Dierre,Yuan et al ,2008) .This sensibility is shown by a decrease of the CL intensity (increase of contrast) or an increase of CL intensity. The variation of CL intensity depends on the chemical nature of surface (Ga face, As face, Zn face ,O face ,N face)(Dierre,Yuan et al ,2008) , on the emission band (energy level in the gap band) (Dierre,Yuan et al ,2008) and on the external treatment supported by the materials (Djemel, 1988) .

Figures 9 and 10 illustrate the impact of electron beam irradiation before and after the treatment by hydrogen (Djemel, 1988).

Different mechanisms are induced by the electron beam. The electron beam irradiation induces an Electron Stimulated Reaction (Bourgoin, Corbett, 1972).The incident electrons depose locally a large energy which causes the bond breaking in the materials and the formation of the reactive sites. These reactive sites can react with the defects from surface or in the bulk .The electron beam, across the locally deposited energy, can enhance the migration of mobile defects and impurities. The interaction , between all defects and the complex defect responsible of luminescence , changes the electronic environment of the last defect .Thus ,these defects can turn into radiative and nonradioactive recombination centres .

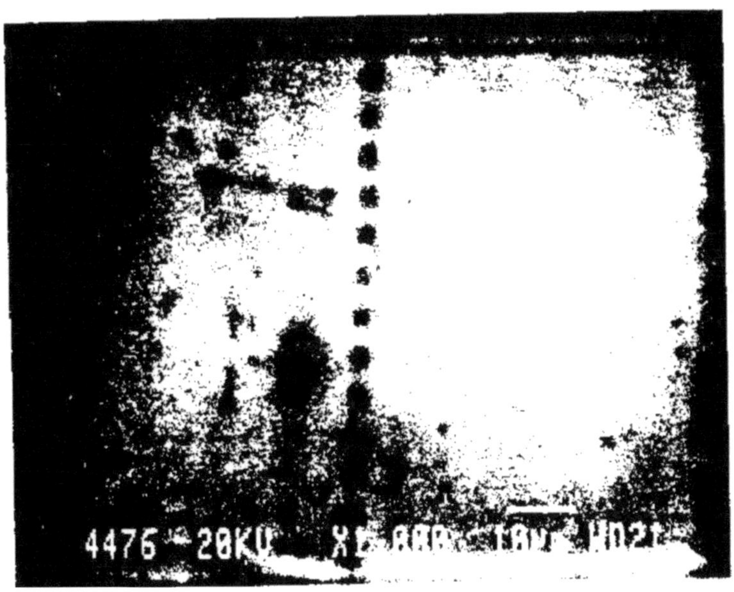

Fig. 9. Impact of electron beam irradiation observed by CL in GaAs before hydrogenation (dark dots) (Djemel, 1988)

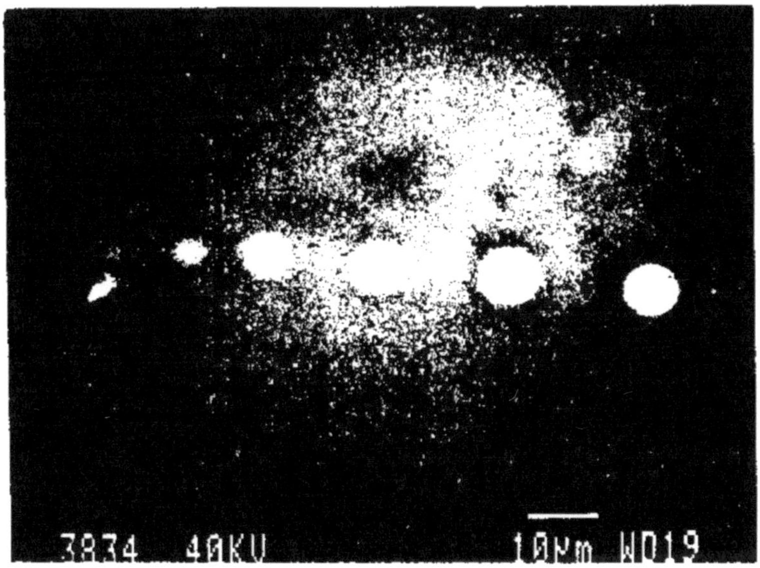

Fig. 10. Impact of electron beam irradiation observed by CL in GaAs after hydrogenation (white dots) (Djemel,1988)

Different mechanisms are induced by the electron beam. The electron beam irradiation induces an Electron Stimulated Reaction (Bourgoin, Corbett, 1972).The incident electrons depose locally a large energy which causes the bond breaking in the materials and the formation of the reactive sites. These reactive sites can react with the defects from surface or in the bulk .The electron beam, across the locally deposited energy, can enhance the migration of mobile defects and impurities. The interaction , between all defects and the complex defect responsible of luminescence , changes the electronic environment of the last defect .Thus ,these defects can turn into radiative and nonradioactive recombination centres.

6. Effect of temperature on CL intensity

Few theoretical and experimental works on the variation of the CL intensity as a function of the temperature have been realized (Jones, Nag et al, 1973; Lei, Leipner et al, 2004; Djemel, Kouissa et al ,2009). These studies concern the temperature variation of CL intensity for the transition conduction band and valence band (Jones, Nag et al, 1973). In this work a large discrepancy exists between the theoretical calculation and the experimntal data. Lei,Leipner et al ,2004 have studied the temperature variation of the luminescence bands for n-GaAs in the temperature range 20K-100K (Lei,Leipner et al , 2004). A new theoretical study on the CL intensity as a function of temperature and taking into account the influence of temperature on all physical parameters is done by Djemel,Kouissa et al ,2009 . Using this model, an improvement in the fitting of experimental data(Lei,Leipner et al , 2004) is shown in (Fig.11) allowing an estimation of capture cross section and the determination of the parameter of the radiative recombination centre.

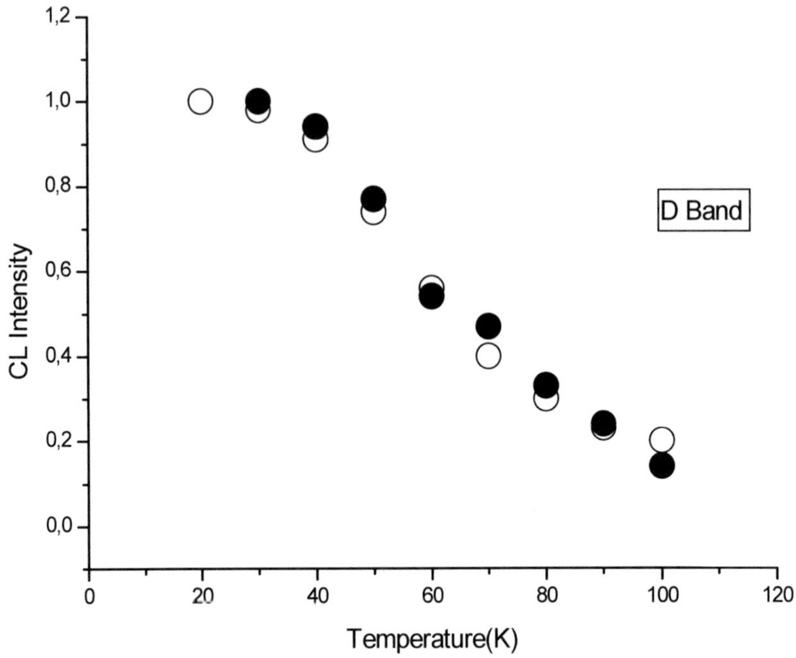

Fig. 11. CL Intensity versus temperature for D band(Et=0.5eV) in GaAs:Si (Lei,Leipner et al , 2004) Numerical results (•) Experimental data (○) (Djemel,Kouissa et al ,2009)

7. Conclusion

In this article we have shown the ability of the cathodoluminescence technique to study defects. We have demonstrated how this technique provides much data which are complementary in nature.

The quantitative study allows the knowledge of the interaction electron-matter in all aspect: energy dissipation of electron beam , generation function , penetration range , creation e-h pair energy ,distribution of generated carriers and their concentration , comparison between the calculated CL intensity and the measured CL intensity ,determination of physics parameters of defects (defects density, energy levels in the band gap, the capture cross section ,recombination velocity) and the parameters of carriers (diffusion length).

The CL micrograph visualisation gives us a general cartography of the active defects and non-active defects .The measurements of luminescence band give valuable information on the energy levels associated to defects and their place in the band gap. The impact of electron beam irradiation allows to determine the thermal stability of the electronic properties of defects ,to study the reactivity between defects and can identify the crystalline face (impact different on Zn face and O face).The measurements of CL intensity as a function of the temperature predict the evolution of many physical parameters with temperature. The comparison between the numerical results and the experimental data of CL intensity allows to extract the data on the radiative recombination centre.

8. References

Bourgoin .J.C, Corbett .J.W, A new mechanism for interstitial migration, Phys. Letter A 38 , 1972, 135-137

Ben Naser.F, Matoussi.A ,Guermazi.S ,Fakhfakh.Z. Cathodoluminescence Investigation of GaAs thin layers , Physics Procedia 2,2009,827-833

Dierre.B,Yuan.X.L,Yao.Y.Z,Yokoyama.M ,Sekiguchi.T.J. Impact of electron beam irradiation on the cathodoluminescence intensity for ZnO and GaAs. Mater.Sci:Mater.Electron ,19 ,2008,S307-S310

Djemel.A . These de Doctorat d'Etat , Université Paris XI Orsay ,Dislocation dans GaAs et Propriétés électroniques par luminescence :Influence des impuretés, 1988

Djemel.A,Castaing.J,Visentin.N,Bonnet.M,Influence of thermal treatments on the electronic activity of dislocations in GaAs observed by cathodoluminescence Semicond.Sci.Technol.5,1990,1221-1224

Djemel.A,.Castaing.J,Chevallier.J,Henoc.P.Hydrogenation of GaAs covered by GaAlAs and subgrain boundary passivation,J.Phys.III,France 2,1992,2301-2307

Djemel.A, Tarento.R.J,Castaing.J,Marfaing.Y,Nouiri A, Electronic Surface Properties of GaAs in CL Experiments ,Phys.Stat.Sol.(a),1998,168,425-432

Djemel.A,Nouiri.A,,Tarento.R.J , Study of suface defects in GaAs by cathodoluminescence: calculation and experiment J.Phys.Condens.Mater 12,2000,10343-10347

Djemel.A,Nouiri.A,Kouissa.S,Tarento.R.J.Cathodoluminescence Calculation of n-GaAs. Surface Analysis and comparaison,Phys.Stat.Sol(a),2002,191,N°1,223-229

Djemel.A, Nouiri.A,Kouissa.S, Tarento.R.J. Characterisation of n-GaAs by cathodoluminescence : Quantitative study and comparison,Current Issues on Multidisciplinary Microscopy Research and Education ,FORMATEX Microscopy Series N°2 Printed in Spain,2004 ,65-71

Djemel.A,Kouissa.S,Tarento.R.J,Temperature dependence of luminescence centre in cathodoluminescence ,Physics Procedia 2,2009,845-851

Everhart.T.E,Hoff.P.H, Determination of kilovolt electron energy dissipation vs penetration Distance in solid materials J.Appl.Phys.Vol.42,1,1971,5837-5846

Hergert.W,Reck.P,Passemann.L,Schreiber.J, Cathodoluminescence measurements using scanning electron microscope for the determination of semiconductors parameters Phys.Stat.Sol,(a),101,1987,661-618

Hergert.W, Hildebrandt.S, Pasemann.L Theoretical Investigations of Combined EBIC, LBIC, CL, and PL Experiments. The Information Depth of the PL Signal Phys.Stat.Sol,(a),102,1987,819-828

Hildebrandt.S, Schreiber.J, Hergert.W,I. Petrov.V.I Determination of the absorption coefficient and the internal luminescence spectrum of GaAs and $GaAs_{1-x}P_x$ (x =0.375, 0.78) from beam voltage dependent measurements of cathodoluminescence spectra in the scanning electron microscope. Phys.Stat.Sol,(a),110,1988,283-291

Jones.G.A.C,Nag.B.R,Gopinath.A,.Temperature variation of cathodoluminescence in direct gap semiconductors,Scanning Electron Microscopy (Part II)Proceeding of the worshop on electron specimen interaction :Theory for SEM IIT Research Institute ,1973,309-316

Lei.H,Leipner.H.S,Bondarenko.V,Schreiber.J.Identification of the 0.95 eV luminescence band in n-GaAs:Si, J.Phys.Condens.Matter,16 ,2004,S279-S285

Oualid.T,Singal.C.H,Dingas.J,Crest.J.P,Amzil.H,. Influence of illumination on the grain boundary recombination velocity in silicon J.Apl.Phys.55,1984,1195-1205

Phang.J.C.H,Pey.K.L,Chan.D.S.H. A simulation model for cathodoluminescence in the scanning electron microscope IEEE Trans.Electron.Devices,39,1992,782-791

Sauncy.T, Palsule.C.P, Holts.H, Gangopadhyay.S,. Lifetime studies of self-actived photoluminescence in heavily silicon-doped GaAs , Phys.Rev.B53,4,1996,1900-1906

Tarento.R.J,Marfaing.Y, Analysis of the recombination velocity and the electron beam induced current and cathodoluminescence contrasts at a dislocation, J.Appl.Phys.71,1992,4997-5003

William.E.W.Evidence for self-actived luminescence in GaAs:The Gallium Vacancy-Donor Centre ,Phys.Rev.168,3,1968,922-928

Wittry.D.B,Kyser.D.K. Measurement of diffusion lengths in direct Gap semiconductors by electron beam excitation ,J.Appl.Phys.Vol.38,1,1967,375-382

Wu.C.J,Wittry.D.B, Investigation of minority-carrier diffusion lengths by electron Bombardment of Schottky barriers.J.Appl.Phys.49,1978,2827-2836

Yacobi.B.G,Holt.D.B,Cathodoluminescence Microscopy of Inorganic Solids Plenium Press, 1990

Yacobi.B.G,Holt.D.B,Cathodoluminescence Scanning Electron Microscopy of semiconductors, J. Appl.Phys.59,4,,1986, R1-R24

Part 2

Application to Semiconductors

5

Cathodoluminescence Properties of Zno Thin Films

M. Addou et al.*
University of Ibn Tofail
Morocco

1. Introduction

The industry has a great need for high performance materials to feature well defined. These needs have prompted the development of methods of study and control of gradually more sophisticated based on the radiation-matter interaction. To identify the properties of materials, we must make a spectral analysis on the emitted photon using a spectrometer combined with a detection system. The analysis of the photon can take place directly in the form of an electrical current, as well as the spectrum received by the measurement system is represented by a function I (λ).

The best radiations are the electrons because they are easy to produce, accelerate and to focus. In this context we speak of cathodoluminescence, which was long used in devices including fluorescent screens. This phenomenon, resulting from the excitation of luminescent materials by electron beam, is leading to photon emission, which is subject to the laws of the transitions. We usually distinguish the intrinsic luminescence (transition band to band, free exciton) and extrinsic luminescence from impurities and defects. The most common application is the television screen (when it is a cathode ray tube). In geology, the cathodeluminescence microscope is used to examine internal structures of geological samples in order to determine, for example, the history of rock formation. Another important application is Cathodoluminescence in image mode. In this mode the cathodoluminescence can view the spatial distribution of responsible levels for a radiative transition observed or locate the non radiative defects such as dislocations, grain boundaries and precipitates. These defects give rise to a generally strong contrast. It is possible to determine the energy levels that are at the root of this contrast. So we used this technique to study the nature and distribution of defects and impurities in materials [1].

Semiconductors exhibit energy gap between the valence band and the conduction band, the order of the electron volt. An incident electron beam of sufficient energy can move electrons from the valence band to the conduction band, thus producing electron-hole pairs; return to the ground state can be achieved through nonradiative transitions or by radiative transitions with photon emission of light. This return is via two mechanisms (direct recombination and

*J. Ebothé, A. El Hichou, A. Bougrine, J.L. Bubendorff, M. Troyon, Z. Sofiani, M. EL Jouad, K. Bahedi and M. Lamrani
University of Ibn Tofail, Morocco

recombination indirect). The CL signal is formed by detecting the photons in the UV, visible and IR.

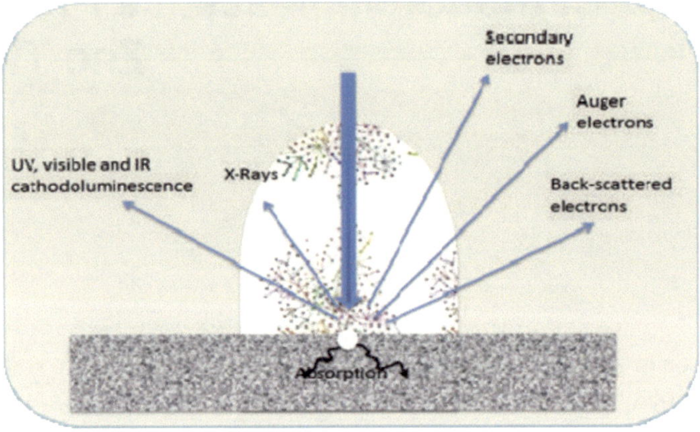

Fig. 1. Cathodoluminescence schema

1. The region of high energy loss (DE> 50eV) corresponding to the excitation of deep levels K, L,... creates the X and Auger radiations.
2. The region of low energy loss (DE <50eV) corresponds to the excitation of the valence and conduction (photon CL).

The most important mechanisms in the analysis of CL in semi-conductors are:

- Electron-solid interaction.

- Dissipation of energy.

- Generation of carriers (e-h).

From the experimental point of view, a study of CL can be simplified to an electron gun, a sample and a detector.

Many parameters are involved and contribute to enhance the cathodoluminecence intensity during the growth process. Temperature effect, doping effect, flow rate deposition effect and the excitation energy effect are strongly among many other parameters, which influence the optical properties of thin films. The investigation on their luminescence remains very limited, especially the cathodoluminescence. However, band-edge photoluminescence in polycrystalline ZnO has not received as much attention as the bulk material because the band-edge photoluminescence is usually weaker and lacks the fine structure of the bulk crystals. Nevertheless, a study of the photoluminescence structure of this material is interesting, because it can provide valuable information on the quality and purity of the material. Polycrystalline films may also have different photoluminescence mechanisms as compared to the bulk material.

A^{II}–B^{VI} compounds, such as transparent and conducting oxides films, have been attracting ample attention as starting material for electro-luminescent devices because of their high visible transmittance and low electrical resistivity in the visible region. Zinc oxide (ZnO) is one of the few metal oxides, which can be used as a transparent conducting material. It has some advantages over other possible materials such as In_2O_3–Sn, $CdSnO_4$, or SnO_2, due to its unique combination of interesting properties: non-toxicity, good electrical, optical and piezoelectric behaviour, high stability in a hydrogen plasma atmosphere and its low price [2,3]. So, ZnO is a good candidate to substitute indium tin oxide (In_2O_3:Sn) and tin oxide (SnO_2), in conductive electrodes of amorphous silicon solar cells.

Furthermore, ZnO has the same crystal structure as GaN: wurtzite crystal structure and direct wide band gap. It is closely lattice matched to GaN. ZnO therefore offers potential as a substrate material on which high quality GaN may be grown. The room temperature band gap of ZnO is 3.3 eV with emission in ultra-violet (UV) region. An outstanding feature of ZnO is its large excitonic binding energy of 60 meV leading to the existence and extreme stability of excitons at room temperature and/or even higher temperatures [4-7]. These characteristics have generated a wide series of applications for example, as gas sensors [8], surface acoustic devices [9], transparent electrodes [10] and solar cells [3] among others. The preparation of zinc oxide thin films has been the subject of continuous research. Many techniques are used for preparing this transparent conductive ZnO such as: RF sputtering [11], evaporation [12], chemical vapour deposition [13], ion beam sputtering [14] and spray pyrolysis [15-18]. Among these methods spray pyrolysis has attracted considerable attention due to its simplicity and large scale with low-cost fabrication. Additionally, by using this technique one can produce large area coatings without the need of ultra high vacuum.

Different process parameters of the intrinsic and doped ZnO films by spray technique have been published [19-22]. The investigation on their luminescence remains very limited, especially the cathodoluminescence. However, band-edge photoluminescence in polycrystalline ZnO has not received as much attention as the bulk material because the band-edge photoluminescence is usually weaker and lacks the fine structure of the bulk crystals [23,24]. Nevertheless, a study of the photoluminescence structure of this material is interesting, because it can provide valuable information on the quality and purity of the material [25]. Polycrystalline films may also have different photoluminescence mechanisms as compared to the bulk material.

This Chapter will be divided on three parts. The first one describe the effects of different process parameters such as substrate temperature, air flow rate and precursors of undoped ZnO films prepared by spray pyrolysis on glass substrate. Several cathodoluminescence bands and the corresponding emission processes are identified. It is also shown that SP is an adapted technique to achieve ZnO films with a quality comparable with that of transparent conducting oxide thin films prepared by other techniques.

The second one concerns some results of a systematic investigation of the crystallinity, the surface morphology, optical properties and the cathodoluminescence measurements of the undoped and Sn-doped ZnO thin films deposited by SP. Several CL bands and the corresponding emission process are identified.

The last section of this chapter is dealing with, the luminescence of the investigated ZnO samples is examined accounting for the effects of two main parameters: the Er dopant concentration and the incident electron beam energy. A general sight of the CL material behavior is first reported, which precedes a more refined study including the spectral peak's integration and the near-field imaging aspects.

2. Effect of deposition temperature (Ts), air flow rate (f) and precursors on cathodoluminescence properties of ZnO thin films

Earlier work indicated that ZnO exhibited three bands of luminescence are centered around 382, 510 and 640 nm, labeled near UV, blue-green and red bands, respectively [26-28]. Our undoped ZnO films displayed the same three cathodoluminescence bands. However, the relative intensities and the disappearance of the blue-green and red emissions depend strongly on the precursor and deposition conditions such as flow rate and/or substrate temperature and doping nature.

2.1 Variation of substrate temperature

Fig. 2 shows the cathodoluminescence of ZnO thin films, at E=5 keV electron beam energy with a beam current of about 1 nA, deposited at different substrate temperatures and different flow rates. The substrate temperatures (Ts) were varied from 350 to 500°Cwith an interval of 50°C, and with a flow off=5 ml/min. When the substrate temperature increases, the surface of our films is entirely covered by grains and condensed.

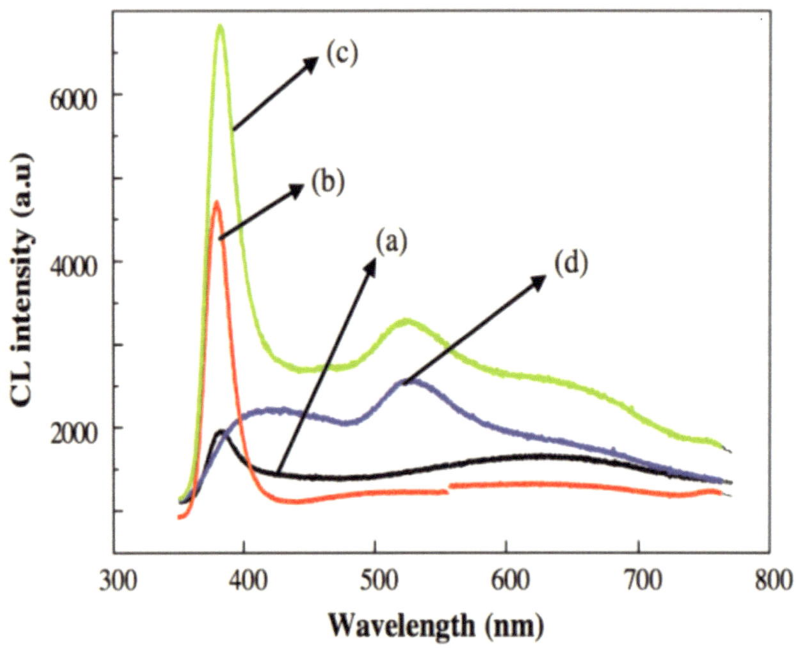

Fig. 2. CL Spectra of ZnO sprayed at flow rate f=5ml/min deposited on different temperature: (a) : T= 350°C, (b): T= 400°C, (c) : T= 450°C, and (d) : T= 500°C.

As indicated, the luminescence intensity depends strongly on the deposition temperature. Extinction of the blue-green emission (centered around 510 nm) is observed at substrate temperature of 350 and 400°C, whereas the near UV emission at 382 nm becomes more dominant than other transitions (blue-green and red emissions) at 450°C. The blue-green emission (510 nm) appears above substrate temperature 450°C but the red emission (640 nm) appears at different substrate temperature. At Ts=500°C, the UV transition shifts to higher wavelength and becomes comparable in cathodoluminescence intensity with blue-green emission. The maximum value of cathodoluminescence intensity for three bands is obtained at T=450°C.

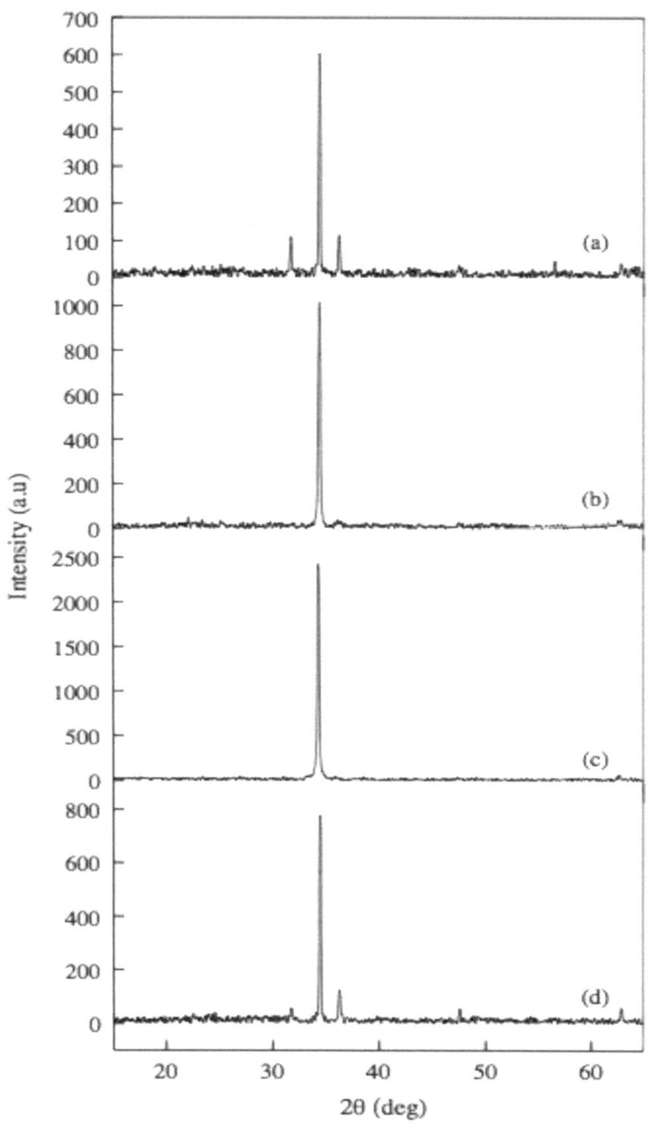

Fig. 3. XRD of ZnO sprayed at flow rate f=5ml/min deposited on different temperature: (a) : T= 350°C, (b): T= 400°C, (c) : T= 450°C, and (d) : T= 500°C.

Before discussing the cathodoluminescence observations of the films in more detail, it is useful to consider their crystal structure and morphology. The XRD of ZnO films, deposited at different substrate temperatures and at flow rate 5 ml/min, indicated that they possess a hexagonal close packed structure (Fig. 3). Moreover, it is the only main peak obtained with all films indicating that the increasing of substrate temperature does not change the preferred textural growth orientation. The [0 0 2] direction corresponds to the c-axis of the crystal lattice that it's normal to the deposition substrate plane. The XRD intensity depends strongly on the deposition temperature. The narrow range of deposition temperature permitted for maximum XRD intensity is illustrated by the pronounced peak at Ts=450°C.

The mean crystallite size D was calculated from the (0 0 2) diffraction peak using Scherrer's formula [29]. Values of D are listed in Table 1 for films prepared under various substrate temperatures. The mean crystal diameter was around 30 nm for the sample deposited at Ts=450 °C which was larger value as compared to all samples. We conclude that the films, which present strong cathodoluminescence intensity in UV emission, have a strong XRD intensity and large grain size, therefore a good crystallinity. We confirmed our results by using SEM images. Fig.4 clearly shows that there is a change in the surface morphology of ZnO films due to a change in the substrate temperature. It is evident that the porous structure occurred throughout the films deposited at Ts=350 °C (Fig. 4(a)). All films deposited at Ts>350 °C attain a microstructure and had a close-packed morphology. The films deposited at Ts=450 °C consists of hexagonal-like grains of approximately 300 nm size. The mean crystal diameters obtained using Scherrer's formula are all case substantially smaller than the dimension of grains observed by SEM image, indicating these grains are probably an aggregation of crystallites. The enhancement of cathodoluminescence intensity at substrate temperature 450°C could be due to a large grain size therefore a better crystallinity. The best luminescence was achieved with samples grown at T=450°C.

Substrate temperatures: T_s (°C)	350	400	450	500
Grain size: D (nm)	27.5	30	31.31	28
CL intensity of UV emission (arb.units)	2000	4700	6800	2100

Table 1. Grain size values at different substrate temperatures calculated using Sherrer's formula (f=5 ml/min)

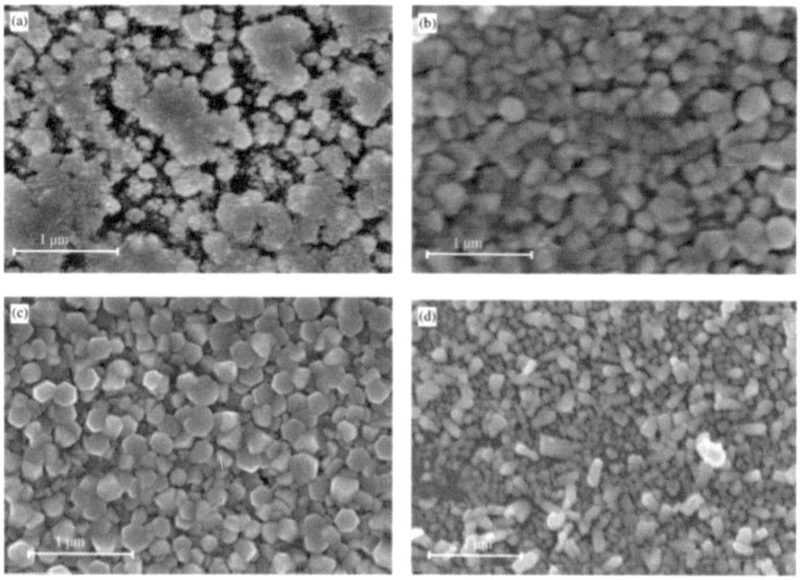

Fig. 4. Morphology of ZnO sprayed at flow rate f=5ml/min deposited on different temperature: (a): T=350°C, (b): T= 400°C, (c) : T= 450°C, and (d) : T= 500°C.

2.2 Variation of airflow rate

The structural and luminescence properties of ZnO thin films are also investigated by varying the flow rate in the region 2.5pfp7.5 ml/min, which here is equivalent to the varying growth rate range of 0.2pro1 mm/s. This formation occurs on a deposited substrate made of soda glass whose temperature is fixed at Ts=450°C. Note that no cracks are observed on large scan area for all samples (SEM images Fig. 6). The films are continuous and in fact consist of grains. The cathodoluminescence spectra of these films (E=5 keV electron beam energy with a beam current of about 1 nA) at different airflow rate and at Ts=450°C are shown in Fig.5. For f=2.5 ml/min, one large band centered at 400 nm have been observed. When the spray rate enhanced, three emissions have been appeared with a dominance of the blue-green emission (510 nm) for f=3.75 ml/min and a strong intensity of the UV transition (382 nm) for f=5 ml/min. Furthermore, the appearance of the UV emission corresponds to the improved crystal quality. At high flow rate f46.25 ml/min, a degradation of cristallinity leads to decrease of the intensity related to UV transition and disappearance of both emissions situated in blue-green and red ranges.

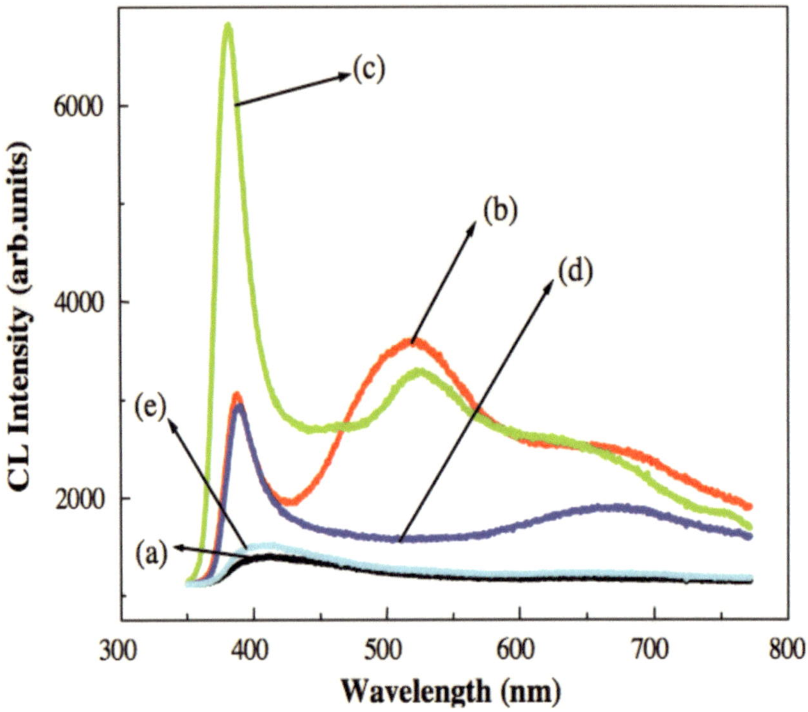

Fig. 5. CL Spectra of ZnO deposited at 450°C for different flow rate : (a) : f= 2,5ml/min, (b) : f= 3.75ml/min, (c) : f= 5ml/min, (d) : f= 6.25ml/min and (e) : f= 7.5ml/min

In Table 2, we have reported the mean grain size of ZnO films at different flow rate (statistical analysis of more than 100 grains on SEM images, Fig. 6). The some behavior of our samples at different substrate temperatures has been observed. Indeed, the low cathodoluminescence intensity of the UV transition is observed for films indicate small grain sizes and porous structure. We can conclude, that good crystallinity and the best luminescence are achieved with samples grown at f=5 ml/min. The decrease in surface grain size with the flow rate beyond f=5 ml/min appears to be mainly the effect of the non-incorporation of Zn particles on the material surface. The optimum flow rate and substrate temperature are f=5 ml/ min and Ts=450 °C, respectively. The films produced exhibit optical characteristics comparable to films grown by more sophisticated techniques.

Air flow rate *f* (ml/min)	2.5	3.75	5	6.25	7.5
Grain size *D* (nm)	200	275	300	260	225
CL intensity of UV emission (arb.units)	1400	3100	6900	3000	1500

Table 2. Grain size calculated from SEM images at different air flow rates (T=450 °C)

Fig. 6. SEM micrographs of ZnO films sprayed at substrate temperature Ts=450 °C and different air flow rate: (a) f=2.5ml/min, (b) f=5 ml/min, (c) f=7.5 ml/min and substrate temperature: Ts=450 °C.

2.3 Cathodoluminescence properties and variation of precursor

Cathodoluminescence spectra obtained from samples deposited at optimum condition Ts=450 °C and f=5ml/min is shown in Fig. 2(c). The three main emissions had peaks at 382, 520 and 672 nm. The luminescent at 382 nm corresponds to the band gap transition of ZnO. The presence of this band is an indicator of the good crystallinity.

The second peak centered at 520 nm is the characteristic of blue-green emission, which is also typical for ZnO material. Despite the numerous reports on the photoluminescence and cathodoluminescence of undoped ZnO, the luminescent centre responsible for this emission is not yet clearly identified. Several assumptions are proposed. Dingle [30] ascribes this emission peak to a substitution of Zn^{2+} by Cu^{2+} in the crystal lattice. According to Vanheusden et al. [31] and Egelhaaf et al. [23] this peak corresponds to a defect-related luminescence (deep-level luminescence), due to the oxygen vacancies in ZnO. This defect-related luminescence is explained by radiative transitions between shallow donors (oxygen vacancies) and deep acceptors (Zn vacancies). In this case, the acceptor level is located 2.5 eV below the conduction band edge [23,32], while the donor level is known as a shallow level at 0.05–0.19 eV, leading to an emission band centered around 508–540 nm. Furthermore, Minami et al. [33] proposed that the blue-green emission in this material might be associated to a transition within a self-activated centre formed by a double-ionized zinc vacancy $(V_{Zn})^{-2}$ and a single-ionized interstitial Zn^{+} at the one and /or two nearest-neighbor interstitial sites. Nevertheless, the composition studies of our films by XPS and EDX analysis do not reveal presence of copper and thus the first explanation must be rejected. Therefore, one of the two other hypotheses may explain the blue-green emission that we observe.

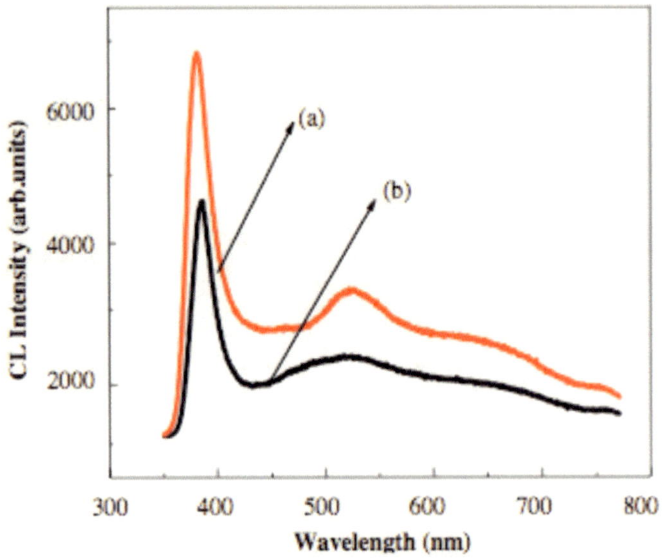

Fig. 7. Cathodoluminescence spectra of ZnO for both precursors (Ts=450 °C and flow rate f=5 ml/min): (a) $ZnCl_2$ and (b) AcZn

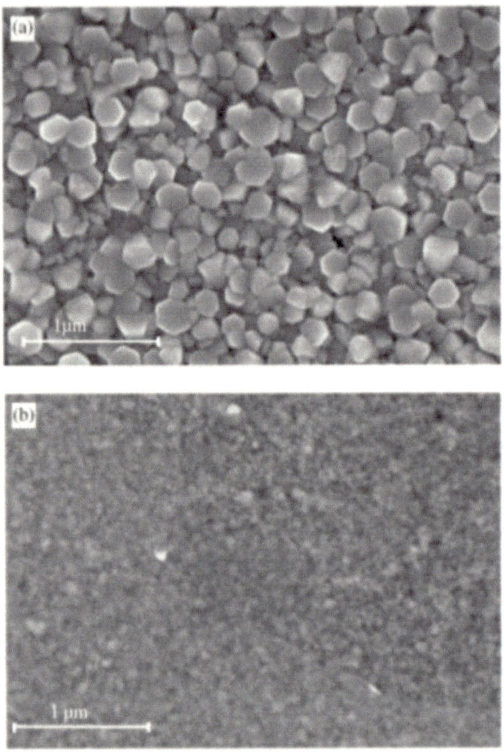

Fig. 8. SEM micrographs of ZnO at different precursors (Ts=450 °C and flow rate f=5 ml/min: (a) ZnCl2 and (b) AcZn.

Previous works have also attributed the blue-green emission to a luminescent centre formed by the association of a doubly ionized zinc acceptor vacancy defect and of halogen impurities such as Cl atom of precursor. Fig.7 shows the cathodoluminescence spectra of undoped ZnO thin films preparing using zinc chloride ($ZnCl_2$) and zinc acetate (AcZn) as precursors. The presence of blue-green emission for both precursors is not attributed to the presence of chlorine atom, but confirms the attribution of blue-green emission to a self-activated centre. Furthermore, the SEM image of both samples shows there is a change in the surface morphology due to the presence of Ac or Cl atoms (Fig.8). The films prepared from precursor $ZnCl_2$ presents a better cathodoluminescence intensity, which confirms our choice for precursor, and the good cristallinity related to a dense structure and crystallites size relatively important. Cathodoluminescence intensity is directly correlated to an improvement of the films crystallinity.

The last peak at 670 nm, which appears markedly broad for this low beam energy (5 keV), may be due to the defect related deep emission, as shown by photoluminescence [34,35]. Lohnert et al. [36] pointed out that the intensity of broad luminescence observed from about 450 to 700 nm depends on the oxygen partial pressure in the sintering process of ZnO ceramics. This also suggests the hypothesis of a transition within a V_o^x neutral and single-ionized oxygen V_o^-.

2.4 Summary

The dependence of cathodoluminescence properties of spray-deposited ZnO films on process parameters such as substrate temperature, airflow rate and precursors, has been studied in detail. We have found that the luminescence intensity depend strongly on process parameters. The optimum values of substrate temperature and airflow rate have been determined to give the best luminescent ZnO films.

In the process of this investigation, it has been observed that the films are polycrystalline and have a preferred orientation in the [0 0 2] direction. The best films in terms of cathodoluminescence consisted of close-packed grains. These grains were an aggregation of many crystallites having a mean diameter around 30 nm.

The cathodoluminescence analysis of the sample deposited at optimum conditions (Ts=450 °C; f=5 ml/min) shows three bands: a near ultra-violet emission at 382 nm, a blue-green emission at 520 nm and a weak red one at 672nm.

3. Cathodoluminescence of undoped and tin-doped ZnO thin films

Cathodoluminescence (CL) spectroscopy, X-ray diffraction and spectrophotometery have been studied of undoped and tin (Sn)-doped ZnO films prepared by spray pyrolysis (SP) technique of zinc chloride and tin chloride. The luminescence films had a polycrystalline hexagonal wurtzite type structure. At room temperature, the cathodoluminescence (CL) spectra of the undoped and doped ZnO films exhibit the common near ultra-violet (UV) band-gap peak at λ = 382 nm but they differ as regards their visible emissions. The undoped ZnO emits an intensive blue-green light (λ = 520 nm) and a red emission (λ = 672 nm). The presence of tin gives rise to a new light emission corresponding to λ = 463 nm and the extension of blue-green light typical of the intrinsic behavior of the material. CL imaging of undoped ZnO films show that the luminescence is located at defined sites giving rise to a grain-like structure inherent to the surface morphology. The presence of tin in the material leads to great luminescent spots, attributed to large grain sizes.

3.1 Experimental procedure

The luminescent films were prepared by spray pyrolysis using air as the carrier gas atmospheric pressure. The apparatus used for deposition is described and schematized elsewhere [37/21]. The spraying solution was prepared from a mixture of 0.05 M zinc chloride ($ZnCl_2$) and deionised water. A small amount of chloridric acid HCl was added in the solution to prevent the formation of zinc hydroxide ($Zn(OH)_2$). The nozzle is directed towards a substrate with a spraying flow rate of 5 ml/min. This substrate is made of bare glass sheets whose temperature is maintained constant at T = 450°C, that is known to be the optimal condition for this material formation. The same film thickness of 0.4 mm, estimated from the deposition time, is retained. A good agreement is obtained with the value deduced from the cross sectional images obtained by transmission electron microscopy. Tin doping was achieved by adding $SnCl_2$ to spraying in a concentration of 6 at.%. The compositionnal analysis of the films as performed with an EDAX 9100 analyzer indicates that the 6% tin percentage of the spraying bath is quite recovered in the Sn-doped ZnO samples.

The crystalline structure was investigated by X-ray diffraction using CuK radiation. Optical transmittance measurements were performed with a Shimadzu 3101 PC UV-VIS-NIR spectrophotometer. The cathodoluminescent (CL) spectra and images were measured at room temperature, using a LEO-GEMINI 982 scanning electron microscopy (SEM), operated from 1 to 30 kV and equipped with a field emission gun. Our CL system, in association with a spectrometer (Triax 190 from JOBIN YVON), allows monochromatic CL imaging as well as acquisition of CL spectra on very localized spots of a sample. The mechanical details of our homemade SFM microscope, implemented inside the SEM, are reported in [37]. In this paper, the photons enter directly a multimode optical fiber (numerical aperture = 0.48) leading to far-field images or spectra. The CL images are acquired with a fixed electron beam, the sample being scanned under the fiber. The spectrometer has a spectral resolution of 0.5 nm and is equipped with an N2-coolled CCD detector. Monochromatic CL images can be obtained by selecting a defined wavelength and by sending the light to a photomultiplier (PM, R4220P from Hamamatsu). The signal is then amplified with an electrometer and images of 256 × 256 pixels at a line frequency of about 0.1 Hz are generated. Panchromatic images are also obtained by directing the light towards the PM without passing through the spectrometer.

3.2 Results and discussion

X-ray diffraction patterns for undoped and Sn-doped ZnO thin films prepared at optimum parameters are shown in Fig.9. The diffraction (0 0 2) peaks located at the angular position $2\theta = 34.4°$ (undoped ZnO) and $2\theta = 34.38°$ (Sn-doped ZnO) are characteristics of the hexagonal close packed structure of this material. Moreover, it is the only main peak obtained with both films indicating that the presence of tin in ZnO keeps unchanged the preferred textural growth orientation. The (0 0 2) direction corresponds to the crystal lattice that is normal to the deposition substrate plane. It should be pointed out that the peak's intensity of the Sn-doped ZnO films in Fig.9 (b) is higher that the one of the undoped sample of Fig.9 (a), it is full width at half maximum (FWHM) being broaded. One deduced that the doped sample has a better crystallinity. Probably, there is the substitution of oxygen by tin anions in the material sites, leading to a partial healing of the crystal network.

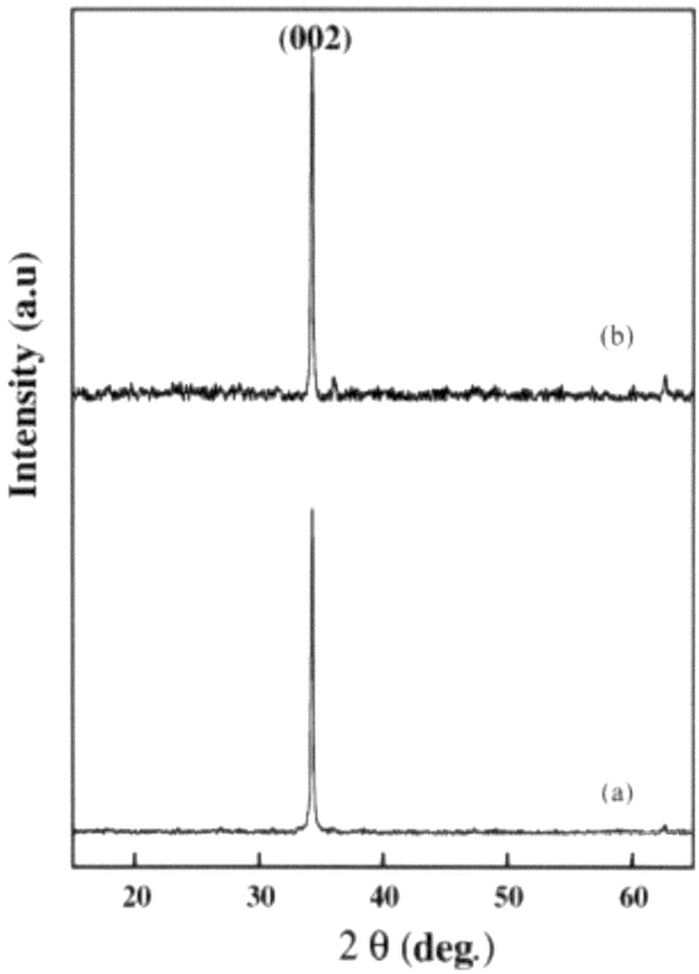

Fig. 9. X-ray diffractograms of: (a) undoped and (b) Sn-doped ZnO films.

Fig.10 consists of SEM showing the effect of doping on the surface morphology. We can see that the morphology evolves considerably after doping our samples. The undoped film in Fig.10 (a), depicts a microstructure consisted of hexagonal-like grain of approximately 200 nm size, the substrate surface being entirely covered for the film thickness retained. The last remark remains valid for the Sn-doped sample image of Fig.10 (b). Where as Sn-doped film presents a plate-like grains with polycristalline-like irregular grains. The size of the grains is larger (350 nm) and not uniform. Therefore, the crystallinity of Sn-doped ZnO films was significantly improved with Sn-doping. This is in a good agreement with X-ray diffraction results.

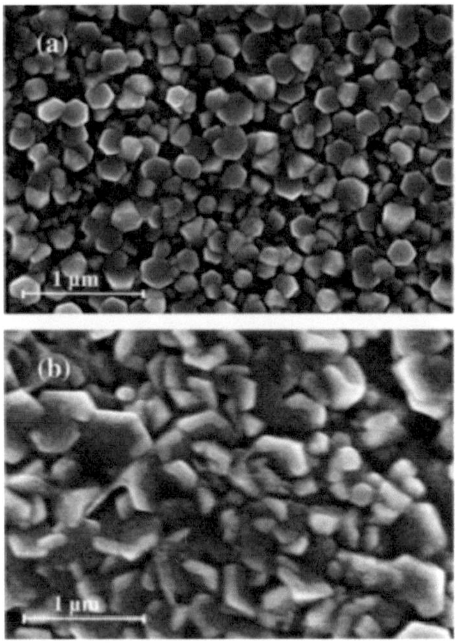

Fig. 10. SEM micrographs of: (a) undoped and (b) Sn-doped ZnO films.

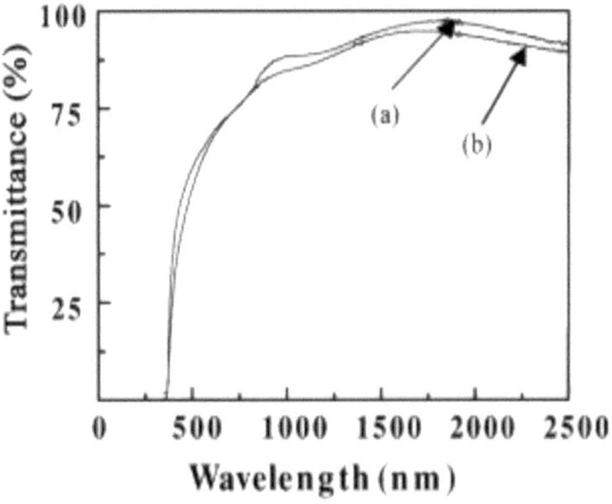

Fig. 11. Optical transmission spectra of: (a) undoped and (b) Sn-dopedZnO films.

The optical properties of undoped and Sn-doped films are restricted to their transmission spectra (Fig.11). The films were highly transparent above 85% in the visible wavelength range. For wavelengths in the absorption edge region, it is observed that it is shifted towards lower wavelengths. This effect has been already observed and discussed by other authors [38], significant an increasing in band-gap value. The results from the previous figures show that the ZnO film with good crystallinity and purity can be obtained by SP.

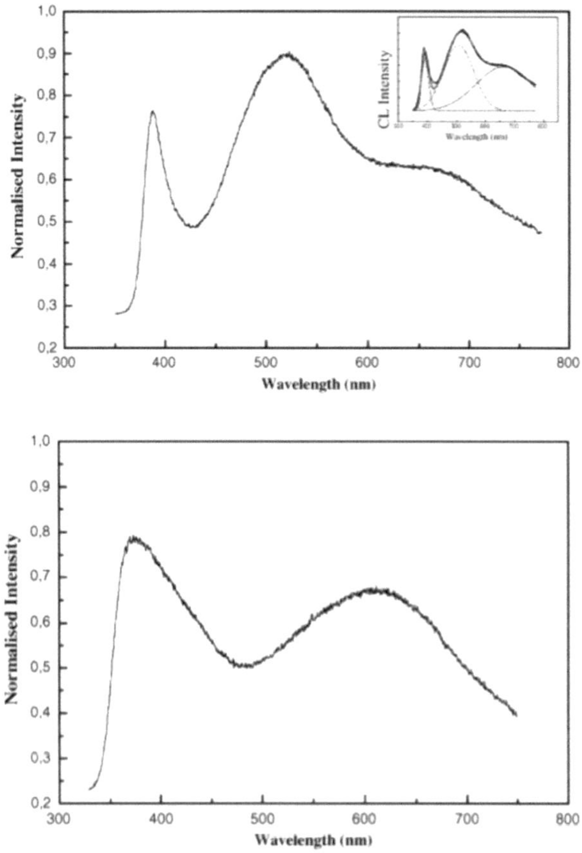

Fig. 12. Cathodoluminescence spectra of: (a) undoped ZnO and (b) Sn-doped ZnO films at E=5 keV electron beam energy with a beam current of about 1 nA.

Cathodoluminescence studies exhibit interesting features. In Figure 12, we present the cathodoluminescence emission spectra of undoped (Fig.12 (a) and Sn-doped ZnO thin film (Fig.12 (b)) at the electron beam energy of 5 keV. Early work indicated that ZnO exhibited three photoluminescence bands centered around 380, 520, and 650 nm, labeled near-ultraviolet (UV), blue-green and red bands [39,27,28]. Our undoped ZnO films displayed the same three cathodoluminescence emission, showing in Fig.12 (a). The first peak localized at λ = 382 nm (Eλ = 3.25 eV) corresponds to the band-gap transition of this material. It is clear that the appearance of the band-edge correspond to the good quality and crystallinity of our films and confirms the previous X-ray diffraction observations. The second peak of the undoped sample is the blue-green emission located at λ = 520 nm (Eλ = 2.38 eV) that is here typical of ZnO material. The luminescent center responsible for this emission is not clearly identified. It is ascribed by Dingle [30] to a substitution of Zn^{2+} by Cu^{2+} in the crystal lattice. Recently, it has been proposed that the blue green emission in this material might be associated to a transition within a self-activated center formed by a double-ionized zinc vacancy V_{zn}^{2-} and the single-ionized interstitial Zn^{+} at the one and/or two nearest-neighbor interstitial sites [39,33]. According to Vanheusden et al. [31], the green–yellow luminescence (deep-level luminescence) is related to the oxygen vacancies in the grown ZnO crystals.

Egelaaf and Oelkrug [45] reported that those defect-related luminescence are caused by radiative transitions between shallow donors (related to oxygen vacancies) and deep acceptors (Zn vacancies). The acceptor level (Zn vacancies) is located 2.5 eV below the conduction band-edge [40,32], while the donor level is known as shallow as 0.05–0.19 eV. In our undoped ZnO samples, no copper impurities were detected using electron probe microanalysis (EPMA) and still we observe the blue–green emission. Therefore, it is likely that the blue-green emission observed in our samples might be better explained by the second hypothesis. The third peak at λ = 672 nm ($E\lambda$ = 1.84 eV), which appears markedly broad for this low beam energy, may due to the defect related deep emissions. Lohnert and Kubalek [41] pointed out that the intensity of broad luminescence observed from about 450 to 700 nm depend on the oxygen partial pressure in the sintering process of ZnO ceramics. This also suggests the hypothesis of a transition within Vo x neutral and single-ionized oxygen V_o^-.

The emission spectrum for the Sn-doped films, using electron beam energy of 5 keV, is a broad band composed of two peaks at 372 and 672 nm (Fig.12 (b)). The incorporation of Sn apparently results in a competitive phenomenon that overshadows the blue–green emission. This behavior could be originated by a more favorable transmission from the localized state generated by the Sn or could be due to a reduction of the self-activated centers responsible for the blue–green emission by occupation of the Zn vacancies by Sn ions. This figure confirms the shift of absorption edge observed by optical results.

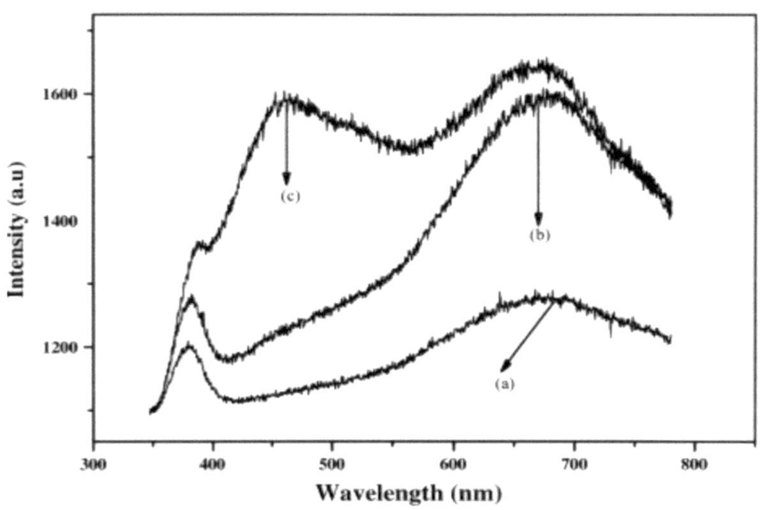

Fig. 13. Cathodoluminescence emission of Sn-doped ZnO films at different electron beam energies: (a) E = 7 keV, (b) E = 10 keV and (c) E = 15 keV.

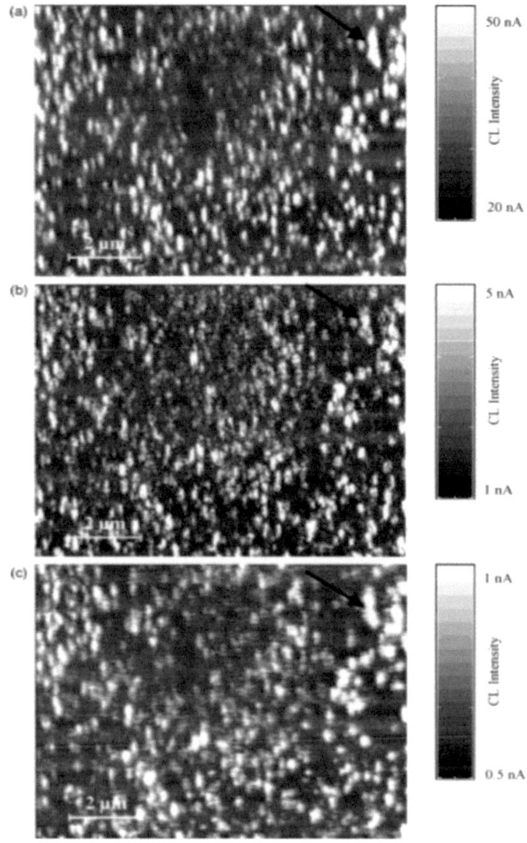

Fig. 14. Cathodoluminescence images of undoped ZnO thin film taken at E = 3 keV: (a) polychromatic CL image, (b) monochromatic CL image taken at λ = 382 ± 8 nm, (c) monochromatic CL image taken at λ = 520 ± 8 nm. The full arrow in each image is a marker for the investigated region of the film surface area.

Fig.13 represents the CL study of the Sn-doped sample at different electron beam energies (E = 7, 10, and 15 keV) proposed for investigating the material homogeneity versus the penetration depth of the electrons. It is clearly appeared that the predominant emission depends strongly by excitation. Indeed, at low excitation alone the tow peaks situated at 372 and 672 nm are observed. The peak localized at 372 nm is due to a transition band-gap with an increasing in band-gap value witch is due to tin doping. From 10 keV a new peak centered at 465 (Eλ = 2.66 eV) began to appear. This emission, that seems typical of tin element, may be associated with a complex luminescent center like (VZn -SnZn)−. At 15 keV, the peak at 465 nm becomes comparable in intensity with the peak situated at 672 nm. Therefore, the increase of this peak's intensity with E means that the tin concentration is higher in the bulk. This result has been verified by X-ray photoelectron spectroscopy (XPS) measurements after successive etching treatment of the sample surface. A similar investigation of the undoped ZnO film has shown that the intensity of the three peaks proportionally increases with the increase of E, implying that the SP deposition method can lead to the formation of homogeneous samples.

Figs. 14 and 15 show polychromatic and monochromatic images of the undoped and Sn-doped ZnO films obtained at the exciting beam energy E = 3 and 5 keV, respectively. The monochromatic images were recorded by selecting the appropriate wavelength with a spectral bandwidth of 16 nm. The CL images of undoped ZnO films (Fig.14) show that the luminescence is located at defined sites giving rise to a grain-like structure inherent to the sample morphology. In view of the relatively low exciting beam energy E = 3 keV, the lateral resolution of the CL images is noticeable. It is smaller than the mean grain size and approximates 100 nm, indicating that the diffusion length of the undoped ZnO film is of this order of magnitude or even smaller. Both monochromatic images at λ = 382 and 520 nm are not much different indicating that the luminescent centers responsible for the band-edge and the blue–green emissions are approximately located at the same places. Nevertheless, in Fig. 6(b), the monochromatic image corresponding to the band-edge emission is presented. A higher density of emitting sites, with smaller luminescent spots than those of Fig. 6(c).

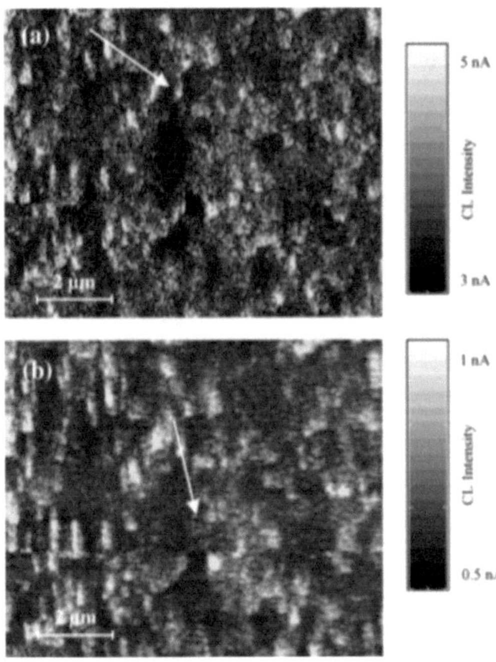

Fig. 15. Cathodoluminescence image of F-doped ZnO thin film taken at 4 keV: (a) polychromatic CL image, (b) monochromatic CL image taken at λ=372±8 nm. The full arrow in each image is a marker for the investigated region of the film surface area.

The CL images of Sn-doped ZnO films (Fig.15) show that the emission is less uniform, appearing more as a granular structure, with a greater luminescent spots than that of the intrinsic ZnO. The grain-like structure cannot be resolved by CL due to the finite carrier-diffusion length. Such a modulation of emission intensity would be attributed to areas with varying grain density, in which the bright regions corresponding to dense grains. Only a few luminescent defects, indicated by arrows on the images, are present. Notes that it is very difficult to take the monochromatic image at λ = 465 nm, because this emission appears at higher exciting beam energy.

In order to study, the effect of Sn doping on the properties of ZnO, we have plotted in Fig.16, the luminescent evolution versus different level doping at a single electron beam energy E = 5 keV. We have remarked a disappearance of bleu green emission observed in undoped ZnO even if level doping and peak intensity becomes more important when level doping increases.

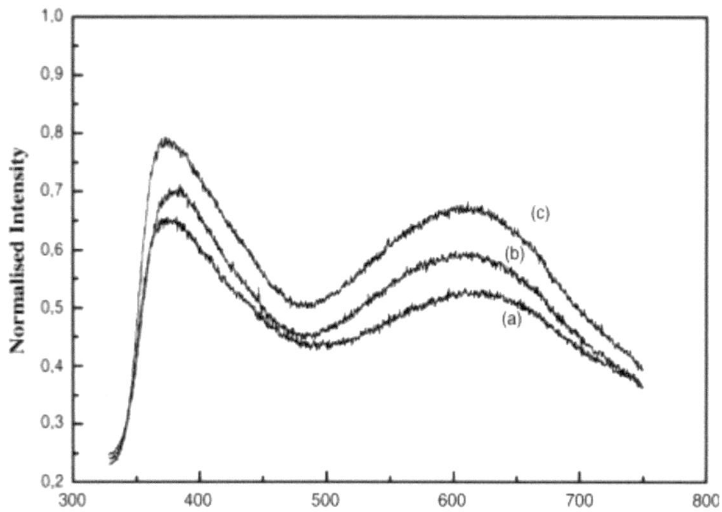

Fig. 16. Cathodoluminescence emission of Sn-doped ZnO films at different level doping: (a) 2 at.%, (b) 4 at.% and (c) 6 at.%.

3.3 Conclusion

Structural, optical and cathodoluminescence characteristics of sprayed ZnO films have been studied here. The effects of tin dopant at the atomic percentage of 6% on these film properties have been analyzed. The hexagonal wurtzite structure of the material is not modified by the presence of the Sn-dopant. The preferred (0 0 2) growth orientation of the films is not affected while an improvement of the material crystallinity is observed with this dopant. The cathodoluminescence analysis of the samples shows that the undoped film presents three bands: a near UV emission at λ = 382 nm in the near UV region, a blue-green emission at λ = 520 nm and a weak red one at λ = 672 nm. Incorporation of tin extinguishes the blue-green band while appears a blue light at λ = 465 nm and increases the value of the band-gap transition. CL imaging analysis shows that the repartition of the emitting centers in the material is intimately connected to the film morphology. The presence of tin in the material leads to great luminescent spots, due to large grain sizes.

4. Cathodoluminescence of undoped and Erbium-doped ZnO thin films

4.1 Introduction

The present work proposes an examination of erbium species influence on the visible and near UV luminescence of thermally sprayed ZnO-Er films. As shown elsewhere, the conditions used for this large scale and cheap deposition technique greatly affect the emitting characteristics of the undoped specimen. Hence, our first objective is the right monitoring of the formation process for this matrix material from which the doping effect can be safely investigated. A particular attention is paid to the structure and composition evolution that mostly remain the basic origin of changes in materials physical properties. The use of rare earth elements as lanthanides engenders more complex phenomena resulting from a combined optical effect of both ZnO and the dopant shell as shown for embedded Er^{3+} species investigated nowadays for planar wave guides in telecommunication applications. In that case, ZnO host appears an excellent amplifier of the Er3+ intra-4f transition signal engendering an infrared line of 1.55 μm wavelength. This somewhat relegates from the time being to minor importance the role of these species in the near UV and visible emissions of ZnO thin films. Despite of it, some works are done here and there in this direction. The study of the lifetime decay in the green fluorescence observed with thermally annealed (Er^{3+}/Si^{4+})-ZnO nanocrystalline films was proposed by Kohls et al. [42] The room temperature blue luminescence of Er-doped ZnO thin films was also investigated by Zhang et al. [43] from samples prepared by simultaneous evaporation of ZnO and Er materials. It clearly appears in most of these reports that only expensive doped Er-ZnO films are involved in this research area. The cathodoluminescent characteristics of the samples in the near ultraviolet and visible region depict a complete extinction of the visible emitted bands (λ =445, 526, and 665 nm) at 1 at. % Er content. Their deactivation below this concentration is explained by a compensation of oxygen defects in the material due to the oxygen-rich medium of the deposition bath. Their reactivation beyond this particular concentration is ascribed to the increase of the Er^{3+} ion shells whose internal radiative transitions lead to a recovering of these visible emitted bands. The radiative mechanism of the transitions from the $^4F_{9/2}$ excited states to then $^4I_{15/2}$ ground state, responsible for the λ =665 nm emission, is predominant in that case. The respective normalized intensity of the violet λ =445 nm and green λ =526 nm emitted bands exhibits a maximal value for 3 at. % Er content, reaching a stabilized regime from about 5 at. % Er.

4.2 Experimental and characterization

Varying Er concentration in ZnO matrix, as proposed in the present work, requires some precautions due to the possible formation of a ternary compound. In actual fact, the physical solubility of Er in ZnO has not yet been clearly evaluated. Due to the high atomic weight of this dopant, this solubility is expected to be somewhat reduced. This necessitates a chemical and structural examination of our investigated samples in the limits of the selected doping concentration interval going till 10 at. %. In these conditions, the proposal of cathodoluminescent study could be done in the light of any change in the material composition and crystal phase.

4.3 Microstructure and composition of ZnO samples

Figure 17 reports the results of x-ray analysis obtained from three representative investigated specimens. As shown, one sees that the material depicts in every case a (002) diffraction peak corresponding to $2\theta = 34.5°$ angular position, which is a characteristic of the hexagonal close packed structure of ZnO $c_0 / a_0 = 1.6024$. This result implies that the related (002) direction, normal to the substrate plane, is the preferential growth orientation in every case. Figure 17(a) clearly reveals that this is the only main orientation of the undoped sample. One sees in Figs.17 (b) and 17(c) that the increase of erbium concentration progressively modifies the film's textural feature with the appearance of two supplementary peaks of the same lattice at $2\theta = 31.8°$ and $2\theta = 36.3°$ that are, respectively, related to the 100 and 101 directions. It should be pointed out that this textural change occurs without any shift in the angular position of 002 diffraction peak and hence with no strain in ZnO matrix structure. A refined analysis of these results reveals that the dependence of the individual intensity of both peaks on the Er content quite differs.

Figure 18 shows that the intensity of the (100) peak always remains higher than the (101) one, regardless of the Er content. However, the (100) peak is more sensitive to the dopant presence particularly for the concentration higher than 5 at. % Er from which the 100 peak becomes markedly enhanced while only a slight increase of the (101) peak's intensity is observed. It should be noted that no typical peak of erbium or Er_2O_3 materials appears in the x-ray diffraction pattern of the films even for 10 at. % Er content. This ultimately gives an indication on the chemical nature of the investigated ZnO films and hence ensures on the real role of Er species in the material CL characteristics.

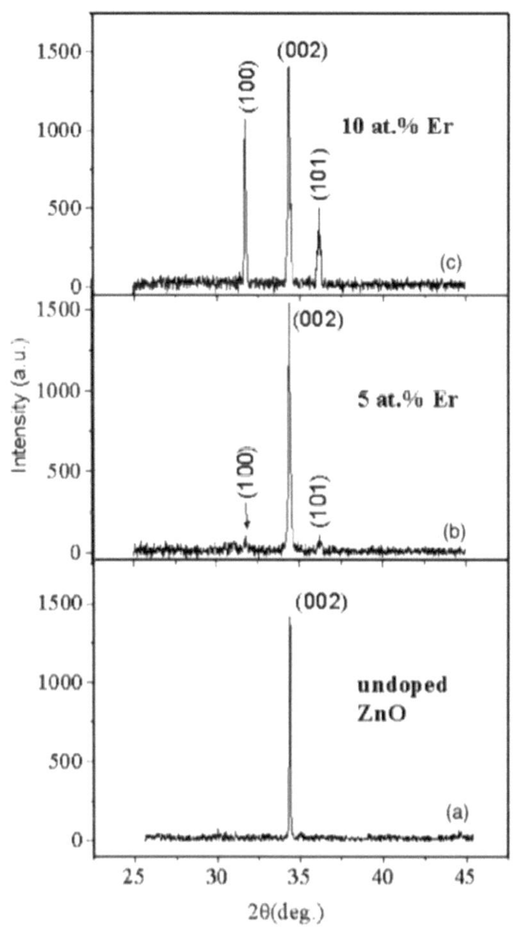

Fig. 17. X-ray diffractograms of ZnO Films: (a) Undoped ZnO (b) Er-doped ZnO 5 at. %, (c) 10 at. %.

The study of the samples composition leading to the results of Table 3 denotes the almost entire recovering of erbium proportion of the spraying bath in Er–ZnO samples, which raises the problem of the dopant insertion mode. Due to the difference in the valence charge [Zn(II) and Er(III)] and the ionic radius, [$R_i(Er^{3+})\cong0.09$ nm and $R_i(Zn^{+2})\cong0.074$ nm], Er is more likely to be interstitial than substitutional in ZnO matrix. On the other hand, the increase of Er content in the doped samples goes together with a decrease of Zn /O ratio. That indicates an increase of oxygen defects in ZnO lattice that is expected to favor the building of Er–O complexes of some interest in the study of the IR emissions of the material.

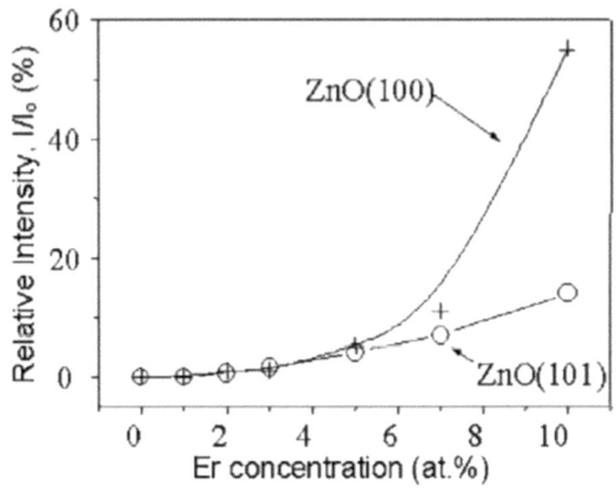

Fig. 18. Relative integrated intensity I/Io of the 100 and 101 x-ray peaks measured from the intensity Io of the 002 peak with respect to Er content in the investigated samples.

Bath composition $[Zn^{+2}, Er^{+3}]$ (*M*)	Zn (at. %)	O (at. %)	Er (at. %)	Zn/O	Cl (at. %)	S (at. %)	C, Si, Be (at. %)
[0.05,0.00]	47.80	47.50	0.00	1.006	2.20	1.60	0.89
[0.049,0.001]	45.57	47.30	2.23	0.963	1.93	1.95	1.02
[0.047,0.003]	42.93	46.25	4.87	0.928	1.85	2.90	1.20
[0.045,0.005]	37.65	46.70	10.15	0.815	1.73	2.61	1.11

Table 3. Compositional feature of the investigated Er doped ZnO films deposited a constant flow rate value f = 5 ml min_{-1}.

The presence of dopant greatly affects the films surface morphology as depicted in the SEM images of Fig.19. If the undoped sample image in the inset of Fig.19 (a) exhibits regular hexagonal-like grains; the insertion of Er species in ZnO matrix engenders variously shaped crystallites, as appearing in Figs. 19a–19c. Besides, one sees that the sample's mean grain size D determined by the Debye-Scherrer broadering method from the main (002) diffraction peak 21 is greatly affected by the Er content. The corresponding D =*f* (Er at. %) curve of Fig. 4 shows that D tremendously increases from 1% concentration onwards, the D value of the undoped sample being slightly higher than that obtained at this particular concentration.

Fig. 19. SEM images of Er-doped ZnO films collected under the incident electron beam energy Eo = 4 keV: (a) for 1 at. %, (b) for 5 at. %, and (c) for 10 at. % Er concentrations. The inset of image (a) depicts the undoped ZnO film's micrograph at the same scale, as proposed in Ref. [44].

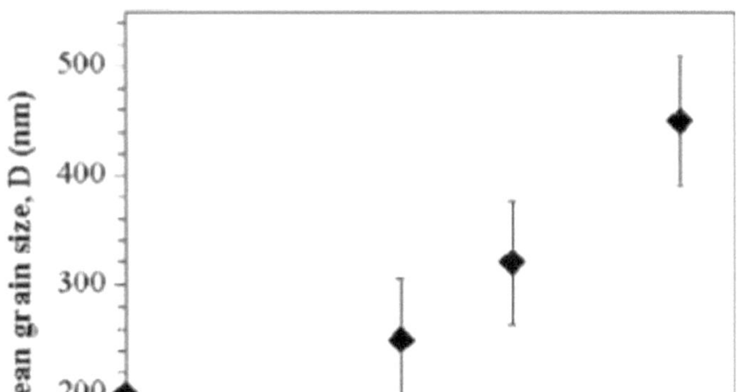

Fig. 20. Dependence of the mean grain size (D) extracted from the SEM images on the sample Er concentration.

4.4 CL– Spectral study of the investigated films

4.4.1 Qualitative CL– Spectral study of the investigated films

The changes observed in the general profile of the CL spectrum linked to the investigated samples give the first indications on the effect of the selected parameters. Referring to the undoped sample result and the common electron beam energy E_0=5 KeV, one can see in Fig.21 that the evolution of the sample CL-spectral profile somewhat deviates from the regular insertion of Er in ZnO matrix. In actual fact, it clearly appears that the spectrum of the Er 1 at. %_-doped ZnO specimen exhibits only the near UV emission peak of 382 nm wavelength. This greatly differs from the result of the other samples for which at least one supplementary peak appears in the visible region. Since these supplementary emission peaks exist for both the undoped and the Er 2 at. % ZnO samples, one infers the occurrence of a specific phenomenon in the material for 1 at. % Er. This phenomenon reflects a deactivation and a reactivation of the peaks, respectively, centered at 440, 520, and 650 nm, below and beyond this particular Er concentration. These peaks are frequently known to be characteristic of the undoped ZnO CL spectrum [45]. All of them are commonly assigned to structural defects, their individual mechanism being not yet clearly identified. The green emission of 520 nm wavelength is currently associated with either Zn or O defect or excess [31].

The result in Table 3 regarding this sample confirms the as expected non stoichiometry feature since it corresponds to the ratio Zn/O=1.006. The observed oxygen defect of the material can be presumably compensated during the film formation by the oxygen-rich phase of the Er_2O_3 deposition solution, the related Er content 1 at. % being marginal. This compensation, directly depending on the defects rate of the undoped sample, is expected to engender deactivation of the green emission as observed here before Er 1 at. % concentration. The increase of Er in ZnO matrix favors the selective excitations of Er^{+3} ions indirectly due to the electron-hole _e-h_ intermediate process via ZnO host, as explained in Ref [46]. This appears as the main cause of the spectral results for Er content higher than 1 at. % for which the inner shell transitions of the Er^{3+} ions engender the emitted peaks. In that case, practically the same visible bands than the undoped ZnO sample are obtained due to the radiative mechanisms from different excitated states to the ground state $I_{15/2}$. 4F5/2 excitated state originates the peak centered at 455 nm, while $^2H_{11/2}$ and $^4F_{9/2}$, respectively, lead to the 526 and 665 nm emissions.

Assuming that the effect of the electron beam energy on the investigated sample is similar regardless of the doping concentration, the study is here proposed from a Er 5 at. % doped specimen. In that case, Fig.22 reveals two main observations: (i) some of the typical CL peaks appear only with the increase of Eo value, which is clearly depicted comparing spectra (a) and (f). The modification obtained can be explained by a deeper penetration of the incident electrons in the material that increases the rate of all the shell electronic transitions of Er^{+3} described above as being responsible for the main peaks obtained in the visible region. We point out that the near UV emission of 385 nm characteristic of the excitonic band edge of pure ZnO material is free of doping, and hence is only affected by this general beam electron penetration that should be the origin of the increase of its peak's intensity. (ii) The results obtained shows that E_o differently affects the as-obtained CL peaks. In actual fact, it clearly appears that the transition from the ($H_{11/2}$) excited states linked to the 526 nm emission peak is predominantly involved in the E_o increase even though those from the ($F_{5/2}$) and ($F_{9/2}$) states, respectively, related to 455 and 665 nm, also remain important. The regular increase of all these peaks intensity contrasts with the appearance of specific emissions linked to particular excitation energy values observed in the photoluminescent results of Er doped ZnO thin films reported by Komuro et al [47].

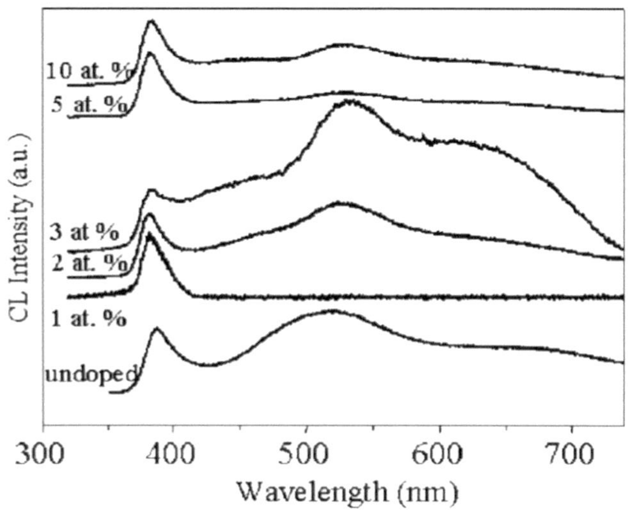

Fig. 21. Dependence of the CL-spectral characteristics of the undoped and Er-doped ZnO films on the sample's Er concentration obtained at the incident electron beam energy E= 10 keV.

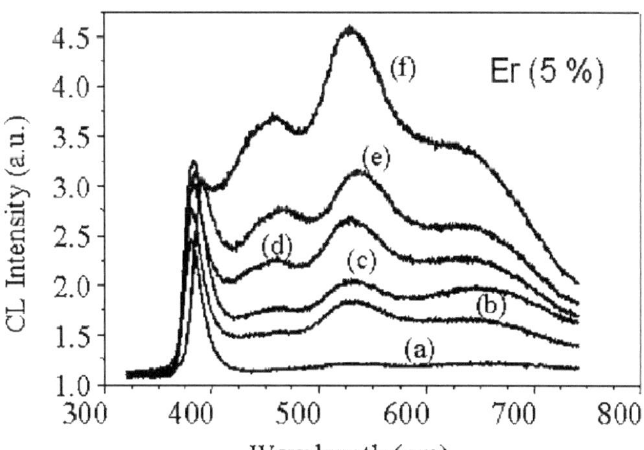

Fig. 22. Dependence of the CL spectral characteristics of Er(5 at. %)-doped ZnO sample on the incident electron beam energies: (a) for Eo = 4 keV, (b) for Eo = 5 keV, (c) for Eo = 7 keV, (d) for Eo = 10 keV, (e) for Eo = 15 keV, and (f) for Eo = 20 keV.

We point out that the broadening of the CL spectral peaks in Figs. 21 and 22 is likely to be connected with electronphonon interactions. This is a characteristic of disordered media in the range of the ZnO band gap (1.5–3.5 eV), particularly under the Er doping effect. [48]

4.4.2 Integration and normalized intensity of the CL – Spectral peaks

An accurate study of a luminescent spectrum requires taking into account the possible peak's overlap phenomenon. This is currently overcame by a spectral modeling leading to a better peak's integration that remains necessary for the assessment of their relative intensity. The application of Gaussian model, as shown in Fig.23, appears to be more appropriate here. The reported result shows that the typical four CL peak feature of the Er doped ZnO material obtained with spectra Figs. 21 and 22 remains unchanged in the modelling despite the shape modification caused by the incident electron beam energy. With the near UV emission being free of the doping process, only the evolution of the three visible CL peaks mentioned above as centered at 445, 526, and 665 nm wavelengths, respectively, can be studied with respect to Er content in ZnO material. Besides, their intensity can lead to examine the effect of the incident electron beam energy for every Er-doped ZnO sample in the region higher than 1 at. %. The normalization of their respective intensity from the one of the near UV value leads to a better assessment of the doping and the incident energy influences.

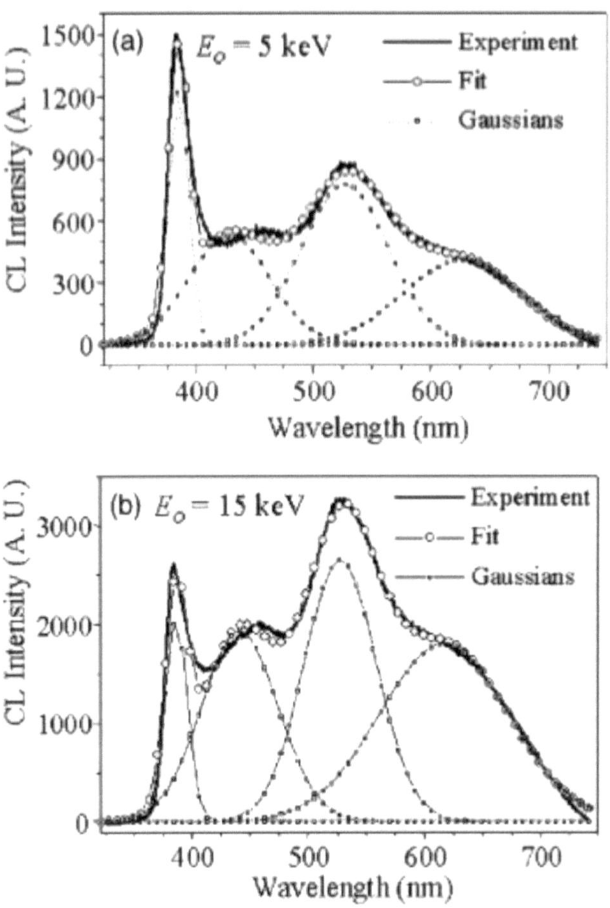

Fig. 23. Gaussian decomposition of CL spectra related to the Er (5 at. %)-doped ZnO sample for two different incident electron beam energies: (a) for E= 5 keV and (b) for E= 15 keV.

The results of Fig.24 show that, for a constant incident electron beam E_o=10 KeV, the increase of Er content in ZnO matrix leads to different peaks intensity profiles. However, a common increasing trend is observed till 3 at. % Er content. This concentration corresponds to a maximum for the intensities of the violet 445 nm and green 526 nm emissions. In both cases, a decrease is observed from this Er percentage towards a plateau neatly appearing at nearly 5 at. %. No decrease is obtained for the red emission 665 nm whose plateau onset is 3 at. % with a higher intensity value. One infers from these results that the radiative mechanism from the excited state $F_{9/2}$ to the ground state $I_{15/2}$ is quite predominant in the investigated Er concentration region. The lowering in the rate of the radiative mechanisms from the $F_{5/2}$ and $H_{11/2}$ excited states to $I_{15/2}$ linked to the 445 and 526 nm peak's intensities might have a link to be clarify with either the reported textural change of Fig.17 occurring at Er concentration higher than 3 at. % or the material composition. Since this material texture remains unchanged below this concentration, the regular increase of the three CL peaks intensity in this region is likely to be the fact of a regular increase of all the radiative mechanisms involved going together with the increase of the active Er species in the material.

The similarity of the electron beam energy effect on the investigated samples assumed above is here confirmed in Fig.25 from the study of samples having three different Er contents. It appears in every case that the increase of E_o leads to that of the normalized intensity of the green emission peak of 526 nm. We point out that the same increasing trend is observed with the violet and red emissions of 445 and 665 nm. As already mentioned, this result has a direct connection with the irradiated volume of the material due to further Er ions involved. A saturation regime is expected at higher E_o value, the related current being mostly affected by the sample Er content and possibly by the textural change observed between 2 and 10 at. %.

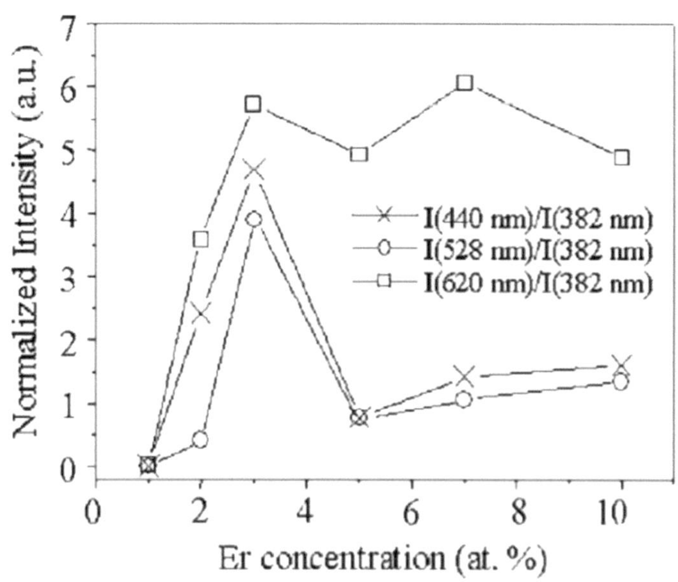

Fig. 24. Normalization (from the UV band) of the integrated intensity related to the main emission peaks and its dependence on the sample Er concentration for E= 10 keV.

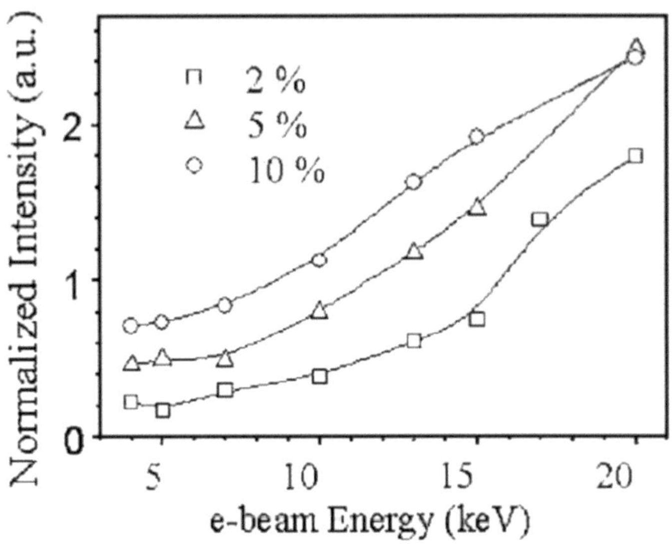

Fig. 25. Normalization (from the UV band) of the integrated intensity related to the green emission peak (λ = 526 nm) and its dependence on the incident electron beam energy for three different samples Er concentration.

4.4.3 CL imaging and repartition of the emitting centers

The study is here performed with a Er 5 at. % doped ZnO sample under an incident beam energy E_o=5 KeV being agreed that the results are similar regardless of the doping concentration and E_o value. The CL images are collected at a required wavelength having a 16 nm spectral bandwidth. Figure 10 depicts those obtained from each of the main emitted wavelengths of the material, the dark mark surrounded by a dashed line in some pictures just serves for the positioning of the investigated sample zone. Figure 10-a related to the UV band edge λ =382 nm clearly shows that the luminescent emitting centers of this particular wavelength are regularly dispersed all over the material, merging with the material surface grains. Being free of doping, such a repartition of emitting centers is easily explained since only ZnO matrix is here concerned. It could be noted that the CL image as obtained is the neatest one due to the predominant intensity of the related peak as it appears in Fig.21. The emitting centers' image of the green wavelength λ =526 nm proposed in Fig.26-b shows that the grain like feature is once again recovered. One infers that Er species are well localized in the surface grains. The result in Fig. 26-c obtained for the red line λ =665 nm greatly contrasts with the two precedent investigated wavelengths since the emitting centers are not clearly distinguished, the intensity signal being quite negligible in Fig. 21. The relationship between the Er species repartition and the emitting centers is thereby difficult to be established. Probably, a better result could be expected for this wavelength at higher incident energy values.

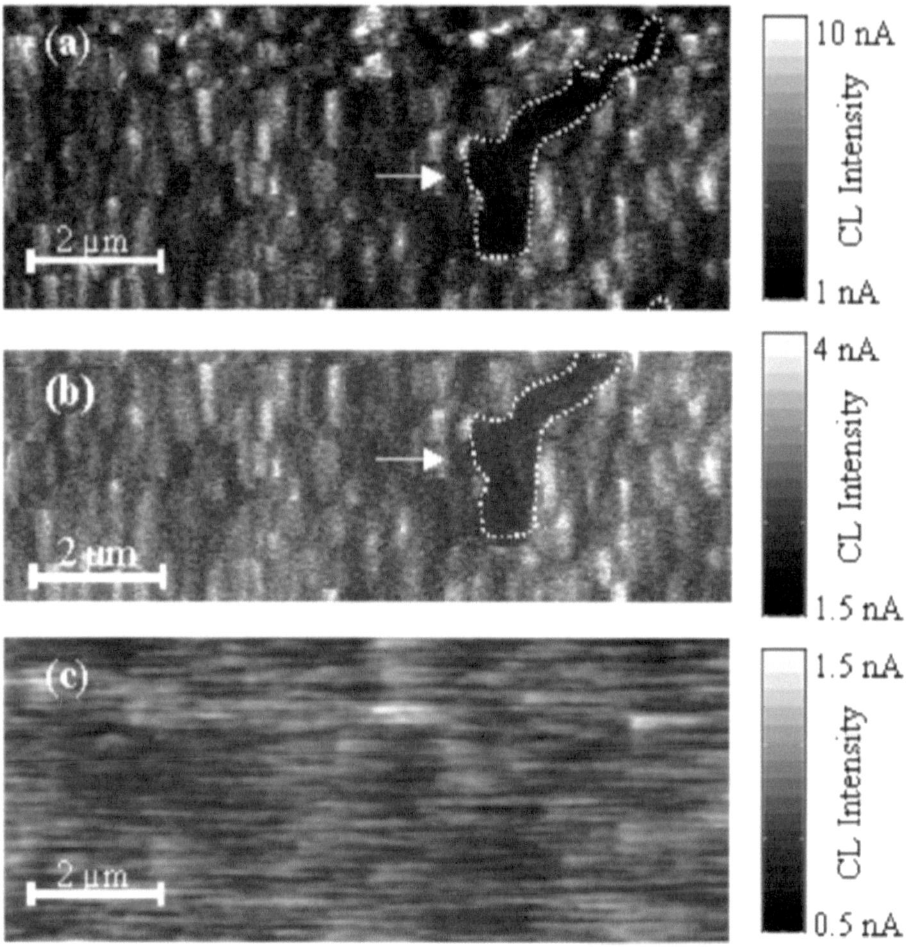

Fig. 26. Monochromatic CL images of the Er (5 at. %)-doped ZnO films collected at E_0 = 5 keV, with a focalized spot of 5 nm diameter: (a) for λ=382 ± 8 nm, (b) for λ=526 ± 8 nm, and (c) for λ=665 ± 8 nm wavelengths.

5. Summary and conclusion

It comes out of the present work that the role of heavy elements as lanthanides in the doping of ZnO material has not yet been quite explored. Dealing here with Er species inserted in this material matrix, we find that the increase of erbium concentration leads to some textural change in the growth of Er–ZnO film. A single 002 preferred growth orientation is obtained at lower Er contents, while the 100 and 101 additional ones become neatly perceptible from about 5 at. % Er concentration. Restricting our luminescence study in the near UV and visible regions, we show that Er concentration greatly affects the CL-spectral feature of the material. All the visible emitted bands entirely disappear for 1 at. % Er concentration. In the investigated wavelength region, the normalization of the main visible peaks intensity from that of the UV peak shows that the radiative transitions in the Er^{+3} ion shell from the F9/2 excited states to I15/2 ground state, responsible for the 665 nm emitted band, are always predominant in the material. Both, the green λ=526 nm and the violet λ=445 nm emitted bands exhibit a maximal normalized intensity for 3 at. % Er concentration that is followed by a lowering and stabilization reached beyond 5 at. % Er content.

6. Conclusion

The dependence of cathodoluminescence properties of spray-deposited ZnO films on process parameters such as substrate temperature, airflow rate and precursors, has been studied in detail. We have found that the luminescence intensity depend strongly on process parameters. The optimum values of substrate temperature and airflow rate have been determined to give the best luminescent ZnO films.

The produced films at optimum condition exhibit luminescence characteristics comparable to films grown by more sophisticated techniques. These films may be useful for specific applications as transparent n-type windows in solar cells or for sensors devices where large surface areas are needed and may also be an important candidate for UV diode lasers applications.

It seems here evident that the particular spectral feature of the single emitted near UV band obtained for 1 at. % Er content in ZnO matrix can be of prime interest in the technology of low cost light emitting diodes LEDs.

7. References

[1] Cowley. J.M. (1969)- Image contrast in a transmission scanning electron microscope. Appl. Phys. Lett., 15, 58.
[2] Z.C. Jin, J. Hamberg, C.G. Granqvist, J. Appl. Phys. 64(1988) 5117.
[3] J.B. Yoo, A.L. Fahrenbruch, R.H. Bube, J. Appl. Phys. 68(1990) 4694.
[4] Y. Chen, D.M. Bagnall, H. Foh, K. Park, K. Hiraga,Z. Zhu, T. Yao, J. Appl. Phys. 84 (1998) 3912.
[5] E. Jeff, Nause, III-Vs Review 12 (1999) 28.
[6] Y. Chen, D.M. Bagnall, T. Yao, Mater. Sci. Eng. B 75(2000) 190.
[7] Y.C. Kong, D.P. Yu, B. Zhang, W. Fang, S.Q. Feng,Appl. Phys. Lett. 78 (2001) 407.
[8] R. Menner, B. Dimmler, R.H. Maunch, H.W. Shock,J. Crystal Growth 86 (1988) 906.
[9] C.Y. Lee, Y.K. Su, S.L. Chen, J. Crystal Growth 96 (1989)785.
[10] A. Sarkar, S. Chaudhuri, A.K. Pal, Phys. Stat. Sol. (A) 119(1990) K21.
[11] T. Minamai, T. Yamamoto, T. Miyata, Thin Solid Films366 (2000) 63.
[12] K. Haga, F. Katahira, H. Watanabe, Thin Solid Films 344(1999) 145.
[13] Y. Natsume, H. Sakata, T. Hirayama, H. Yanigida,J. Appl. Phys. 72 (1992) 4203.
[14] M. Ruth, J. Tuttle, J. Goral, R. Noufi, J. Crystal Growth96 (1989) 363.
[15] A. Ortiz, C. Falcony, J. Hernandez, M. Garcia,J.C. Alonso, Thin Solid Films 293 (1997) 103.
[16] A. Maldonado, R. Asomoza, J. Canetas-Ortega,E.P. Zironi, R. Hernandez, O. Solorza-Feria,Sol. Energ. Mater. Sol. Cells 57 (1999) 331.
[17] S.A. Studenikin, M. Cocivera, W. Kelner, H. Pascher,J. Lumin. 91 (2000) 223.

[18] M. de la, L. Olvera, A. Maldonado, R. Asomoza,O. Solorza, D.R. Acosta, Thin Solid Films 394 (2001) 242.
[19] N. Benhamdane, W.A. Murad, R.H. Misho, M. Ziane,Z. Kebbab, Mater. Chem. Phys. 48 (1997) 119.
[20] S.A. Studenikin, N. Golego, M. Cocivera, J. Appl. Phys.83 (1998) 2104.
[21] M. Addou, A. Moumin, B. El idrissi, M. Regragui, A.Bougrine, A. Kachouane, J. Chem. Phys. 96 (2) (1999) 232.
[22] F. Paraguay, D.J. Morales, W. Estrada, L.E. Andrade,M. Miki-Yoshida, Thin Solid Films 366 (2000) 16.
[23] H.J. Egelhaaf, D. Oelkrug, J. Crystal Growth 161 (1996)190.
[24] T. Sekiguchi, N. Ohashi, Y. Terada, Jpn. J. Appl. Phys.Part 2 Lett. 36 (1997).
[25] P. Fons, K. Iwata, S. Niki, A. Yamada, K. Matsubara,J. Crystal Growth 201/202 (1999) 627.
[26] H. Nanto, T. Minami, S. Takata, Phys. Status Solidi A 65(1981) K131.
[27] J. Zhong, A.H. Kitai, P. Mascher. J. Electrochem. Soc. 140(1993) 3644.
[28] K. Vanheusden, W.L. Warren, C.H. Seager, D.R. Tallant,J.A. Voigt, B.E. Gnade, J. Appl. Phys. 79 (1996) 7983.
[29] P. Scherrer, Goettinger Nachr 2 (1918) 98.
[30] R. Dingle, Phys. Rev. Lett. 23 (1969) 579.
[31] K. Vavheusden, C.H. Seager, W.L. Warren, D.R. Tallant,J.A. Voigt, Appl. Phys. Lett. 68 (1996) 403.
[32] E.G. Bylander, J. Appl. Phys. 49 (1978) 1188.
[33] T. Minami, H. Nanto, S. Takata, J. Lumin. 24/25 (1981)63.
[34] C. Falcony, A. Ortiz, M. Garcia, J.S. Helman, J. Appl.Phys. 63 (1998) 7.
[35] A. Mitra, R.K. Thareja, V. Ganesan, A. Gupa, P.K.Sahoo, V.N. Kulkarni, Appl. Surf. Sci. 174 (2001) 232.
[36] K. Lohnert, E. Kubalek, Microscopy of SemiconductingMaterials, Institute Of physics Publishing, Bristol, 1983,
[37] M. Troyon, D. Pastre, J.P. Jouart, J.L. Beaudoin, Ultramicroscopy75 (1) (1998) 15.
[38] I. Shin, C.V. Qin, J. Appl. Phys. 58 (1985) 2400.
[39] S.A. Studenikin, N. Golego, M. Cocivera, J. Appl. Phys. 84 (1998) 4.
[40] H.J. Egelaaf, D. Oelkrug, J. Cryst. Growth 161 (1996) 190.
[41] K. Lohnert, E. Kubalek, Microscopy of Semiconducting Materials,IOP, Bristol, (1983) 303.
[42] M. Kolhs, M. Bonanni, L. Spanhel, D. Su, and M. Giersig, Appl. Phys. Lett. 81, (2002) 3858
[43] X. T. Zhang, Y. C. Liu, J. G. Ma, Y. M. Lu, D. Z. Shen, W. Xu, G. Z. Zhong, and X. W. Fan, Thin Solid Films 413, (2002) 257 .
[44] A. El Hichou, A. Bougrine, J.-L. Bubendorff, J. Ebothé, M. Addou, and M. Troyon, Semicond. Sci. Technol. 17, (2002) 607
[45] C. Shi, Z. Fu, C. Guo, X. Ye, Y. Wei, J. Deng, J. Shi, and G. Zhang, J. Electron Spectrosc. Relat. Phenom. 101–103, (1999) 629.
[46] X. T. Zhang, Y. C. Liu, J. G. Ma, Y. M. Lu, D. Z. Shen, W. Xu, G. Z. Zhong, and X. W. Fan, Thin Solid Films 413, (2002) 257.

[47] S. Komuro, T. Katsumata, T. Morikawa, X. Zhao, H. Isshiki, and Y. Aoyagi, J. Appl. Phys. 88, (2000) 7129.
[48] I. V. Kityk, J. Wasylak, D. Dorosz, J. Kucharski, S. Benet, and H. Kaddouri, Opt. Laser Technol. 33, (2001) 511.

6

Multicolor Luminescence from Semiconductor Nanocrystal Composites Tunable in an Electric Field

G.N. Panin
[1]*Department of Physics, Quantum-Functional Semiconductor Research Center Dongguk University, Seoul,*
[2]*Institute of Microelectronics Technology & High Purity Materials, Russian Academy of Sciences, Chernogolovka,*
[1]*Korea*
[2]*Russia*

1. Introduction

Semiconductor nanostructures have interesting electro-optic properties due to quantum confinement of electrons, strong exciton binding energies, and the possibility to tailor the bandgap and radiative emission rates. An electric field normal to the quantum well, for example, shifts the absorption edge to lower energies (red-shift) and increases the refractive index below the absorption edge allowing to control the intensity of a light beam through electroabsorption. Such nanostructures can be conveniently used for the direct modulation of light, since they show much larger electro-optic effects than bulk semiconductors. Moreover, the optical properties of semiconductor nanostructures can be controlled by an electric field due to different emission rates from radiative centers in different charge states.

In the Chapter, the effect of an external electric field on cathodoluminescence from semiconductor nanocrystals and nanocomposites with different radiative emission rates is described, giving special emphasis to ZnO nanocrystals in MgO and polymer matrix.

In the next Section we shortly review the preparation of ZnO nanocrystals and nanocomposites, including a chemical solution deposition (CSD) of core/shell nanocrystals, as well as their structural and optical properties.

In Section 3 of the chapter, the effect of an external electric field on cathodoluminescence from the nanocrystal/polymer structure will be studied, giving special emphasis to the doped ZnO nanocrystals in Poly(4,4′- diphenylene diphenylvinylene) (PDPV) matrix and PBET/ITO structures. Electric field-induced color switching of cathodoluminescence from nanocrystals/polymer is reviewed. The assumed mechanism of electric field-tunable cathodoluminescence implies the presence of radiative recombination channels, which are sensitive to the electric field through the band bending at the crystal surface. The experiments on a reversible quenching of the UV near-band-edge emission under visible illumination confirm an appearance of the recombination channel after recharging oxygen-vacancy states in the surface depletion zone. It has been shown that the UV near-band-edge emission can be modulated at frequencies of hundreds of hertz. Limiting performance bounds for potential future devices fabricated from nanocrystals with different radiative emission rates are considered.

2. Structural and optical properties of ZnO (MgO) nanocrystals and nanocomposites

Zinc oxide is a II-VI compound semiconductor with a wide direct band gap of 3.3 eV at room temperature (RT), which makes it attractive for optoelectronic devices in the near UV region (Kang et al., 2006; Look, 2001; Özgür et al. 2005). ZnO has a large exciton binding energy of about 60 meV, which promises efficient RT exciton emission. The unique physical and electronic properties of ZnO, such as large bond strength and large damage threshold for laser irradiation, (Minne et al., 1995; Yamamoto et al., 1999) and its high exciton binding energy allow to fabricate an excitonic laser operating at room temperature (Bagnall et al., 1997; Ohtomo et al., 2000; Yu et al., 1997). A lower pumping threshold for laser operation is expected when exciton-related recombination rather than electron–hole plasma recombination occurs. The use of a ZnO-based quantum structure with a higher exciton binding energy could decrease significantly the current threshold, which is the key point for current injection laser. The modified density of states in the ZnO-based low-dimensional quantum structure can confine both excitons and photons, thus achieving high-efficiency stimulated emission (Sun et al., 2000). The RT cathodoluminescence (CL) spectrum of ZnO along with the exciton emission band at near 380 nm contains a deep-level defect related emission band at 500–650 nm. The deep-level emission band, as usual, is considered consisting of the green (500–520 nm), yellow-orange (560– 600 nm), and red (650 nm) bands. The green emission band is frequently observed and the early studies it was attributed to copper impurities (Özgür et al., 2005). However at present, oxygen vacancies have been assumed to be the most likely candidate for recombination centers involved in the green luminescence of ZnO (Cheng et al., 2006; Leiter et al., 2001; Studenikin et al., 1998b; VanDijken et al., 2000a; Vanheusden et al., 1996a, 1996b). UV (exciton) and visible (defect) emissions are in competition with each other and it is presumed that the deep level emission centers are preferable channels of the electron-hole recombination. Such behavior of the luminescence bands gives rise to interesting physical phenomena, which creates an ability to control the intensity of the UV emission, and could be also of practical relevance. In practice the dependence of the excitonic band on external and internal perturbations is widely used to modulate its intensity. The modulations of the excitonic PL by means of acoustic (Takagaki et al., 2003), terahertz waves (Klik et al., 2005) and visible light (Chen et al. 1996; Shih H. Y. et al., 2011; Kurbanov et al., 2008a; Kurbanov et al., 2008b) as well as the excitonic PL and CL by electrical field (Hagn M., et al., 1995; Panin et al., 2005a; Schmeller et al., 1994; Zhang S. K., et al., 2001; Zimmermann et. al., 1997; Zimmermann et al., 1998; Zimmermann et al., 1999) were reported. The opportunity to control the characteristics, in particular, reflectance, of the sample under study by the external periodic perturbations serves as a basis for the modulation spectroscopy and allow to study electron states and features of the semiconductor band structure (Aigouy et al., 1997; Motyka et al., 2006; Rowland et al., 1998). In ZnO nanocrystals the electron paramagnetic-resonance signal assigned to singly ionized oxygen vacancy (V_0^+) was observed to be photosensitive for photon energies down to as low as 2.3 eV (Vanheusden et al., 1996a, 1996b). Upon illumination, the V_0^+ density was observed to grow. Moreover, it has been found that the intensity of the green emission in ZnO powder correlates well with the paramagnetic single-ionized oxygen-vacancy (V_0^+) density. Nevertheless, the effect of the subband excitation on emission from ZnO is still unclear and remains unexplored in spite of the observed correlation between the density of V_0^+ and the green emission intensity and the sensitivity of V_0^+ to a visible light.

MgO (E_g = 7.3-7.7 eV) (Roessler & Walker, 1967; Johnson, 1954) solved in ZnO can produce a large band gap $Mg_xZn_{1-x}O$ alloy, which is well suitable for quantum structures operating in the ultraviolet spectral range (Ohtomo et al., 1998; Sharma et al., 1999). The radiative recombination of electron–hole pairs in terms of the quantum confined Stark effect was observed in ZnO/$Mg_xZn_{1-x}O$ structures (Makino et al. 2002). According to the phase diagram of ZnO/MgO binary systems (Raghavan et al., 1991), the thermodynamic solid solubility of MgO in a ZnO matrix is less than 4 mol%. The crystal structure of ZnO (hexagonal, a = 3.24 Å and c = 5.20 Å) is vastly different from that of MgO (cubic, a = 4.24 Å). However, since the ionic radius of Mg^{2+} (0.57Å) is almost the same as that of Zn^{2+} (0.60 Å) (Shannon, 1976), Zn^{2+} can be replaced by Mg^{2+} in the ZnO matrix. The synthesis of cubic $Mg_xZn_{1-x}O$ nanocrystals under pressure was reported (Baranov et al., 2005a; Baranov et al., 2010). Stable cubic $Mg_xZn_{1-x}O$ solid solutions *(*$0.33 < x < 0.68$*)* with continuous band gap up to 3.74 eV have been synthesized under high pressures and high temperatures.

2.1 The growth techniques

The growth techniques of NCs are very important since they influence their structure, shape and distribution of nanocrystal sizes, stoichiometry, structure of the surfaces or interfaces, and their optical properties. Therefore, before studying in the next sections the electric field induced optical properties of NCs, we want first to briefly look to the chemical growth techniques, paying special attention to the techniques of self-assembling (Fu et al, 2003; Fu et al., 2005; Panin et al., 2003b; Panin et al 2004d; Panin et al., 2005c; Panin et al., 2007d; Kurbanov et al., 2008c) as well as structural and luminescent properties of as-grown and annealed NCs. Nanostructuring by *physical techniques* based on lithography and etching does not usually produce NCs of sizes small enough for the detection of size effects, even if one uses electron beam techniques. The advantage of these techniques is their compatibility with microelectronics techniques. However, the drawback top / bottom of lithographic techniques in their low productivity and poor optical quality of structures produced by etching. The *chemical techniques* are the ones most used for the preparation of NCs. We will mention a few of them which have been used in preparation of ZnO-based composites.

2.1.1 Colloidal synthesis of ZnO nanoparticles

Colloidal synthesis, by the reduction of metal salts in solutions with organic ligands, are frequently used to produce metal nanoclusters (e.g. gold). As for II-VI semiconductors, nanoparticles of ZnO, CdSe, CdS, etc. are produced from reagents containing the nanocrystals constituents. One reagent contains the metal ions (e.g. Zn^{2+}) and the other provides the oxygen (O^{2-}) or the chalcogenide (e.g. Se^{2-}). The size of the nanocrystals is controlled by the temperature of the solutions and the concentrations of the reagents and the stabilizers. Synthesis of ZnO nanoparticles with sizes of 2–8 nm is fairly detailed, (Baranov et al., 2008; Sakohara et al., 1998; Van Dijken et al., 2001; Wong et al., 2001; Meulenkamp, 1998; Pesika et al., 2003; Spanhel & Anderson, 1991; Tokumoto et al., 2003; Hosono et al., 2004) and the conditions for nanoparticle manufacturing by precipitation of zinc acetate with alkali from alcoholic solutions are also well studied. In particular, particle growth rate curves were obtained and the effects of several factors were determined, such as the water proportion in the solution, the nature of the solvent, the starting solution concentration, and precipitation temperature.

A typical absorption spectrum from suspended ZnO nanoparticles is shown in Fig 2.1 The simplest way to estimate the particle size in colloidal ZnO solutions is to determine the transition position in optical absorption spectra. The absorption onset of ZnO nanocrystals appears at 337nm, which is blueshifted relative to 382 nm absorption onset of bulk ZnO. The spectrum shows that these uncapped ZnO nanoparticles were well dispersed in ethanol, and that the band gap absorption was much more effective compared to the absorption by surface defect states (Chen et al., 1997).

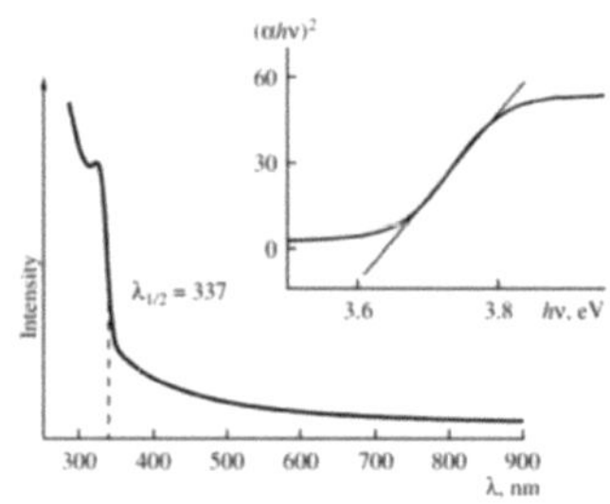

Fig. 2.1. Optical absorption spectra of colloidal ZnO particles prepared in ethanol. In the inset, the same data in other coordinates. (Baranov et al., 2008)

The inset in Fig. 2.1 illustrates the experimental data in the $(\alpha hv)^2$ – hv coordinates. The existence of a linear segment on the $(\alpha hv)^2$ versus hv curve verifies the existence of a direct electron transition which defined the particle adsorption edge. Extrapolation of this segment to axis hv gave a value of 3.6 eV for the band gap of suspended ZnO nanoparticles. Compared to the band gap of micron-sized ZnO (E_g = 3.3 eV), the band gap of ZnO nanoparticles was increased by the quantum confinement effect. The method which we used is to equate E_g with the wavelength at which the absorption is 50% of that at the excitonic peak (or shoulder), so called $\lambda_{1/2}$. To convert measured values of $\lambda_{1/2}$ into particle size an expression (1) can be used:

$$1240/\lambda_{1/2} = a + b/D^2 - c/D \tag{1}$$

where $\lambda_{1/2}$ in nm, diameter D in Å and the values a= 3.556, b =799.9, and c = 22.64, which gives a good description of the experimentally found size dependence for 25 < D < 65 Å (Meulenkamp, 1998). The mean particle size calculated using expression 1 was ~3.0 nm. Comparison of the data derived from the absorption spectra with TEM data and theoretical calculations demonstrates good convergence.

ZnO nanoparticles synthesized from aqueous solution (Kurbanov et al., 2007a; Wang et al., 2005) and deposited on a substrate show usually the larger size (100-1000 nm) and cone-shape (Fig. 2.2 (a)).

Fig. 2.2. Typical SEM images of ZnO nanocrystals (a) as-prepared and (b) annealed in air at 500 °C. The inset shows the specklike defects revealed after annealing at higher magnification. (Kurbanov, et al., 2009a)

As-prepared ZnO nanocrystals display a strong UV luminescence accompanied with a weak yellow-orange emission at around 565 nm. With increasing the excitation power a violet PL band at ~400 nm starts to appear and it dominates over the UV near-band-gap emission at high excitation power. While the UV-PL band intensity displays saturation behavior, the violet band intensity grows superlinearly. The violet luminescence most likely arises from defects associated with zinc vacancy or zinc vacancy related complexes.

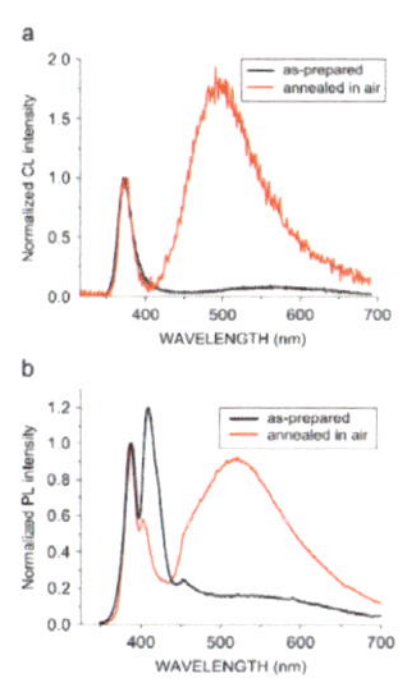

Fig. 2.3. Normalized (a) CL and (b) PL spectra of ZnO nanoparticles: (black) as- prepared and (red) after annealing in air at 500 ºC for 1h. (Kurbanov et al., 2009b)

Figs. 2.3 (a) and (b) show the normalized room temperature CL and PL spectra of the ZnO nanoparticles before and after annealing in air at 500 ºC for 1h, respectively. The annealing resulted in a strong increase in intensity of the green band centered at 505 and 520 nm in the CL and PL spectra, respectively. This band became dominant emission band in the visible region. The decrease in intensity of the violet band at ~400 nm after annealing is clearly observed in the spectra obtained at the high laser power excitation (80 mW) (Fig. 2.3 (b)). The UV emission from nanocrystals is ascribed to near-band- edge luminescence of ZnO. The variation of the UV peak energy as a function of temperature described by the Varshni equation also confirms this statement (Kurbanov et al., 2009b). Because the ZnO nanocrystals were prepared without intentional doping, the violet, green and yellow-orange emissions are attributed to native defect centers and their complexes. Namely, the yellow-orange emission band around 560–580 nm from the as-prepared ZnO nanocrystals has been assigned to oxygen interstitial related defects. The green band peaked at 505–520 nm appearing after the high-temperature annealing and being dominant in the visible emission is attributed to oxygen vacancy related centers. The observed differences in the green band positions in CL and PL spectra could be attributed to specific features of photon and electron beam excitation, particularly, to a different penetration depth of photons and electrons. The first principles investigations show that under oxygen-rich conditions the zinc vacancy and oxygen interstitials are the dominant defect types and they have low formation enthalpies than other defects (Erhart et al., 2006; Kohan et al., 2000). Based on these calculations, it is reasonable to suggest that the violet emission from ZnO nanocrystals grown in thermal equilibrium conditions is originated from Zn vacancy related defects or their complexes. Moreover, as was mentioned above, in most cases the violet emission band is accompanied with the yellow-orange band, attributed to oxygen interstitials (Kumar et al., 2006; Hua et al., 2007; Jeong et al., 2003; Wu et al., 2001; Zhao et al., 2005;). The first principles studies predict that for Zn vacancy, the transition level (0 /-1) lies at ~0.45 eV (Kohan et al., 2000) or ~0.3 eV (Erhart et al., 2006) above the valence band and the transition level (-1/-2) is located at ~0.8 eV (Kohan et al., 2000) or 0.7eV (Erhart et al., 2006) above the valence band. Thus a transition between the conduction band and the single charged Zn vacancy acceptor level would give rise to luminescence at 405 nm (3.06 eV), in reasonable agreement with the predicted transition energy.

A postgrown annealing treatment significantly improved the UV emission efficiency and resulted in the clear appearance of a low temperature emission band around 3.31 eV (so-called A-line). Spatially and wavelength resolved CL measurements revealed a spotlike distribution of the A-line emission on a nanocrystal surface with a strong correlation between the emission around 3.31 eV and the specklike defects that appeared on the nanocrystal surface after annealing. At low temperatures, bound exciton emissions from ZnO nanocrystals are the dominant radiative recombination channels. The energy range between 3.375 and 3.367 eV is normally assigned to ionized donor bound excitons, while the neutral donor bound excitons are positioned between 3.366 and 3.36 eV. The free exciton emission is usually observed at 3.377 eV. According to the conventional classification for the bulk ZnO crystals, the lines situated at lower energies, between 3.33 and 3.31 eV, are attributed to the two-electron satellite (TES) recombination of the donor bound excitons (Kang et al., 2006; Meyer et al., 2004; Teke et al., 2004). However, in recent years many research groups reported on an observation of a strong emission line in the TES region the so called A-line. The position of the A-line varies from 3.31 to 3.333 eV depending on the growing methods of ZnO. The A-line's intensity is strongly enhanced in ZnO powders when compared to bulk ZnO (Fallert et al., 2007). The origin of the A-line emission is currently under discussion. Possible assignments of the A-line include: neutral excitons bound to nitrogen impurities (Look et al., 2002; Yang et al., 2006), donor-acceptor pair recombination (Zhang et al., 2003), free-to neutral- acceptor transitions (Cao et al., 2007), two-electron transition of exciton bound to neutral donor (Studenikin et al., 2000), the first longitudinal optical (LO) phonon replica of free excitons (Hirai et al., 2008) or excitons bound to surface or structural defects (Fallert et al., 2007; Fonoberov et al., 2006; He et al., 2007; Meyer et al., 2004; Teke et al., 2004). Recently, on a basis of cathodoluminescence measurements, it was suggested that this line is composed of two overlapped bands related to point defects and the first LO phonon replica of the free exciton (Mass et al., 2008).

Wavelength resolved CL images as well as, spatially resolved CL spectra from ZnO nanocrystals show a direct correlation between the A-line emission at 3.311 eV and the annealing induced specklike defects (Kurbanov et al., 2009a). The specklike defects appear on all faces of the annealed nanocrystals and lead to a spotlike distribution of the emission. Figure 2.3 (b) shows a SEM image of the annealed ZnO nanocrystals. The annealing of the nanocrystals leads to the formation of specklike defects on their surface, which appear as dark spots and are not observed on the as-prepared nanocrystal faces.

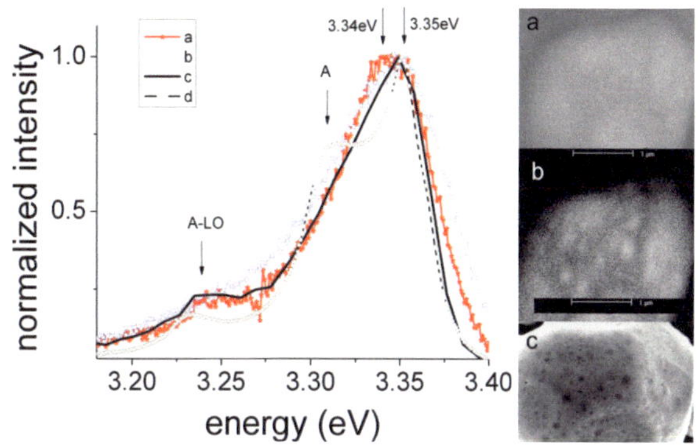

Fig. 2.4. (Left) Spatially resolved CL (a, b) and integral CL(c) and PL (d) spectra at 90 K. Spatially resolved CL measurements were performed on ZnO nanocrystal face: specklike defect region (a) and defect-free region (b); (Right) Wavelength resolved CL images of the ZnO nanocrystal face obtained at 90 K for (a) 3.358 eV (DX-line) and (b) 3.313 eV (A-line) energies as well as a SE image of the ZnO nanocrystal face with specklike defects (c).

The spatially resolved CL spectra, as well as, the integral PL and CL spectra obtained from a large number of ZnO nanocrystals at 90 K are presented in Fig. 2.4. It can be seen that in contrast to the PL spectrum the DX and A-lines in the integral CL spectrum are not well resolved; whereas, the A-LO line at around 3.24 eV appears clearly [Fig. 2.4, curve (c)] The spatially resolved CL spectrum from the nanocrystal surface region without specklike defects peaks at 3.35 eV; it is broadened and displays no A-LO line [Fig 2.4, curve (b)]. A spectral maximum of the CL spectrum from the specklike defect region is redshifted and located at 3.34 eV [Fig. 2.4 (a).] Moreover, it shows the pronounced A-LO peak at around 3.24 eV. Obviously, the A and A-LO lines appear only in the CL spectra collected locally from the regions with the specklike defects, or a large number of nanocrystals with the specklike defects. Fig. 2.5 shows the wavelength resolved CL images taken at the DX (3.358 eV) and A-line (3.311 eV) energies at 90 K. The CL image monitored at the bound exciton energy exhibits an uniform distribution and no extraordinary features are observed [Fig. 2.4 (a)]. In contrast, the emission detected at the A-line energy is spotlike and localized [Fig. 2.4 (b)]. The distribution of the spots coincide with position of the specklike defects in the secondary electron image [Fig. 2.4 (c)].

2.1.2 ZnO nanorods

Numerous techniques have recently been developed to synthesize quasi-1D ZnO nanostructures, (Baranov et al. 2004; Liu et al., 2004; Xia et al., 2003; Yang et al., 2002; Yao et al., 2002; Yin et al., 2004; Wang et al., 2004; Zhang et al., 2003) including growth with or without catalysts by a vapour-liquid- solid epitaxial (VLS) mechanism (Yang et al., 2002; Yao et al., 2002; Wang et al., 2004), microwave plasma growth (Zhang et al., 2003) and solution methods (Liu et al., 2004; Yin et al., 2004). An important issue in the fabrication of nanostructures is the control of the shape and size of nanomaterials, both of which affect their optical and electrical properties. ZnO nanorods grown from a NaCl (NaCl-Li_2CO_3) salt mixture were described by (Baranov et al., 2005b; Baranov et al. 2004). The salt composition and the size of the Zn-containing precursor particles play a key role in the synthesis and define their size and properties.

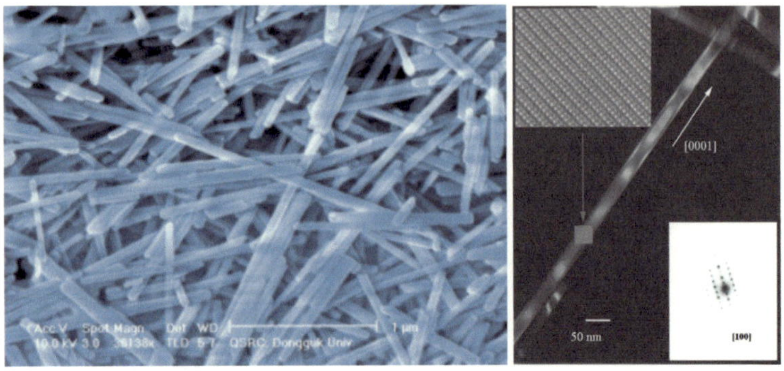

Fig. 2.5. (Left) HRSEM images of ZnO nanorods grown from the NaCl–Li_2CO_3 salt mixture at 700° C for 1 h and (Right) TEM images of ZnO nanorods grown from the NaCl mixture at 700° C for 1 h. Insets—selected area of electron diffraction and the ZnO atom-scale resolution image (HRTEM).

Figures 2.5 show HRSEM and TEM images of ZnO nanorods synthesized from the NaCl–Li_2CO_3 salt mixture as well as their selected area diffraction analysis. The presence of Li_2CO_3 allows to decrease the diameters of ZnO nanorods. The product prepared at 600° C for 3 h consists of quite uniform nanorods with diameters ranging from 8 to 40 nm and of length from 200 to 500 nm. As the temperature increased to 800° C, the diameters of the nanorods increased drastically and ranged from 200 to 500 nm. The results indicate that individual nanorods are single crystalline and possess the wurtzite structure (see the inset in figure 2.5, right).

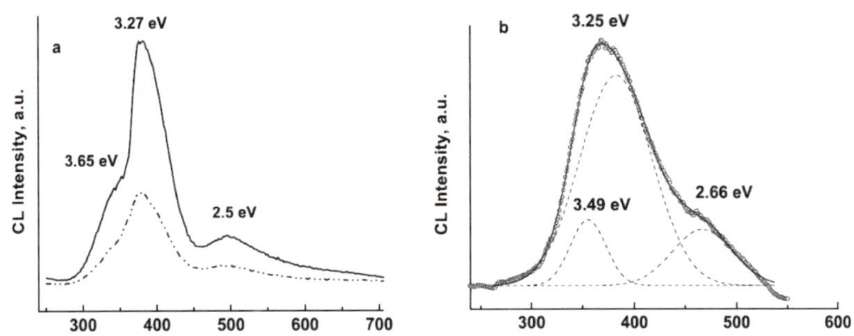

Fig. 2.6. (a) CL spectra taken at room temperature (V = 15 kV) from ZnO nanorods grown from (a) the NaCl mixture at 500° C (dash-dot line) and 700° C (solid line), (b) the NaCl–Li_2CO_3 salt mixture at 700° C (solid line) (the dashed lines are Gaussian multi-fit approximations) (Baranov et al., 2005b).

The CL spectra for the ZnO nanorods are shown in figure 2.6. Two peaks at around 3.25 eV and 2.5 eV and a high-energy shoulder at around 3.65 eV are observed. The high-energy bands 3.65 eV and 3.25 eV correspond to the near band-gap emission (NBE) of small and large rods, respectively. The high energy shoulder in the spectrum indicates some dispersion of small size rods. Strong quantum confinement effect could be attributed to the real size limitation by an potential barrier formed due to the charge surface states or the exciton dead layer (Pekar, 1958; Combescot et al., 2001), which increases with nanocrystal size (Fonoberov et al., 2004). The UV shift of PL from nanocrystals which are large in comparison with the exciton Bohr radius in ZnO (~1 nm (Fonoberov et al., 2004; Gill et al., 2002)) has been reported. The UV NBE shift of about 40 meV was observed for surprisingly large (24 nm) ZnO nanocrystals (Hur et al., 2005).

2.1.3 CVD grown ZnO nanorods and tetrapods

Chemical vapor deposition (CVD) and vapor transport (VT) techniques are widely used for the growth of nanorods and tetrapods. The effect of growth conditions on their morphology, size and properties has been well studied (Lyapina et al., 2008) and high efficient light emitting diodes based on ZnO nanorods were reported (Lee et al., 2011).

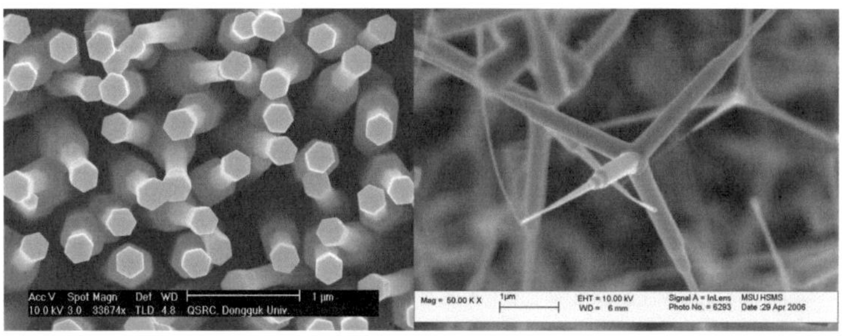

Fig. 2.7. SEM micrographs of ZnO nanorods and tetrapods grown on Si substrates.

Figure 2.7 shows typical ZnO nanostructures grown on a Si substrate. The supersaturation of the zinc vapor plays a key role in determining the morphology of the structures. The nanorods have a hexagonal shape and its size can be well controlled by the zinc vapor. The tetrapods are not bonded to the substrate surface, which implies that they were formed in the gas phase. The lower layer consists of small tetrapods with legs ~ 2–3 μm in length and ~ 350 nm in diameter. In the upper layer, the tetrapods have elongated legs on the order of 5μm in length, with a diameter of 650 nm at the junction and 100 nm at the tip. The rods of the tetrapods have long whiskers, up to 10 μm in length, with a sharp transition from a leg thickness of 500 to 80–100 nm.

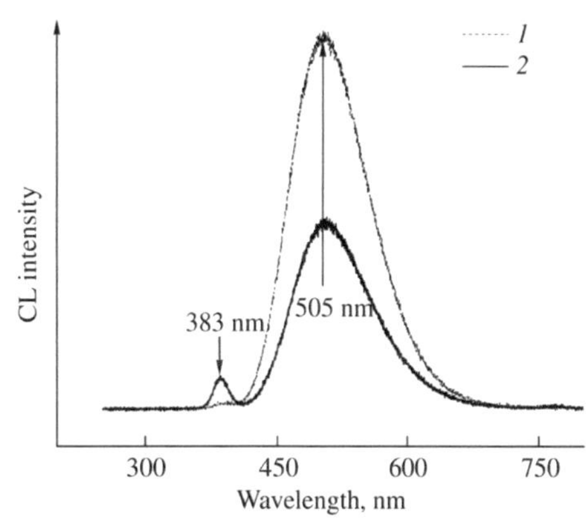

Fig. 2.8. CL spectra of tetrapods grown in (*1*) oxygen-deficient and (*2*) oxygen-enriched conditions (Lyapina et al., 2008).

Figure 2.8 shows the RT CL spectra of tetrapods obtained in different conditions. The UV peak at 383 nm is due to free excitons. The green band centered at 505 nm is markedly stronger. The intensity ratio between the UV and green emissions is larger in the samples obtained at the highest oxygen flow rate and reduced zinc vapor concentration.

2.1.4 ZnO/MgO nanocrystals and nanocomposites

The techniques for growing nanocrystals inside a matrix are very developed, partly as a consequence of the industrial production of color filters and photochromic glasses based on copper halides, such as CuCl, CuBr, CuI, etc. Nanocrystals of II-VI compounds (ZnO, CdS, CdSe, ZnSe, etc.), starting from supersaturated viscous solutions, are also frequently incorporated in glass matrices for applications in optical filters due to the ease in controlling the dot size. ZnO/MgO composites can be prepared by using aqueous solution and methanol solution techniques (Panin et al., 2004c; Panin et al., 2005b). Fig. 2.9 (Left) shows SEM images of the ZnO/MgO particles prepared by these techniques. The sizes of the individual ZnO and MgO particles prepared from the aqueous solution are approximately 100 and 500 nm, respectively, while the size of the particles prepared from the methanol solution is about 10 nm.

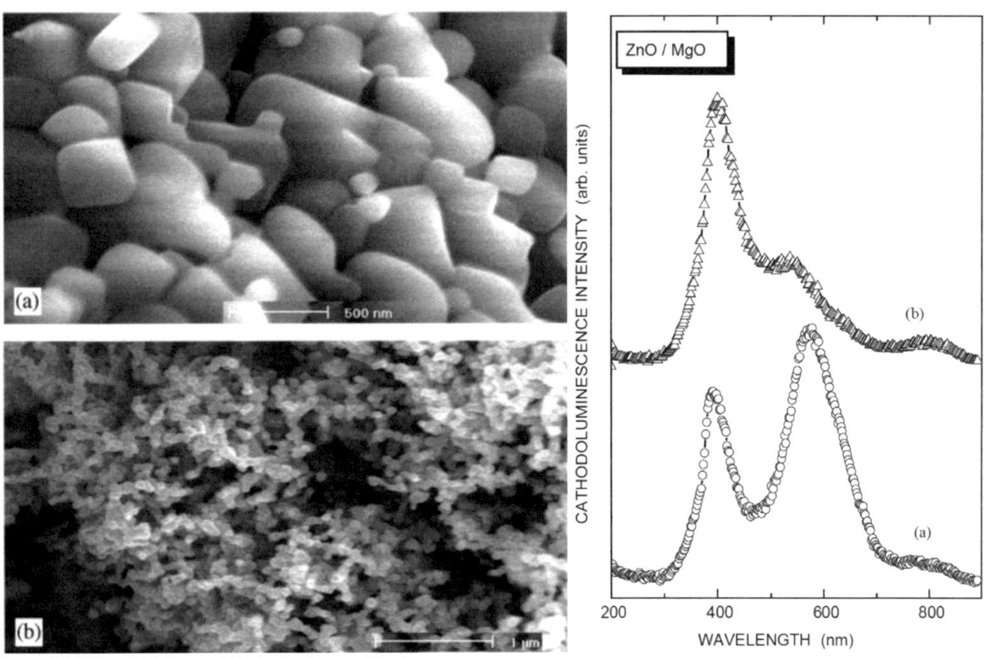

Fig. 2.9. (Left) Scanning electron microscopy images of ZnO/MgO nanoparticles prepared by using (a) aqueous solution and (b) methanol solution techniques. (Right) RT CL spectra of the ZnO/MgO particles annealed in (a) N_2 and (b) O_2 atmospheres.

(right panel) shows CL spectra of the large-size ZnO/MgO particles annealed in an oxygen or nitrogen environments. Annealing in oxygen environment inhibits the green luminescence and enhances the near-band-gap emission.

2.1.5 ZnO/MgO core-shell nanocrystals

ZnO/MgO core-shell nanocrystals were synthesized by chemical solution deposition (CSD) technique using a colloidal zinc oxide solution and alcoholic solution of magnesium acetate as described in (Baranov et al., 2008).

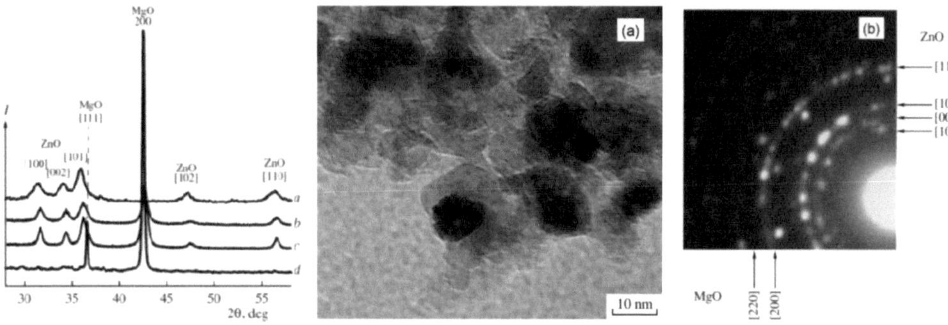

Fig. 2.10. XRD patterns from powdered nanoparticles of (*a*) ZnO annealed at 400° C and (*b*), (*c*), and (*d*) ZnO/MgO composites, annealed at 400, 500 and 600° C, respectively. (Right) (a) TEM micrograph and (b) electron diffraction pattern of the ZnO/MgO nanocomposite annealed at 500° C.

X-ray powder diffraction and electron diffraction show that, although the synthesis temperature was rather low, nanoparticles have a crystal structure. The XRD pattern of ZnO sample (Fig. 2.10, curve (a)) consistent with the hexagonal phase of ZnO crystals; the unit cell parameters insignificantly differ from literature data: a = 3.249 Å and c = 5.206 Å. The coherence length for zinc oxide determined from diffraction peak broadening agrees well with the particle size derived by digitizing TEM micrographs (Fig. 2.10, (image) (a)) and is within 5 – 12 nm. X-ray powder diffraction and TEM data for samples annealed at 400, 500, and 600°C indicate that the optimal temperature range for ZnO/MgO composite manufacturing is 400–500°C. The X-ray diffraction patterns contain both peaks associated with the cubic MgO structure (with Miller indices 111 and 200) and those associated with zinc oxide (wurtzite; indices 100, 002, 101, 102, and 110) (Fig. 2.10, curves *b* and *c*). These peaks are considerably broadened, but there is no anisotropy in directions 100 and 002; that is, particles have near-spherical shapes, which are also seen in TEM micrographs (Fig. 2.10, image (a)). The ZnO coherence length (crystallite size) in the composite with MgO varies as a function of synthesis parameters more widely than for pure zinc oxide; crystallite sizes range from 8 to 14 nm. The unit cell parameters calculated for MgO and ZnO phases almost coincide with the literature values; this means that, there is no noticeable interaction between magnesium and zinc oxides with solid solution formation at the low annealing temperatures. It confirms also by the electron diffraction pattern for the nanocomposite annealed at 500°C which contains both indices attributed to the cubic MgO structure ([220], [200]) and those associated with wurtzite zinc oxide ([100], [002], [101], and [110]) (Fig. 2.10, image (b)). A different situation is observed for samples annealed at 600°C (Fig. 2.10, curve *d*) or higher temperatures. In this case, wurtzite phases vanish and there are only peaks of newly formed solid solution $Mg_{1-x}Zn_xO$; the init cell parameters of the cubic phase increase systematically, which corresponds to a partial substitution of zinc ions (with a larger ionic radius) for magnesium ions.

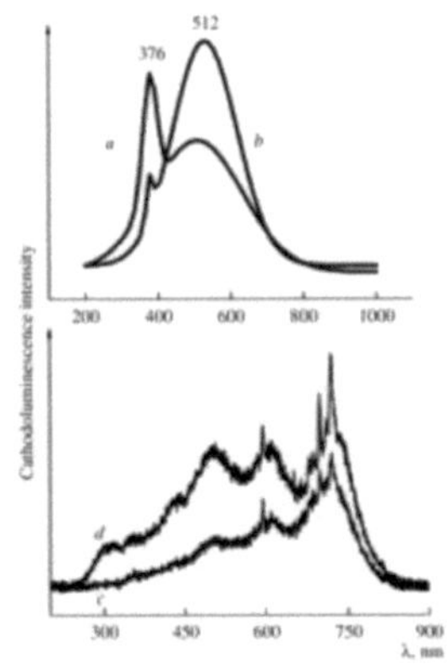

Fig. 2.11. Cathodoluminescence spectra of the ZnO/MgO nanocomposites annealed at (*a*) 400, (b) 500, (c) 600, and (d) 700° C (Baranov et al., 2008).

Cathodoluminescence spectra of annealed samples are shown in figure 2.11. The spectra of the samples annealed at 400 and 500°C show two peaks at 376 and 512 nm which are attributed to the free exciton luminescence and the oxygen-vacancy-related green emission from ZnO nanoparticles. The strong blue shift of the excitonic luminescence could be due to the quantum confinement effect in ZnO nanocrystals. The relationship between nanocrystal size and band gap can be obtained using a number of models (Efros & Rosen, 2000; Andersen et al., 2002; Hyberstsen, 1994). The effective mass model for spherical particles with a Coulomb interaction term (Brus, 1984; Brus, 1986) where the band gap E*[eV] can be approximated by Eq. 2 (Pesika et al., 2003):

$$E^*=E_g^{bulk}+[(\hbar^2\pi^2)/2er^2][(1/m_e m_0)+(1/m_h m_0)]-1.8e/4\pi\varepsilon\varepsilon_0 r, \quad (2)$$

where E_g^{bulk}=3.3 eV is the ZnO bulk band gap, r is the particle radius, m_e=0.26 is the effective mass of the electrons, m_h=0.59 is the effective mass of the holes, m_0 is the free electron mass, ε=8.5 is relative permittivity of ZnO, ε_0 is the permittivity of the free space, ħ is Plank's constant divided by 2π, and e is the charge of the electron. The size of the ZnO particles estimated from Eq. 2 was 8 nm that well agreed with TEM and X-ray diffraction data for the nanocomposites studied. The relative intensity of the green luminescence increases with an increase of annealing temperature (Fig. 2.11, curve b). This behavior of luminescence could arise from an increase in density of single-ionized oxygen vacancies after annealing as a result of an increase of the electron concentration in the ZnO particles. The spectra of the samples annealed at 600 and 700°C are shown in figure 2.11 (c, d). For the sample annealed at 600°C, edge luminescence is not observed but several peaks in green, yellow and red spectral regions are appeared. After annealing at 700°C, a peak in the deep UV region (at 307 nm) appears, confirming the formation of a solid $Mg_{1-x}Zn_xO$ solution, for which the UV shift of edge luminescence is characteristic. The peaks at around 600 and 700 nm can be attributed to Mg- or oxygen-related defects in the solid solution which together with oxygen vacancy defects reduces transparency of the nanocomposite in the visible range. The hexagonal ZnO (core) and cubic MgO (shell) phases in ZnO/MgO nanocomposites are thermally stable and high transparent up to 500°C. The annealing at temperatures 600 -700°C led to Mg doping and formation of the $Mg_{1-x}Zn_xO$ solid solution with the radiative defect centers, emitted in green, yellow and red spectral regions.

The luminescent properties of nanocomposite consisting of ZnO nanotetrapods (NT) and MgO particles were also investigated. The nanocomposites were prepared from ZnO tetrapods (300-5000 nm in length and 20-500 nm in diameter) grown by CVD method (Lyapina et al., 2008) and capped by MgO nanoparticles (3-50 nm) which were obtained by thermolysis of jellylike product containing mixture of magnesium acetate and sodium hydroxide alcohol solution (Baranov et al., 2008). The nanocomposite was annealed at 500°C in air. The low temperature annealing was used to prevent the diffusion of Zn^{2+} cations to MgO lattice and vice versa Mg^{2+} cations to ZnO lattice and keep the cubic MgO and hexagonal ZnO structures (Baranov et al., 2008; Panin et al., 2005b).

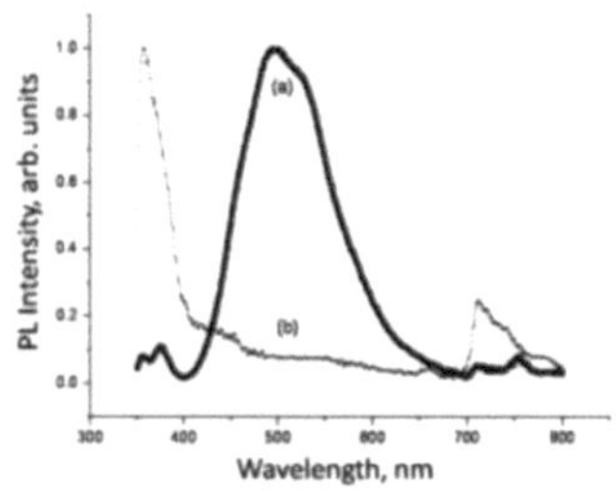

Fig. 2.12. The normalized PL spectra of (a) ZnO nanotetrapods (solid line) and (b) ZnO nanotetrapods covered by MgO (dot line).

Fig. 2.12 shows PL spectra of ZnO tetrapods and the tetrapods capped with MgO particles. The annealed tetrapods demonstrate high intensity of the green luminescence attributed to the oxygen-vacancy-related defects. The tetrapods annealed with capping MgO nanoparticles show however the suppressed green emission and the relatively-enhanced excitonic luminescence (Fig. 2.12, curve (b)). To study the electronic and optical properties of the samples on nano-scale through the lens detector (TLD) and CL measurements were carried out (Ryu et al., 2002; Panin et al., 2003a). Figure 2.13 (a) shows an HRSEM image of the tetrapod and (b) profiles of CL intensity and the TLD signal obtained from the structure by scanning an electron beam along the A-B line. The enhanced TLD signal indicates the electron depletion of a ZnO surface with formation of the space charge region. The intensity of CL from the depletion zone is suppressed significantly (about 25 %).

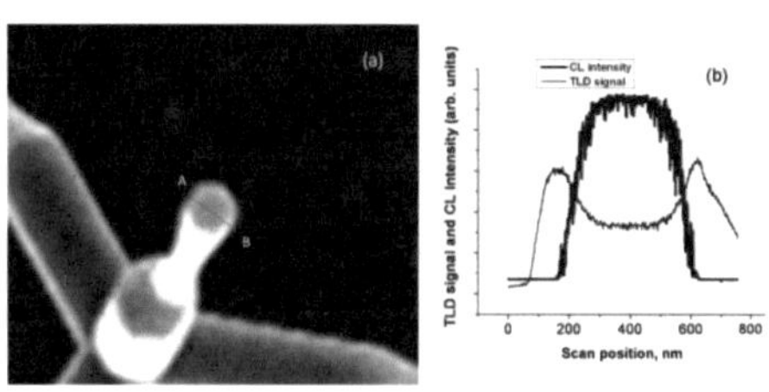

Fig. 2.13. (a) An HRSEM image of the nanotetrapod structure and (b) profiles of the CL intensity and the TLD signal from the structure obtained by an electron beam scanning (along the A-B line in (a)).

The green luminescence from ZnO nanocrystals are generally explained by the formation of single ionized oxygen vacancies (Vanheusden et al., 1996b). The mechanism of the observed suppression of the green luminescence from the nanocomposite structure could be explained by the electron depletion of ZnO nanocrystals due to the band bending on the interface between ZnO and MgO capping agent. In the part of this depletion region where the Fermi level E_F passes below the V_0^+ / V_0^{++} energy level, all oxygen vacancies are in the nonradiative V_0^{++} state and the green luminescence from the nanocrystals is suppressed (Panin et al., 2008). In case of uncapped tetrapods the electron density is relatively high and the green emission from such samples is well appeared.

3. Tuneable luminescence from nanocrystals and composites

3.1 ZnO Nanocrystals capped by Rh6G

ZnO-based polymer and molecular composites have demonstrated the attractive optical and mechanical properties. In particular, it is found that capping of ZnO nanocrystals with polymer, organic dye, etc. can effectively passivate the surface defects and decrease the surface-related visible emission (Borgohain et al., 1998; Du et al., 2006; Harada et al., 2003; Tong et al., 2004; Yang et al., 2001). The remarkable increase of the near-band-edge emission to the visible emission intensity ratio (NBE/VB) in ZnO nanocrystals coated with Rhodamine 6G (Rh6G) organic dye was reported by (Kurbanov et al., 2007b). These results show that capping with Rh6G leads not only to passivation of the ZnO nanocrystals surface but also to formation of dye monomers mostly, due to advanced surface of ZnO nanocrystals.

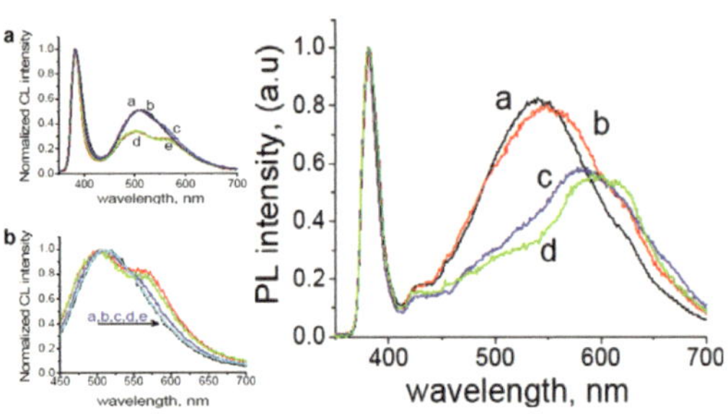

Fig. 3.1. RT normalized CL (Left) and PL (Right) spectra for ZnO nanocrystals impregnated with the various Rh6G concentration solutions: (Left) (a) uncapped, (b) 10^{-6} mol%, (c) 10^{-5} mol%, (d) 10^{-4} mol% and (e) 10^{-3} mol% and (Right) (a) uncapped, (b) 10^{-4} mol%, (c) 10^{-3} mol% and (d) 10^{-2} mol%. The spectra were normalized to (a) UV and (b) visible emission bands.

Fig. 3.1 shows the RT CL and PL spectra of ZnO nanocrystals impregnated with Rh6G solution of various concentrations. Both the CL and the PL spectra of the uncapped nanocrystals show two emission peaks - the sharp peak at 382 nm and a broad peak at ~510–530 nm. The luminescence at 382 nm is attributed to the near-band-edge emission (free exciton) and the broad emission peak at 510–530 nm is contributed to deep level defects, probably ionized charge states of oxygen vacancies (Studenikin et al., 1998b; VanDijken et al., 2000a; Van Dijken et al., 2001; Vanheusden et al., 1996a, 1996b). After capping with dye the emission spectra of ZnO nanocrystals display a new band at 560–570 nm. Its intensity increases with increasing dye concentration and it can be attributed to Rh6G luminescence. Moreover, at high dye concentrations a second band at 625 nm, which is also related to dye molecules, appears (Fig. 3.1 (right panel) d). However, the visible emission intensity around 510 nm decreases with increasing dye concentration and at higher dye concentrations, the emission arisen from Rh6G becomes dominant over the visible luminescence of ZnO. The capping gave rise to increasing the intensity ratio of the UV emission to the visible emission of the ZnO nanocrystals. After capping, the (NBE/VB) intensity ratio for the CL spectrum increases from 1.96 to 3, and that for the PL spectrum increases from 1.22 to 3.6. Improvement of the (NBE/VB) intensity ratio indicates that dangling bonds and defect states in the surface layer of ZnO nanocrystals are significantly passivated by dye. The visible luminescence of ZnO nanocrystals located around 510-520 nm (''green'' emission) is affected by the surface (Matsumoto et al., 2002; Norberg & Gamelin, 2005; Ramakrishna & Ghosh, 2003; Shalish et al. 2004; VanDijken et al., 2000a; Van Dijken et al., 2001). A direct relationship between the size and the excitonic emission intensity in ZnO crystals in the 40–2500 nm size range was reported (Matsumoto et al., 2002; Shalish et al. 2004; Van Dijken et al., 2001). Moreover the ''green'' emission intensity was reported to display a direct correlation with surface hydroxide concentrations (Liu et al., 1992; Norberg & Gamelin, 2005; Van Dijken et al., 2000b). On the other hand, it has been found that surface OH groups can form H-bridges with dye molecules through their OH groups and/or nitrogen (Ying et al., 1993; Wood & Rabinovich, 1989). The evaporation of the residual ethanol solvent is accompanied with the increased interaction between dyes and hydroxides. Statistically neighboring molecules are bonded to the hydroxide groups, resulting in stabilization of dye molecules at the ZnO surface. The obtained results indicate that Rh6G molecules present as monomers at ZnO nanocrystals at lower dye concentrations. Fig. 3.2 shows PL spectra for the ethanol solutions containing Rh6G of 10^{-5}, 10^{-4}, and 10^{-3} mol%. The Rh6G solutions display only one PL band at ~560 nm resulting from monomers (Innocenzi et al., 1996). With increasing the dye concentration, this band exhibits redshift, which is attributed to formation of dimers and increasing their amount (Bojarski, 2002; Innocenzi et al., 1996). The aggregate formation process is clearly displayed in the PL spectrum of the dye deposited onto a flat quartz glass surface (Fig. 3.2 (curve d)). The PL spectrum contains peaks near 665, 625 and 530 nm. Using the results (Bojarski, 2002) we assume that the PL bands at 625 and 665 nm are formed due to dimer and aggregate emissions. The broad PL peak at 530 nm is related to the emission of the quartz glass. PL of organic dye on a quartz glass surface undergoes a strong quenching effect owing to strong aggregation and as a result, the dye emission intensity on the quartz glass surface is many orders of magnitude less then that in the ethanol solutions as well as on the ZnO nanocrystal surface.

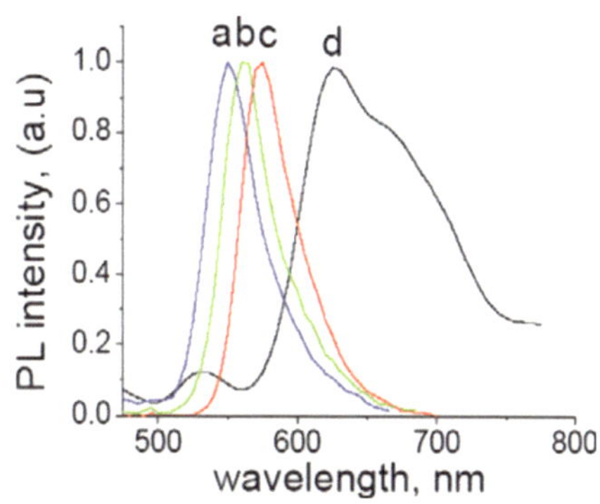

Fig. 3.2. Normalized PL spectra of Rh6G in ethanol solution for the different concentrations: (a) 10^{-5} mol%, (b) 10^{-4} mol%, (c) 10^{-3} mol% and (d) Rh6G deposited onto a flat quartz glass surface (Kurbanov et al., 2007b).

Based on these results, we can infer that the emission band at 625 nm appearing in the ZnO nanocrystal PL spectra at higher dye concentrations is caused by dimers and the spectrum redshift caused by increasing the dye concentration is evidence of the dimer formation process. However, the dye emission from the capped ZnO nanocrystals displays basically monomer-like behavior. The large surface to- volume ratio of nanocrystals contributes to isolating the dye molecules from each other that reduces the formation of dimers or complexes, especially at low dye concentrations. Similar behavior of materials with advanced surface structure to prevent the dye dimerization process has been observed in artificial opal (Kurbanov et al., 2007c), porous aluminum (Levitsky et al., 2002) and mesa structured titanium dioxide (Vogel et al., 2004) impregnated with dye. The dye covering the ZnO surface absorbs also the emission at around 500 nm. It is known that Rh6G has strong absorption bands near 500–530 nm and in the UV region (Bojarski, 2002; Ethiraja et al., 2005; Innocenzi et al., 1996). The findings indicate that the excitation of Rh6G at the surface of ZnO nanocrystals during our experiment may occur through two channels, which are an energy transfer from the ZnO deep level emission centers and a direct laser excitation. A significant increase in the suppression of the green luminescence appears at higher dye concentrations when the dimer related PL band at 625 nm has appeared. The Rh6G dimers have two absorption bands, higher intensity one at the short wavelength side of the monomer band (around 500 nm) and lower intensity one overlapping with the monomeric one at 560– 570 nm, which reabsorbs the emission of monomers. An increase of the dimer concentration might also lead to an increase in the absorption of the visible emission from ZnO nanocrystals and a decrease in the luminescence intensity from monomers. The surface passivation of ZnO nanocrystals due to the organic dye capping improved their UV-to-visible luminescence intensity ratio. Furthermore, a large surface area of nanocrystals helps to suppress the aggregate formation process of dye molecules, and Rh6G on ZnO nanocrystal surface displays a monomer-like emission, position of which depends on the dye concentration. The results indicate that ZnO nanocrystals capped with Rh6G organic dye hold promise for potential electro-optic applications.

3.2 CL from PBET structure

Organic semiconductors have the potential to provide a compact, low-cost light-emitting and laser diodes over a broad range of wavelength throughout the UV-visible spectrum. The large exciton binding energy in the conjugated polymers (Van der Horst et al., 2002) and efficient excitonic recombination can be the basis of high quantum yield devices operating at room temperature. Injecting electrons and holes into the charge states of the conjugated polymers leads to electroluminescence (Burroughes, 1990). These carriers should be balanced at the junction of the emitting layer to yield the maximum exciton formation. Light emission of poly [3-(2-benzotriazolo) ethylthiophene] (PBET) at a different electron beam current density has been investigated by the CL spectroscopy (Panin et al., 2004a). An electron beam was used to inject directly electrons and holes in the polymer and induced an electric field.

PBET was synthesized by (Ahn et al., 2001), to increase quantum efficiency by introducing benzotriazole, an electron-withdrawing moiety, to the thiophene. Singlet excitons are usually generated in polymer either by photoexcitation or by electrical injection of negative and positive charge carriers to the lowest unoccupied molecular orbital (LUMO) and the highest occupied molecular orbital (HOMO), respectively. Due to large exciton binding energy and the lattice relaxation (Bredas et al., 1996) energy of photons emitted by polythiophene is usually significantly lower than the HOMO–LUMO band gap, which is around 3.58 eV (Van der Horst et al 2002). Both photoluminescence and electroluminescence spectra of the polymer show yellow-orange luminescence (Ahn et al., 2002).

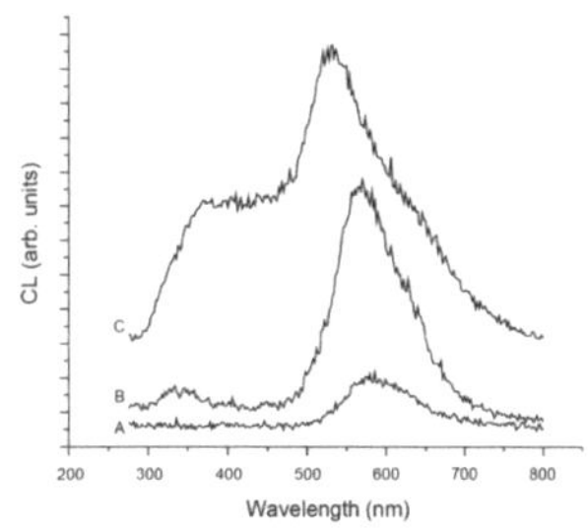

Fig. 3.3. CL spectra of PBET/ITO structure obtained at 10 kV accelerating voltage of incident electrons and at different electron beam current densities: (A) 0.4 A/cm^2, (B) 30 A/cm^2 and (C) 2 kA/cm^2 (Panin et al., 2004a).

Room temperature CL spectra of the PBET/ITO structure under various electron beam excitations are shown in Fig. 3.3. Under the low-excitation conditions with an incident beam current density of 0.4 A/cm^2 the spectrum (Fig. 3.4, A) is similar to the photoluminescence spectrum previously reported (Ahn et al., 2002). The peak centered at 580 nm with a shoulder at 620 nm were attributed to the π–π^* transition of the conjugated thiophene segments. Under higher excitation with a beam current density of 30 A/cm^2 (Fig. 3.3, B) three peaks centered at 340, 565 and 620 nm are revealed. Under the highest excitation used in our experiment with a beam current density of 2 kA/cm^2 the broad ultraviolet-blue bands are detected. In addition, a blue shift of green luminescence in the spectrum up to 529 nm and a red shift of the UV peak to 370 nm are observed. The peak at 347 nm could be attributed to the direct interband radiative transition. The absorption spectrum of the PBET samples has a strong absorption band at 282 nm (Ahn et al., 2002).

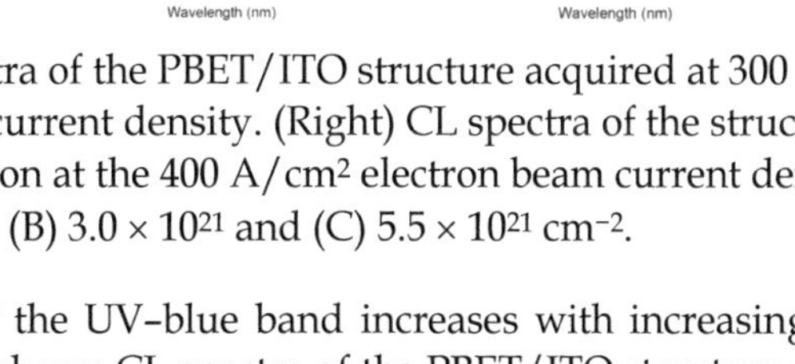

Fig. 3.4. (Left) CL spectra of the PBET/ITO structure acquired at 300 and 80 K at a 400 A/cm^2 electron beam current density. (Right) CL spectra of the structure obtained after the electron beam irradiation at the 400 A/cm^2 electron beam current density for different fluences: (A) 1.5×10^{21}, (B) 3.0×10^{21} and (C) 5.5×10^{21} cm^{-2}.

A relative intensity of the UV-blue band increases with increasing electron beam current density. Fig. 3.4 (Left) shows CL spectra of the PBET/ITO structure obtained under the high electron beam excitation at different temperatures. The UV band centered at 350 nm along with green-orange luminescence bands are clearly observed. The relative intensity of the bands depends strongly on temperature. At 80 K the polymer structure shows the high intensity of the UV luminescence, which is substantially higher than that of the green-orange luminescence. Some red shift of a 620 nm peak is also observed. Note that no shifts for the UV band and no blue band appearances are detected under such excitation conditions. To investigate the origin of the blue band we measured the luminescence of the polymer structure after the electron beam irradiation with the high electron beam current density for various fluences and temperatures. As the fluence of the electron beam irradiation increases, the intensity of the broad violet-blue luminescence at 370–470 nm increases, while an intensity of the green-orange luminescence decreases (Fig. 3.4 (Right)). The blue-violet band is detected under room temperature excitation, but it has not appeared under the highest electron beam current density excitation at 80 K. The results indicate that the structural/electron change in the polythiophene derivative may occur under the room temperature electron beam irradiation with a high fluence. The most important observation is the appearance of UV and blue-violet emissions centered at 347 and 420 nm. The wide spectra with violet and UV emissions have been previously demonstrated in insulating polymers at high electric fields (Chayet et al. 1997; Massines et al., 1997). The high-energy photon emission in these systems was attributed to the so-called "impact ionization," which occurs when the injected and thermally generated carriers are accelerated by a strong electric field to cause collision excitations and subsequent radiative recombination. Theoretical models (Artbauer, 1996) suggest that the number of these collisions increases with electric field, which can be arisen from the electron beam induced charging. Our experimental results indicate that an increase of the electron beam current density results in charging of the polymer and an induced electric field which increases the UV CL intensity. The charging effect with formation of the built-in electric field in the polymer structure was clearly observed in a secondary electron mode and could be controlled by the electron beam current adjustment. The UV luminescence suggests the recombination of "hot" nonequilibrium carriers in the presence of the strong electric field. Impact excitation induced by the electric field produces a large number of excited carriers in excess of equilibrium ("hot" carriers). These excited states decay thus approaching their number at equilibrium via several competing mechanisms. Formation of singlet excitons, which decay radiatively emitting green luminescence, is one such possible mechanism. However, the strong electric field inhibits this type of decay by dissociation of the excitons, thereby enhancing the direct HOMO-LUMO interband radiative transition with UV emission. Under such conditions direct radiative transition is favorable over exciton recombination. The saturation of the green luminescence at high current density is a further evidence for the partial inhibition of exciton recombination. The appearance of the blue luminescence at high current densities could be due to the electron beam induced phase transformation in PBET at room temperature. When the polymer backbone conformations are coupled with that of the sidegroups the first-order phase transition is occasionally realized. Such polymers should show the bistable states, which are separated by a potential barrier in free energy. The phase conversion between the two states in some temperature region is thermally forbidden but may be possible by electron beam excitation. In the hysteretic temperature region of the phase transition the switching of the electronic phase was previously demonstrated by pulsed photoexcitation of electron-hole pairs in polythiophene (Hosaka et al., 1999). The photo-induced change in the absorption and luminescence spectra arising from the structural change, which is generally termed photochromism, has been observed for many photochromic molecules with photoactive π-electron units and thin films of the semiconducting π-conjugated polymer (Koshihara et al. 1992). The structural change induced by the photoexcitation is considered to be minimal with least steric hindrance but significantly affects the π-conjugation length or the electronic structure. The photoexcited electron-hole pairs or their lattice relaxed analogs (polarons) are responsible as precursory excitations for the observed photoinduced phase transition. Origin of the blue luminescence from the polymer might be attributed to the electron beam induced change in the ring torsion angle by the interchain interaction in a substituted benzotriazole derivative segment.

Stimulated emission from the PBET/ITO structure in a broad range of wavelengths are observed at high current density (Panin et al., 2004b).

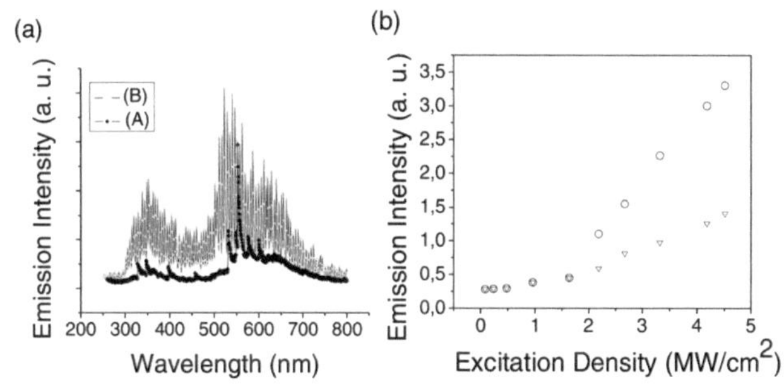

Fig. 3.5. a) RT spectra of stimulated emission (A) and lasing (B) from PBET obtained under electron beam excitation with different power densities: 1.5 MW/cm^2 (A) and 2 MW/cm^2 (B) and b) the intensity of emission as a function of the beam excitation density. (Panin et al., 2004b)

Figure 3.5 (a) shows RT spectra of stimulated emission (curve A) and multimode lasing in broad range of wavelengths (curve B) from the device structure obtained under electron beam excitation with high power densities. When the electron beam current density increases, the narrow emission lines appeared and the FWHM strongly decreases. A finely structured spectrum with features as narrow as 0.5 nm is observed at 2 MW/cm^2 power density. The narrow lines appear above a threshold power density and indicate stimulated emission. The dependence of the emission intensity on the pumping is shown in Fig. 3.5 (b). The broad spontaneous emission depends approximately linearly on excitation density. In contrast, stimulated emission shows the changeover to superlinear behavior. From the superlinear dependence we estimate an 1.3 MW/cm^2 threshold power density. The narrow lines appearing in the spectrum under high excitation are attributed to laser modes. These modes could be arising from self-assembled cavities formed in the polymer film due to recurrent light scattering (Cao et al., 1999) by domain boundaries. Small (10–20 nm) nanostructures in a fractal-like organization and elongated self-assembled aggregates of these structures forming extended domains (100–2000 nm) have been previously demonstrated in these polymer films using scanning tunneling microscopy and near-field optical measurements (Micheletto et al., 2001). It was shown that these structures are the source of the electroluminescence emission pattern which was observed due to domain boundaries scattering. Laser emission from these resonators results in a small number of discrete narrow peaks in the emission spectrum. As the pump power increases further, the gain increases and it exceeds the loss in the lossier cavities. Laser oscillation in those random cavities adds more discrete peaks to the emission spectrum.

3.3 ZnO Nanocrystal- PDPV composite

The combination of conjugated polymer and inorganic semiconductor nanoparticles is an attractive field in fabrication of light-emitting devices. Such nanocomposites allow combining the efficient luminescence from inorganic semiconductor with good mechanical and optical properties of polymer films (Friend et al., 1999; Vanheusden et al., 1996b). Poly(4,4'-diphenylene diphenylvinylene), (PDPV) (Eg=3.0 eV) (Feast et al., 1995) is a conjugated polymer for electroluminescence (EL) in the green region (Cacialli et al., 1997). The oxide semiconductor ZnO is a promising material for EL in the blue-UV region (Lee et al., 2011; Mordkovich et al., 2003; Panin et al., 2004c;). The luminescence efficiency of ZnO nanoparticles can be higher than that of ZnO epitaxial films and the particle-size dependent band gap allows controlling the excitation spectrum. ZnO nanocrystals doped by Mg, W, Fe, Mn, Y, Eu, V etc. show the high efficiency in multicolor low-energy CL and are prospective to use in advanced flat panel displays, solid state lighting applications and smart solid state electronics (Baranov et al., 2006; Mordkovich et al., 2003; Panin et al., 2007a; Panin et al., 2007b; Panin et al., 2007c).

PDPV (cis/trans~50:50) was prepared by condensation polymerization of 4,4'-dibensoyl-biphenyl following the procedure described elsewhere (Feast et al., 1995). The particles were prepared by dissolving the mixture of analytic-grade $ZnCO_3$ and $(MgCO_3)4Mg(OH)25H_2O$ (Aldrich) in a mole Zn/Mg ½ ratio in nitric acid and then precipitating by adding a $2M(NH_4)2CO_3$ solution. The precipitate was decomposed in a vacuum furnace at 320 °C and finally oxidized at 550 °C in air. The doping effect of ZnO by Mg has been observed after annealing of ZnO/MgO samples at temperatures above 500 °C (Panin et al., 2005b).This procedure leads to the formation of 12-60 nm sized Mg doped ZnO crystals (Panin et al., 2004c). PDPV was dissolved in chloroform and ZnO:Mg powder was dispersed in chloroform separately by treatment in an ultrasonic bath for 30 min. The PDPV mixed with the particles was deposited on a Si/SiO_2 substrate with gold electrodes separated by 5 μm by spin coating at 1000 rpm. The same procedure was used to get separate PDPV and ZnO:Mg particle films (Fig. 3.6).

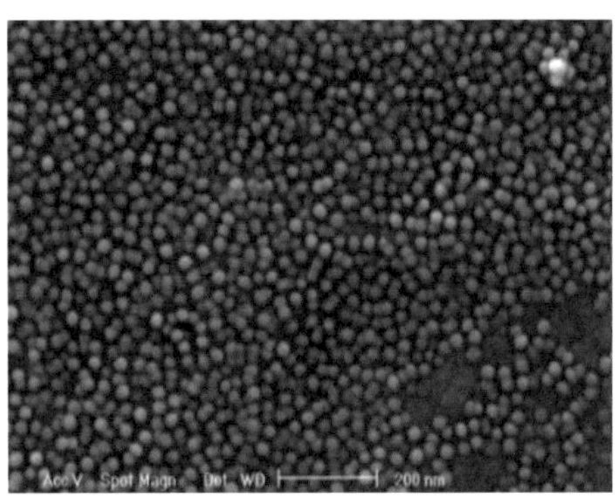

Fig. 3.6. An HRSEM image of the ZnO:Mg particle film (Panin et al., 2008).

Figs. 3.7 show CL spectra of the pristine PDPV film and the pristine ZnO:Mg particle film. PDPV has a well pronounced CL with the maximum emission in the green region, which is consistent with the luminescence results obtained for this polymer (Cacialli et al., 1997; Feast et al., 1995). The CL spectrum of ZnO:Mg nanoparticles reveals a more complicated structure with several peaks at 3.54; 2.83; 2.02 and 1.72 eV, respectively. The intensity of CL at 1.72 eV is higher with respect to other picks, which indicates that the red emission from ZnO:Mg nanoparticles related to impurities or traps (Falcony et al., 1998; Studenikin et al., 1998a) is the dominant one.

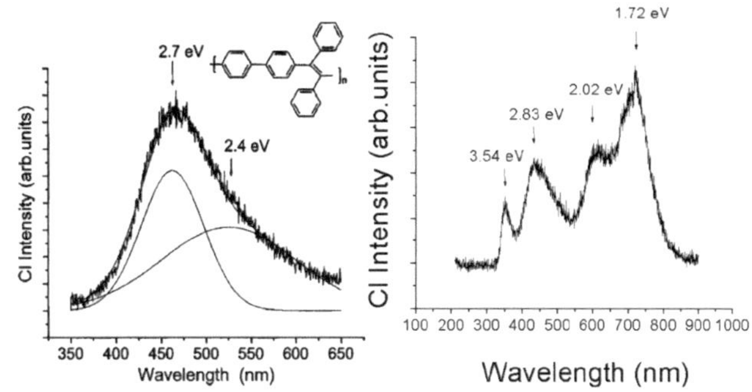

Fig. 3.7. CL spectrum of the PDPV film (Left) (inset shows chemical structure of PDPV) and the pristine ZnO:Mg particles (right).

3.3.1 Switching of cathodoluminescence from ZnO nanoparticle/polymer nanostructure by an electric field

The effect of reversible switching between blue-green and red emission in PDPV mixed with ZnO:Mg nanoparticles by an electric field was reported by (Panin et al., 2005a). Figure 3.8 (a) shows the CL spectrum of the PDPV–ZnO:Mg composite film without application of an electric field. The resulting spectrum combines CL from all the materials involved, which caused the broadening of PDPV spectrum with the pronounced peaks at 3.50, 2.86, and 1.79 eV. The main maximum of this asymmetric spectrum in the blue-green region (2.86 eV) results from pronounced blue-green emission from both PDPV and ZnO:Mg nanoparticles. As can be seen from Fig. 3.8 (b), application of a positive bias (+5 V) to the electrode suppresses the blue-green emission and shifts the emission maximum to the red region. The effect is found to be reversible with respect to the application of the electric field, namely then field was off, the emission maximum returned to the blue-green region.

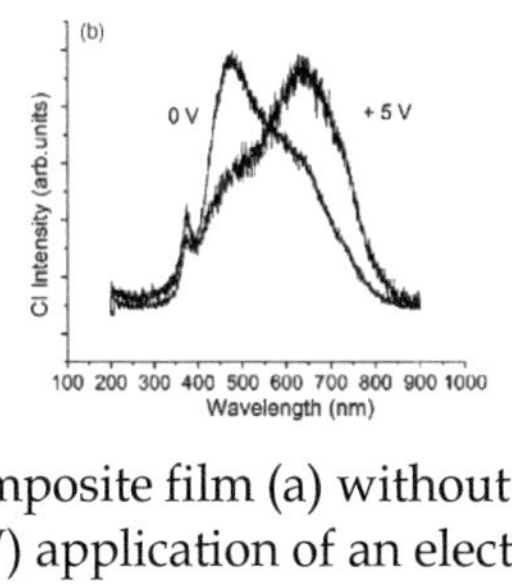

Fig. 3.8. CL of PDPV–ZnO:Mg composite film (a) without electrodes and (b) with electrodes biased with (+5 V) or without (0 V) application of an electric field (Panin et al., 2005a).

3.3.2 The mechanism of the electric field-tuneable luminescence from the ZnO nanoparticle/polymer structure

Before starting the analysis of the luminescence switching mechanism in PDPV–ZnO:Mg composite films, we have to note that the similar effect has been found in PPV films doped with ruthenium dinuclear complex (Welter et al., 2003). The band gap of ruthenium complex is about 0.5 eV smaller than that of the polymer. Therefore, the mechanism for the formation of the excited state in the polymer involves the ruthenium complex in a stepwise electron transfer process. In the case of the PDPV– ZnO:Mg composite the band gap of ZnO is higher than that of the polymer and the latter mechanism of the electric-field switching between blue-green and red emissions does not work (as well as the mechanism proposed for the emission from PPV–carbon nanotube composite (Woo et al., 2000). Fig. 3.9 shows the positions of the molecular orbitals and band structures of the materials used in our composite structure. As can be seen, there is a direct contribution to the green emission from LUMO-HOMO radiative recombination in PDPV and contributions to the blue, green and red emissions from radiative deep levels recombination in ZnO.

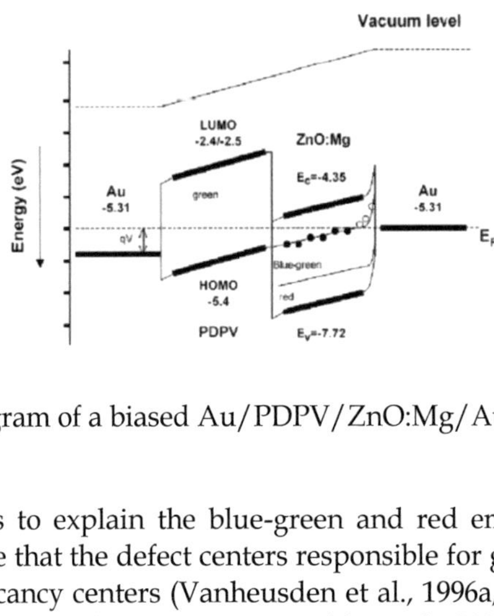

Fig. 3.9. Energy band diagram of a biased Au/PDPV/ZnO:Mg/Au structure (Panin et al., 2005a).

There are several models to explain the blue-green and red emission in ZnO. The most acceptable models assume that the defect centers responsible for green luminescence are the singly ionized oxygen vacancy centers (Vanheusden et al., 1996a, 1996b) or donor–acceptor level transitions (Egelhaaf & Oelkrug, 1996; Reynolds et al., 2001; Studenikin et al., 2002). The acceptor level (Zn vacancy) is located 2.5 eV below the conduction band edge (Bylander et al., 1978; Egelhaaf & Oelkrug, 1996), while the donor level (oxygen vacancy) is known as a shallow level at 0.05–0.19 eV, leading to an emission band centered around 508–540 nm. The blue-green emission in ZnO might also be associated with a transition within a self-activated center formed by a double-ionized zinc vacancy $(V_{Zn})^{-2}$ and the single-ionized interstitial Zn^+ at the one or two nearest-neighbor interstitial sites (Studenikin et al., 1998c). At last the blue-green emission can be ascribed to a substitution of Zn by extrinsic impurities such as Cu or Mg in the crystal lattice (Dingle, 1969; Mordkovich et al., 2003). The blue-green emission in the doped ZnO particles can be attributed to recombination of $V_O^{\bullet}$ electrons with excited holes in the valence band (Fig. 3.10), while the red emission in the particles at 670 nm (E_d =1.84 eV) could originate from the complex defect-related centers such as deep donor and deep acceptor centers associated with V_{Zn} (Fig. 3.10) (Ohtomo et al., 1999; Studenikin et al., 1998b; Mitra et al., 2001).

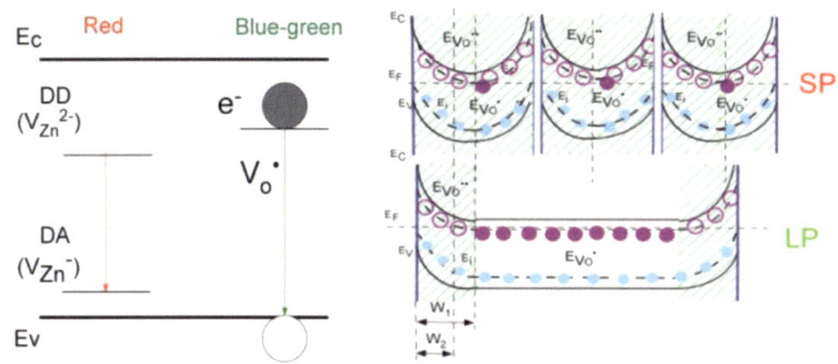

Fig. 3.10. (Left) Electron energy level diagram for red and blue-green radiation centers in ZnO. (Right) The energy-band diagrams of small (SP) and large (LP) particles in cross-sections.

The mechanism for the formation of the excited states in the PDPV-ZnO:Mg composite structure implies the presence of channels of radiative recombination, which can be controlled by electric field. Figure 3.10 (right panel) shows the energy-band diagrams for small (SP) and large (LP) particles in cross-sections. Application of an electric field causes depletion effects, which are known to affect the green luminescence of ZnO crystals (Vanheusden et al., 1996a; 1996b). A high density of electrically active surface states enhances the effect of band bending at a semiconductor surface, which creates an electron depletion region of width W in the particle. In the part of this region where the Fermi level E_F passes below the V_0^+ / V_0^{++} energy level, all oxygen vacancies are in the nonradiative V_0^{++} state. The CL maximum at ~2.86 eV indicates that both radiative recombination mechanisms including oxygen-vacancy related centers in ZnO and direct LUMO-HOMO transfers in PDPV coexist at zero electric field. It should be noted that band bending might be reduced by both polymer ZnO surface passivation and strong electron beam illumination. This effect results in an attraction of minority carriers to the surface, where they become trapped, and converts some of the V_0^{++} centers to the V_0^+ state. As for the red emission related to the recombination via impurity-oxygen involved centers in ZnO:Mg, the radiative recombination mechanism also works and can be recognized as the shoulder of the main spectra at ~650 nm. Complete band bending suppression at both interfaces in the case of the Au-PDPV-ZnO:Mg-Au structure can be obtained by applying a bias voltage to electrodes. PDPV is a *p*-type conductor similar to many other conjugated polymers. It means that application of positive bias to the Au-ZnO-PDPV increases the barrier and the green emission from PDPV is reduced because of charge separation at the polymer-nanocrystal interface (Greenham et al., 1996). Furthermore, the interface band bending creates the electron depletion region (W) at the ZnO particle surfaces and the conversion of the radiative V_0^+ centers to the nonradiative V_0^{++} state. As a result, blue emission becomes dominant. This mechanism is reversible and when the bias becomes zero, the green radiative recombination channel from the V_0^+ related states as well as PDPV emission should become dominant again in agreement with our experimental data. Note that the band bending in our experiment does not affect red emission, which is attributed to the deeper levels. A depletion region width (W) on the surface of ZnO particles depends on the surface state density and the charge carrier density in particles as it can be seen from equation 3 :

$$W=(2\varepsilon_{Zn0}V_{bi}/eN_d)^{1/2} \quad (3)$$

where ε_{Zn0}- is dielectric constatnt of ZnO, V_{bi} – potential barrier, e-electron charge, N_d – donor density. The size of particles and their charge density can control the luminescence spectrum. It is noteworthy that the CL spectrum of the PDPV–ZnO:Mg film contains peaks at ultraviolet-blue, blue-green, and red regions. This multicolor CL and the ability to tune the intensity of the bands by an electric field makes this polymer-nanoparticles composite promising for the creation of flexible tunable light emitters.

3.4 Optical tuning the luminescence of nanocrystals

The possibility of optical control of luminescence from nanocrystals was investigated by (Shih et al., 2011; Kurbanov et al., 2008a, 2008b). The visible light illumination effect on near-band-edge emission intensity from ZnO nanocrystals excited by UV He-Cd laser radiation (325 nm) was reported by (Kurbanov et al., 2008a). A cw Ar^+-laser irradiation (488 nm) in addition to the UV excitation can effectively (up to 75%) reduce the exciton emission from ZnO. This reduction in integrated intensity ($\Delta I/I$)(quenching) of the near-band-edge emission is reversible and has a pronounced dependence on temperature and ratio of Ar^+ and He-Cd laser power densities. The mechanism of the observed quenching effect was proposed to be similar the mechanism for the electric field tuned emission from ZnO nanocrystals in PDPV (Section 3.3). It implies an appearance of the additional recombination channel under a visible light illumination due to the recharging of oxygen-vacancy states in the surface depletion zone. The ZnO nanocrystals investigated in this work have the regular cone form and the size of 100–500 nm (Kurbanov et al., 2007a; Wang et al., 2005). PL studies were carried out on a SPEX spectrometer equipped with a 0.75 m grating monochromator using a 50 mW cw He-Cd laser operating at the wavelength of 325 nm as the excitation source. PL spectra were measured in the backscattering configuration and the registration of the PL signal was carried out by using a conventional lock-in technique with a mechanical chopper. A 12 mW cw Ar^+-laser operating at the wavelength of 488 nm (2.54 eV) was used as an additional illumination source. Both the laser beams were focused on the sample by employing quartz lenses. The beam spot size after the lens was estimated to be about 100 μm in diameter and the Ar^+-laser beam's spot covered the He-Cd-laser's spot. In order to vary the excitation intensity (power density), the spot size was kept constant while the input power was attenuated by the neutral filters. The corresponding maximum intensities of laser beams at 325 and 488 nm were estimated to be about 500 and 120 W/cm^2, respectively. The samples were mounted on the cold finger of a closed-cycle helium cryostat and the sample temperature was controlled in the range from 10 K up to room temperature. Figures 3.11 (a) and (b) display the RT PL spectra collected from ZnO nanocrystals under single (He-Cd laser) (black) and double (He-Cd and Ar^+-lasers) (red) illuminations at different excitation power densities of He-Cd laser. The quenching effect depends strongly on the power-density ratio of the employed lasers (I_{488}/I_{325})-with increasing the (I_{488}/I_{325}) ratio the magnitude of the quenching increases. A decrease in power density of the He-Cd laser by 2 orders of magnitude, at the constant Ar^+-laser intensity of 120 W/cm^2, results in an increase in the integrated intensity quenching ($\Delta I/I$) of the UV emission from 0.07 (7%) to 0.14 (14%). The observed quenching effect is reversible; the PL restores the initial intensity and form instantly after the turning off the Ar^+-laser illumination regardless of excitation power density.

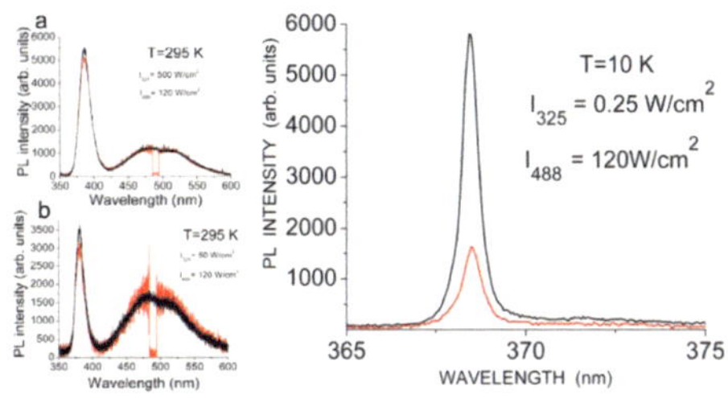

Fig. 3.11. (Left) RT PL spectra collected from CSD ZnO nanocrystals under single (He-Cd laser) (black) and double (He-Cd and Ar^+-lasers) (red) laser illuminations at different excitation power densities of He-Cd laser: (a) I_{325}~500 W/cm^2 and I_{488} ~120 W/cm^2 and (b) I_{325}~50 W/cm^2 and I_{488}~120 W/cm^2. (To protect the photomultiplier tube from the scattered Ar^+-laser radiation around 488 nm the spectrometer's slit was shut down). (Right) PL spectra collected from CSD ZnO nanocrystals under single (He-Cd laser) (black) and double (He-Cd and Ar^+-lasers) (red) laser illuminations at different excitation power densities of He-Cd laser at 10 K: I_{325}~0.25 W/cm^2 and I_{488}~120 W/cm^2.

The UV emission quenching effect was found to depend strongly on temperature and the laser power density. With the decrease in temperature to 10 K the magnitude of the quenching increases and under the He-Cd laser power density of 50 W/cm^2, it reaches to 0.38 (38%). The reducing of the He-Cd laser power density by 2 orders of magnitude at the fixed Ar^+-laser power density increases the quenching value up to 0.55 (55%) (Kurbanov et al., 2008a). The highest effect achieved in these experiments was ~75%. It was obtained at 10 K when the power-density ratio of the Ar^+ and He-Cd lasers was 500 (Fig. 3.11, right panel). The reversible behavior of the quenching effect remains also at low temperatures. However the time required reestablishing the initial PL intensity increases with a decrease in temperature. At 10 K the recovery time takes several minutes. Moreover the recovery time was found to depend on an illumination time by Ar^+-laser. The long term illumination results in prolongation of the recovery time of the PL intensity. The restoration of the PL initial intensity could be significantly accelerated by using a short-term increasing of temperature.

3.4.1 Electro-optical processes in nanocrystals under double excitation

The additional Ar^+-laser illumination of ZnO nanocrystals creates a new partway for excited charge-carrier recombination, which results in quenching of the near-band-edge emission intensity. Obviously, the energy of Ar^+-laser radiation (2.54 eV) is not enough to create new defects in ZnO lattice. However such low-energy photons can effect on charge states of the existing defects. As was identified in previous studies, the major defect related PL band at around 500 nm is originated from oxygen-vacancy center and the quenching effect also was more pronounced in samples with higher relative intensity of this band. Oxygen vacancies in ZnO can occur in three different charge states: the $V_0°$ state which has captured two electrons and is neutral relative to the lattice, the singly ionized V_0^+ state with one electron, and the V_0^{++} state which has no electrons and is doubly positively charged with respect to the lattice. Only the V_0^+ state is radiative and paramagnetic and consequently observable by EPR measurements. It is well known that the ZnO particle surface layer contains an electron depletion region created due to the surface states. The existence of such region results in the band bending at the surface. In the fraction of the depletion region where the Fermi level passes below the V_0^+ / V_0^{++} energy level, all oxygen vacancies will be in the diamagnetic V_0^{++} state while oxygen vacancies in V_0^+ state will exist in the particle core. The first principle studies have shown (Lany & Zunger, (2005); Kohan et al., 2000; Van de Walle, 2001; Zhang et al., 2001) that the oxygen vacancy is a "negative-U" center, i.e., V_0^+ state is deeper (farther from the conduction band) than either V_0^{++} or V_0° for any Fermi-level position. As already mentioned the intensity of the green luminescence in ZnO correlates very well with the paramagnetic single-ionized oxygen-vacancy density (Vanheusden et al., 1996a, 1996b). V_0^+ is found to be photosensitive for photons with an energy of >2 eV and this effect was explained by converting some of the V_0^{++} centers to the paramagnetic V_0^+ state. Considering these results and fact that the photon energy of Ar^+-laser radiation is 2.54 eV, the observed quenching effect of the near-band-edge emission from ZnO nanocrystals is well interpreted in terms of recharging of the oxygen-vacancy centers. Since V_0^{++} centers can be easily formed in the depletion region, consequently their number depends on the depletion region width (see Eq.3, Sec. 3.3.2). As can be seen, the width of the depletion region is inversely proportional to the square root of the free-carrier concentration in the nanocrystal and proportional to the potential at the surface. It is well known that in the case of high excitation, the surface potential depends on excitation intensity (Studenikin et al., 1998a). At high intensities the width of the depletion region is smaller than at low intensities. On the other hand an increase in temperature leads to additional narrowing of the depletion region due to the increase in the free carrier concentration. The simultaneous action of these factors (i.e., the increase in excitation intensity and temperature) results in the strong decrease in the depletion region width, consequently, the concentration of V_0^{++} centers. These speculations are in the good agreement with a strong increase in the quenching effect with a decrease in excitation power density and temperature. The temperature lowering from 295 to 10 K and a decrease in excitation intensity in 3 orders of magnitude (from 500 to 0.5 W/cm^2) resulted in a large enhancement of the quenching effect (from 7% up to 53%) at the fixed Ar^+-laser power density of 120 W/cm^2. Even though the Ar^+-laser illumination quenches well the UV emission, its impact on the visible emission intensity is observed very week. Although at room temperature under the double illumination the visible emission intensity slightly increases, at low temperatures it is rather unremarkable. This circumstance may be originated from different excitation efficiencies of the UV and visible emissions depending on temperature and excitation power density. One can see a decrease in temperature results in an increase in ratio of intensity of the UV emission to that of the visible emission. It increases from 2.25 to 66.5 with the decrease in temperature from 295 to 10 K at low excitation power density. Such behavior indicates that at low temperatures the visible emission is excited with relatively low efficiency than UV emission and the effect of Ar^+-laser irradiation on the visible PL, probably, also decreases. On the other hand, Ar^+-laser illumination could induce mostly the nonradiative states. It is known that there exists another type of single-ionized oxygen-vacancy-related center: V_0^+ complex, (Vanheusden et al., 1997) which can be formed in the presence of nearby interstitial oxygen (Frenkel pair) (Hoffmann & Hahn, 1974). This center unlike the isolated V_0^+ is not a green luminescent site, even though it shows a somewhat similar nature. Moreover the V_0^+ complex may act as a quenching center for radiative recombination in ZnO. The CSD technique used to grow ZnO nanocrystals provides the oxygen rich samples (Kurbanov et al., 2007a). This circumstance allows us to infer that in the ZnO nanocrystals prepared in the oxygen rich conditions, some of V_0^{++} centers under Ar^+-laser illumination creates V_0^+ -complex which does not take part in green emission but acts as nonradiative, quenching center for UV PL. The reversible behavior of the quenching effect and dependence of the recovery time on temperature and exposure time by the visible light confirm the proposed recharging model of the oxygen-vacancy-related centers in the ZnO nanocrystal depletion region. As was reported (Vanheusden 1996a), the EPR signal from V_0^+ decays when illumination was interrupted and at 294 K this decay is instantaneous, but at 150 K it requires several minutes. In our experiments the recovery process of the UV emission intensity could be also accelerated by a short-term rising of temperature. These observations indicate that there is some energetic barrier, escaping of which the system can return into the initial state. The existence of energetic barrier for decay of photogenerated V_0^+ into the other charge states was predicted by Walle (Van de Walle, 2001) based on the first principles investigation. For a further understanding of the mechanism of the observed UV PL band quenching, we investigated the quenching effect dependence on the excitation power density in details. The integrated PL intensity versus the excitation power density under (a) the single (He-Cd laser) and (b) the double (He-Cd-and Ar^+-lasers) illuminations as well as (c) the quenching ($\Delta I/I$) of the UV emission at 10 K are plotted in Fig. 3.12. The both PL intensity dependences display a tendency to saturation at high excitation level. The luminescence intensity I versus excitation power density can be expressed as

$$I = \eta I_0^{\alpha} \tag{4}$$

In this relation I_0 is the power density of the excitation laser radiation, η is the constant of proportionality (some authors define it as an emission efficiency), and the exponent a represents the radiative recombination mechanism. For excitonic recombination, $1<a<2$, for band-gap emission, i.e., electron-hole bimolecular recombination, $a\sim 2$, and a is less than 1 when an impurity is involved in the transitions, as well as for donor-acceptor transitions (Bergman et al., 2004; Jin et al., 1997). Using Eq. (4) to fit the data of the single and double laser illuminations we found $a \sim 0.61$ and $a\sim 0.7$, respectively. The obtained values of the exponent a are less 1 and they could be interpreted as evidence that the UV emission originates from native donor-acceptor transitions or impurities. However all results undoubtedly indicate that UV PL has exciton nature. As is well known, a saturation effect of the excitonic emission under high-power excitation is not rarely observed phenomenon (Bergman et al., 2004; Jin et al., 1997). The saturation of the excitonic emission from GaN powder excited in ambient air at RT was ascribed to a thermally activated nonradiative process due to laser heating of particles (Bergman et al., 2004). A slow increase and saturation of the PL intensity of free excitons in a GaInAsSb/GaAlAsSb single-quantum-well structure under higher excitation intensity observed at 10 K was attributed to the exciton screening effect (Jin et al., 1997). It seems in our experiment that both suggested saturation mechanisms are present.

The visible light illumination increases the exponent α from 0.61 to 0.70 by of approximately 15%. This growth undoubtedly is not result of a decrease in the laser heating. It originates rather from a decrease in exciton relative population due to opening an additional channel of recombination. Although this channel is basically nonradiative, the decrease in exciton relative population can reduce the exciton screening effect and provide a faster increase in the PL intensity with excitation power than that in the case of the single He-Cd laser illumination. It may be expected that use of the observed quenching effect allows effectively modulate the UV emission from ZnO and an employment of high power lasers with wavelength in the 450–550 nm region could provide it.

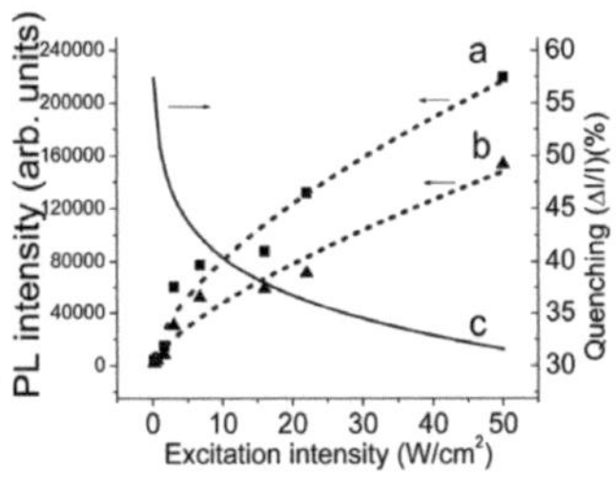

Fig. 3.12. Integrated exciton PL intensity (a) without and (b) under Ar^+-laser illumination as well as (c) the quenching of the PL intensity as a function of He-Cd laser excitation power density at 10 K. The dash lines are the fit to a power law (see the text) (Kurbanov et al., 2008a).

3.4.2 The optical modulation of UV PL from ZnO nanocrystals

The optical modulation of ultraviolet (UV) photoluminescence from ZnO nanocrystals excited by He-Cd laser (325 nm) with visible Ar^+ laser radiation (488 nm) was reported by (Kurbanov et al., 2008b). The effective reversible quenching of the UV luminescence intensity was achieved. The quenching efficiency was found to depend on temperature, the ratio of He-Cd and Ar^+ laser intensities, and the frequency of modulation. The observed quenching effect was used to modulate UV emission by chopped visible Ar^+ laser radiation. A sufficiently large modulation depth was obtained.

Figure 3.13 (solid line) shows the PL spectrum of ZnO nanocrystals obtained at 10K under radiations with the He-Cd laser. The spectrum displays a dominant near-band-edge emission peak at 368.6 nm and two weak emission bands at 373.6 and 382.8 nm as well as a weak visible emission band. The peak around 368 nm is related to the PL emission of a donor bound exciton (Kang et al., 2005; Look, 2001; Özgür et al. 2005). The bands at 373.6 and 382.8 nm, as indicated by arrows in Fig. 3.13, might be attributed to donor–acceptor pair recombination (Tomzig & Helbig, 1976) or exciton phonon replicas (Reynolds & Collins, 1969) or two photon transitions (Studenikin et al., 2000). An additional radiation using the Ar^+ laser leads to the strong quenching of the UV emission up to approximately 55% with some changes in deep-level PL [Fig. 3.13 (dashed line)].

The Ar^+-laser-radiation-induced quenching effect was used to modulate the intensity of the UV emission from ZnO nanocrystals. For this objective, the ZnO nanocrystals at 10K were simultaneously excited using an unmodulated cw He-Cd laser beam and a chopped Ar^+ laser beam. The power densities of the UV and visible lasers were 0.5 and 120W/cm^2, respectively. The UV peak intensity variations at 368.6 nm were observed using a SPEX monochromator and a photomultiplier tube on a Tektronix oscilloscope. The intensity variations of the exciton emissions at different modulation frequencies of the Ar^+ laser are presented in Fig. 3.13 (right panel). The curve (a) (Fig. 3.13. (Right)) represents the dc PL signal from the ZnO nanocrystals obtained under the single cw He-Cd laser excitation. The curves (b), (c), and (d) are PL signals modulated by Ar^+ laser radiation in various frequency regions: low (20 Hz), middle (200 Hz), and high (400 Hz), respectively. With increasing Ar^+ laser beam modulation frequency, both the amplitude of the PL signal and its mean level decrease. The curve (e) represents the dc PL signal obtained when nanocrystals were illuminated with an unmodulated cw Ar^+ laser beam. It was the lowest level that was achieved under Ar^+ laser radiation. The decrease in the modulation depth of the PL intensity with increasing frequency at low temperature is caused by the long recovery time of the quenching effect. As can be seen from Fig. 3.13, the complete modulation {i.e., the change in the PL intensity from the initial level to the lowest level induced by cw Ar laser radiation [curve (e)]} was observed at low (10- 20 Hz) modulation frequencies. It seems reasonable to expect that the complete modulation of the UV emission at RT could be obtained at higher frequencies. However, a high-power Ar^+ laser or a pulse laser operating in the 480 - 500 nm range should be employed owing to the small magnitude of the quenching effect at 300 K.

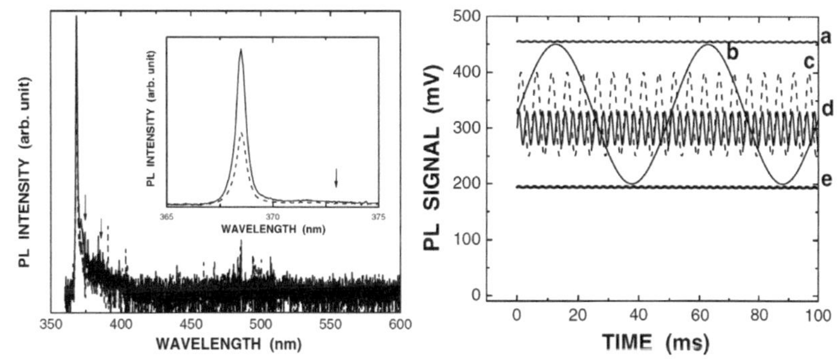

Fig. 3.13. (Left) PL spectra of ZnO nanocrystals obtained at 10K under He-Cd laser excitation (soid line) and combined He-Cd and Ar^+ laser radiations (dashed line). (The intensities of the He-Cd and Ar^+ lasers are approximately 0.50 and 120 W/cm^2, respectively.) The inset shows the PL spectra in the wavelength range between 365 and 375 nm. (Right) Schematics of PL signals from ZnO nanocrystals at 10K observed using oscilloscope. The curve (a) indicates the dc PL signal from ZnO nanocrystals under cw He-Cd laser excitation and the curves (b), (c), and (d) represent the PL signals modulated by Ar^+ laser radiation under cw He-Cd laser excitation at approximately 20, 200, and 450 Hz, respectively. The curve (e) represents the dc PL signal under both cw He-Cd and Ar^+ laser radiations.

4. Conclusion

In this chapter recent progress in multicolor luminescence from semiconductor nanocrystal composites tunable in an electric field have been highlighted. The effect of an external electric field on cathodoluminescence from semiconductor nanocrystals and nanocomposites with different radiative emission rates was described, giving special emphasis to ZnO nanocrystals in MgO and polymer matrix. The structural and optical properties ZnO nanocrystals and nanocomposites, including the core/shell nanocrystals were investigated. An ability to control the relative intensity of the near-band-gap emission (UV, violet-blue) and the deep-level luminescence (green, red) from ZnO nanocrystals and polymer composites by an electron-hole pumping and an electric field was demonstrated. It allows to adjust the luminescence of the composite in a broad visible wavelength range. The effect of an external electric field on cathodoluminescence from the nanocrystal/polymer structure was studied, giving special emphasis to the doped ZnO nanocrystals in Poly(4,4′- diphenylene diphenylvinylene) (PDPV) matrix and PBET/ITO structures. Cathodoluminescence of the polythiophene derivative structure was investigated under various electron beam current excitations. UV and blue bands in the spectrum of the polymer structure at the high electron beam current density are observed. The intensities of these bands increase as an electron beam current density increases while the green-orange luminescence is saturated. The induced electron beam field in the PBET/ITO suggests inhibiting the green luminescence by dissociation of the excitons, thereby enhancing the direct interband radiative transition with UV emission. The room temperature electron beam irradiation with a large fluence at a high current density results in the broad blue luminescence that may be attributed to the electron/structural changes in the polythiophene derivate film under an induced electric field. These effects could be used to obtain white light luminescence from the polythiophene derivative composites. Electric field-induced color switching of cathodoluminescence from ZnO:Mg nanocrystals/PDPV composite from blue-green to red is considered. The assumed mechanism of electric field-tunable cathodoluminescence implies the presence of radiative recombination channels, which are sensitive to the electric field through the band bending at the crystal surface. Optical control luminescence through the impact of visible Ar^+-laser illumination (488 nm) on the UV emission from ZnO nanocrystals excited by He-Cd laser (325 nm) at various excitation intensities and temperatures has been investigated. It was found that a visible light illumination simultaneously with UV excitation results in a decrease in the near-band-edge emission intensity. The experiments on a reversible quenching of the UV near-band-edge emission under Ar^+-laser illumination confirm an appearance of the recombination channel after oxygen-vacancy charging in the ZnO surface depletion zone. The quenching effect of the UV emission observed in ZnO samples is suggested to be due to the recharging of oxygen-vacancy states under a visible light irradiation. The strong quenching of UV luminescence from ZnO nanocrystals under Ar^+-laser illumination (488 nm) and an opportunity to modulate the UV PL intensity by visible light irradiation was demonstrated. The quenching effect depends on the intensity ratio of the visible and UV lasers, temperature, and the relative intensity of the green luminescence band. The highest quenching effect (75%) was achieved at I_{488}/I_{325}~500 at 10 K. It was shown that the UV near-band-edge emission is modulated at frequencies of hundreds of hertz. It is reasonable to expect that the complete modulation of the UV emission at RT could be obtained at higher frequencies when using a high-power Ar^+ laser or a pulse laser operating in the 480 - 500 nm range. In this context the quenching effect may find application in ZnO based optoelectronic devices and optical communication systems.

5. Acknowledgment

This work was supported by a National Research Foundation of Korea (NRF) grant funded by the Ministry of Education, Science and Technology (MEST) No. 2011-0000016 as well as by the Leading Foreign Research Institute Recruitment Program through NRF funded by MEST No. 2010-00218, and a Russian Ministry of Science and Education grant No. 02.740.11.5215. Author is grateful to all colleagues who took part in joint work, especially A.N. Baranov, O.O. Kapitanova, S.S. Kurbanov, A.N. Aleshin, I. A. Khotina, and T.W. Kang.

6. References

Ahn S.H., Czae M.Z., Kim E.R., Lee H., Han S.H., Noh J. & Hara M. (2001). Synthesis and Characterization of Soluble Polythiophene Derivatives Containing Electron-Transporting Moiety. Macromolecules, 34, pp. 2522-2527.

Ahn T., Lee H. & Han S.-H. Effect of annealing of polythiophene derivative for polymer light-emitting diodes. Appl. Phys. Lett. 80, pp. 392-394

Aigouy L., Holden T., Pollak F. H., Ledentsov N. N., Ustinov W. M., Kopev P. S. & Bimberg D. (1997). Contactless electroreflectance study of a vertically coupled quantum dot-based InAs/GaAs laser structure. Appl. Phys. Lett. 70, pp. 3329-3331.

Andersen K.E.; Fong C.Y. & Pickett W.E. (2002). Quantum confinement in CdSe nanocrystallites. J.Non. Cryst. Solids 99, pp.1105-1110.

Artbauer J. (1996). Electric strength of polymers. J. Phys. D 29, pp. 446-456.

Bagnall D.M., Chen Y.F., Zhu Z., Yao T., Koyama S., Shen M.Y & Goto T. (1997). Optically pumped lasing of ZnO at room temperature. Appl. Phys. Lett. 70, pp. 2230-2232.

Baranov A N, Chang C H, Shlyakhtin O A, Panin G N, Kang T W & Oh Y J (2004). In situ study of the ZnO–NaCl system during the growth of ZnO nanorods. Nanotechnology 15, pp. 1613-1619.

Baranov A N, Solozhenko V L, Chateau C, Bocquillon G, Petitet J P, Panin G N, Kang T W, Shpanchenko R V, Antipov E V & Oh Y J (2005a). Cubic MgxZn1−xO wide band gap solid solutions synthesized at high pressures.: J. Phys.:Condens. Matter 17, pp. 3377–3384

Baranov A N, Panin G N, Kang Tae Wong & Oh Young-Jei (2005b). Growth of ZnO nanorods from a salt mixture. Nanotechnology 16, pp. 1918–1923

Baranov A. N.; Panin G. N.; Yoshimura Masahiro & Oh Young-Jei (2006). Growth and magnetic properties of Mn and MnSn doped ZnO nanorods. J. Electroceram., 17, pp. 847-852.

Baranov A. N., Kapitanova O. O., Panin G. N. & Kang T. V. (2008). ZnO/MgO Nanocomposites Generated from Alcoholic Solutions. Russ. J. of Inorg. Chem., 53 (9), pp. 1366-1370.

Baranov A. N.; Kurakevych O. O.; Tafeenko V. A.; Sokolov P. S.; Panin G. N. & Solozhenkopp V. L. (2010). High-pressure synthesis and luminescent properties of cubic ZnO/MgO nanocomposites. J. Appl. Phys. 107, pp. 073519-1-073519-5.

Bergman L., Chen X. B., Morrison J. L., Huso J. & Purdy A. P. (2004). Photoluminescence dynamics in ensembles of wide-band-gap nanocrystallites and powders. J. Appl. Phys. 96, pp. 675-682.

Borgohain K. & Mahamuni S. (1998). Luminescence behaviour of chemically grown ZnO quantum dots. Technol. 13, pp. 1154-1157.

Bojarski P. (2002). Concentration quenching and depolarization of rhodamine 6G in the presence of fluorescent dimers in polyvinyl alcohol films. Chem. Phys. Lett. 278, pp. 225-232.

Bredas J.-L., Cornil J. & Heeger A.J. (1996). The exciton binding energy in luminescent conjugated polymers. Adv. Mater. 8, pp. 447-452.

Bredas J. L., Logdlund M. & Salaneck W. R. (1999). Electroluminescence in conjugated polymers Nature, 397, pp. 121-128.

Bylander E. G. (1978). Surface effects on the low-energy cathodoluminescence of zinc oxide. J. Appl. Phys. 49, pp. 1188-1195.

Burroughes J.M., Burroughes J. H., Bradley D. D. C., Brown A. R., Marks R. N., Mackay K., Friend R. H, Burns P. L. & Holmes A. B. (1990). Nature (London) 347, pp. 539-541.

Brus L.E. (1984). Electron–electron and electron-hole interactions in small semiconductor crystallites: The size dependence of the lowest excited electronic state. J. Chem. Phys. 80. pp. 4403-4409.

Brus L.E. (1986). Electronic wave functions in semiconductor clusters: experiment and theory. J. Phys. Chem. 90, p. 2555-2560.

Cacialli F., Friend R. H., Haylett N., Daik R., Feast W. J., Dos Santos D. A. & Bredas J. L., (1997). Efficient green light emitting diodes from a phenylated derivative of poly(p-Phenylene-Vinylene). Synth. Metals, 84, pp. 643-644.

Cao B. Q., Lorenz M., Rahm A., von Wenckstern H., Czekalla C., Lenzner J., Benndorf G. & Grundmann M. (2007). Phosphorus acceptor doped ZnO nanowires prepared by pulsed-laser deposition. Nanotechnology 18, pp. 455707.

Cao H., Zhao Y. G., Ho S. T., Seelig E. W., Wang Q. H. & Chang R. P. H. (1999). Random Laser Action in Semiconductor Powder. Phys. Rev. Lett. 82, pp. 2278-2281.

Chayet H; Pogreb R. & Davidov D. (1997). Transient UV electroluminescence from poly(p-phenylenevinylene) conjugated polymer induced by strong voltage pulses. Phys. Rev. B 56 (20), pp. 12702-12705.

Chen I. J.; Chen T. T.; Chen Y. F. & T. Y. Lin (1996). Nonradiative traps in InGaN/GaN multiple quantum wells revealed by two wavelength excitation. Appl. Phys. Lett. 89, 142113-142115.

Chen W., Wang Z. G., Lin Z. J. & Lin L. (1997). Absorption and luminescence of the surface states in ZnS nanoparticles J. Appl. Phys. 82, p. 3111-3115.

Cheng W.; Wu P.; Zou X. & Xiao T. (2006). Study on synthesis and blue emission mechanism of ZnO tetrapodlike nanostructures. J. Appl. Phys. 100, pp. 054311(4).

Combescot M, Combestcot R & Roulet B (2001). The exciton dead layer revisited. Eur. Phys. J. B 23 139-151.

Dingle R. (1969). Luminescent Transitions Associated With Divalent Copper Impurities and the Green Emission from Semiconducting Zinc Oxide. Phys. Rev. Lett. 23, pp. 579-581.

Djurisˇic A. B., Leung Y. H., Tam K. H., Ding L., Ge W. K., Chen H. Y. & Gwo S. (2006). Green, yellow, and orange defect emission from ZnO nanostructures: Influence of excitation wavelength. Appl. Phys. Lett. 88, pp. 103107(3).

Du X.W., Fu Y.S., Sun J., Han X. & Liu J. (2006). Complete UV emission of ZnO nanoparticles in a PMMA matrix. Semicond. Sci. Technol. 21, pp. 1202-1206.

Efros A.L. & Rosen M. (2000). The electronic structure of semiconductor nanocrystals. Annu. Rev. Mater. Res, 30 pp. 475-521.

Egelhaaf H. J. & Oelkrug D. (1996). Luminescence and nonradiative deactivation of excited states involving oxygen defect centers in polycrystalline ZnO. J. Cryst. Growth 161, pp. 190-194.

Erhart P., Albe K. and Klein A. (2006).First-principles study of intrinsic point defects in ZnO: Role of band structure, volume relaxation, and finite-size effects. Phys. Rev. B 73, pp. 205203(9).

Ethiraja A.S., Hebalkara N., Kharrazia Sh., Urban J., Sainkar S.R., Kulkarn S.K. (2005). Photoluminescent core-shell particles of organic dye in silica. J. Lumin. 114, pp. 15-23.

Falcony C., Ortiz A., Garcia M. & Helman J. S. (1988).Photoluminescence characteristics of undoped and terbium chloride doped zinc oxide films deposited by spray pyrolysis.J. Appl. Phys. 63, pp. 2378-2388.

Fallert J., Hauschild R., Stelzl F., Urban A., Wissinger M., Zhou H., Klingshirn C. & Kalt H., (2007). Surface-state related luminescence in ZnO nanocrystals. J. Appl. Phys. 101, pp. 073506(4).

Feast W. J., Millichamp I. S., Friend R. H., Horton M. E., Phillips D., Rughooputh S. D. D. V. & Rumbles G. (1985). Optical absorption and luminescence in poly(4,4'-diphenylenediphenylvinylene). Synth. Metals 10, pp. 181-191.

Fonoberov V. A., Alim K. A., Balandin A. A., Xiu F. & Liu J. (2006). Photoluminescence investigation of the carrier recombination processes in ZnO quantum dots and nanocrystals. Phys. Rev. B 73, pp. 165317(9).

Fonoberov V A & Balandin A. A. (2004). Radiative lifetime of excitons in ZnO nanocrystals: The dead-layer effect. Phys. Rev. B 70, pp. 195410(5).

Friend R. H., Gymer R. W., Holmes A. B., Burroughes J. H., Marks R. N., Taliani C., Bradley D. D. C., Dos Santos D. A., Bredas J. L., Logdlund M. & Salaneck W. R. (1999). Electroluminescence in conjugated polymers. Nature, 397, pp. 121-128.

Fu D. J., Panin G. N. & Kang T. W. (2003). GaN Pyramids Prepared by Photo-Assisted Chemical Etching. Journal of the Korean Physical Society, Vol. 42, pp. S611-S613.

Fu Dejun; Park Young Shin; Panin Gennady N. & Kang Tae Won (2005). Formation of Hexagonal GaN Pyramids by Photo Assisted Electroless Chemical Etching. Jap. Journal of Applied Phys, Vol.44, No.11, pp. L342-L344.

Gaspar C., Costa F. & Monteiro T. (2001). Optical characterization of ZnO. J. Mater. Sci.: Mater. Electron. 12, pp. 269-271.

Gill B. & Kavokin A. V. (2002). Giant exciton-light coupling in ZnO quantum dots. Appl. Phys. Lett. 81, pp. 748-750.

Greenham N. C., Peng Xiaogang & Alivisatos A. P. (1996). Charge separation and transport in conjugated-polymer/semiconductor-nanocrystal composites studied by photoluminescence quenching and photoconductivity. Phys. Rev. B 54, pp. 17628-17637.

Hagn M., Zrenner A., Böhm G. & Weimann G. (1995). Electric-field-induced exciton transport in coupled quantum well structures, Appl. Phys. Lett. 67 (2), pp. 232-234.

Harada Y. & Hashimoto S. (2003). Enhancement of band-edge photoluminescence of bulk ZnO single crystals coated with alkali halide. Phys. Rev. B 68, pp. 045421(4).

He H. P., Tang H. P., Z. Z. Ye, Zhu L. P., Zhao B. H., Wang L. & Li X. H. (2007). Temperature-dependent photoluminescence of quasialigned Al-doped ZnO nanorods. Appl. Phys. Lett. 90, pp. 023104(3).

Hirai T., Ohno N., Harada Y., Horii T., Sawada Y. & Itoh T. (2008). Spatially resolved cathodoluminescence spectra of excitons in a ZnO microparticle. Appl. Phys. Lett. 93, pp. 041113(3).

Hoffmann K. & Hahn D. (1974). Electron Spin Resonance of Lattice Defects in Zinc Oxide. Phys. Status Solidi A 24, pp. 637-648.

Hosaka N., Tachibana H., Shiga N., Matsumoto M., & Tokura Y. (1999). Photoinduced Phase Transformation in Polythiophene. Phys. Rev. 82 , pp. 1672-1675.

Hosono Eiji, Fujihara Shinobu, Kimura Toshio & Imai Hiroaki (2004). Non-Basic Solution Routes to Prepare ZnO Nanoparticles. J. Sol-Gel Sci. Technol. 29, pp. 71-79.

Hyberstsen M.S. (1994). Absorption and emission of light in nanoscale silicon structures. Phys. Rev. Lett. 72, p. 1514-1517.

Hua G., Zhang Y., Ye Ch., Wang M. & Zhang L. (2007).Controllable growth of ZnO nanoarrays in aqueous solution and their optical properties. Nanotechnology 18, pp. 145605

Hur T B, Hwang Y H & Kim H. K. (2005). Quantum confinement in Volmer–Weber-type self-assembled ZnO nanocrystals. Appl. Phys. Lett. 86, pp. 193113(3).

Innocenzi P., Kozuka H. & Yoko T. (1996). Dimer-to-monomer transformation of rhodamine 6G in sol—gel silica films. J. Non-Crystal. Sol. 201, pp. 26-36.

Yamamoto A., Kido T., Goto T., Chen Y., Yao T. & Kasuya A. (1999). Dynamics of photoexcited carriers in ZnO epitaxial thin films. Appl. Phys. Lett. 75, pp. 469-471.

Jeong S.H., Kim B.S. & Lee B.T. (2003). Photoluminescence dependence of ZnO films grown on Si(100) by radio-frequency magnetron sputtering on the growth ambient. Appl. Phys. Lett. 82, pp. 2625-2627.

Jin Sh., Zheng Y. & Li A. (1997). Characterization of photoluminescence intensity and efficiency of free excitons in semiconductor quantum well structures. J. Appl. Phys. 82, pp. 3870-3873.

Johnson P.D. (1954). Some Optical Properties of MgO in the Vacuum Ultraviolet. Phys. Rev. 94, pp. 845-846.

Yu P., Tang Z.K., Wong G.K.L., Kawasaki M., Ohtomo A., Koinuma H. & Segawa Y. (1997). Ultraviolet spontaneous and stimulated emissions from ZnO microcrystallite thin films at room temperature. Solid State Commun. 103, pp. 459-463.

Kang T. W.; Yuldashev Sh. & Panin G. N. (2006). Electrical and optical properties of ZnO thin films and nanostructures. Chapter 4, In Handbook of Semiconductor Nanostructures and Nanodevices, edited by A. A. Balandin and K. L. Wang, American Scientific, Los Angeles.

Kasai P. H. (1963). Electron Spin Resonance Studies of Donors and Acceptors in ZnO. Phys. Rev. 130, pp. 989-995.

Klik M. A. J.; Gregorkiewicz T.; Yassievich I. N.; Ivanov V. Yu. & Godlewski M. (2005). Terahertz modulation of the blue photoluminescence in ZnSe. Phys. Rev. B 72, pp. 125205(5).

Kohan A. F., Ceder G., Morgan D. & Van de Walle Chris G. (2000). First-principles study of native point defects in ZnO. Phys. Rev. B 61, pp. 15019-15027.

Kroger E. A. & Vink H. J. (1954). The Origin of the Fluorescence in Self-Activated ZnS, CdS, and ZnO. J. Chem. Phys. 22, pp. 250-252.

Koshihara S., Tokura Y., Takeda K. & Koda T. (1992). Reversible photoinduced phase transitions in single crystals of polydiacetylenes. Phys. Rev. Lett. 68, pp. 1148-1151.

Kumar B., Gonga H., Vicknesh S., Chua S.J. & Tripathy S. (2006). Luminescence properties of ZnO layers grown on Si-on-insulator substrates. Appl. Phys. Lett. 89, pp. 141901(3).

Kurbanov S., Panin G., Kim T. W. & T. W. Kang (2007a). Thermo- and Photo-annealing of ZnO Nanocrystals. Jpn. J. Appl. Phys., Part 1 46, pp. 4172-4174

Kurbanov S.S., Panin G.N., Kim T.W. & Kang T.W. (2007b). Luminescence of ZnO nanocrystals capped with an organic dye. Optics Communications 276, pp. 127–130.

Kurbanov S.S., Panin G. N., Park Y. S., Kang T. W. & Kim T. W. (2007c). Photo- and Cathodoluminescence Studies of ZnO-Filled Opal Nanocomposites. Journal of the Korean Physical Society, Vol. 50, No. 3, pp. 617-621.

Kurbanov S. S., Panin G. N., Kim T. W. & Kang T. W. (2008a). Impact of visible light illumination on ultraviolet emission from ZnO nanocrystals. Phys. Rev. B 78, pp. 045311-1-045311-6

Kurbanov Saidislam, Panin Gennady, Kang Tae Won & Kim Tae Whan (2008b). Modulation of Excitonic Emission from ZnO Nanocrystals by Visible Light Illumination. Jap. J. Appl. Phys. Vol. 47, No. 5, pp. 3760–3762.

Kurbanov S.S.; Panin G.N.; Kim T.W. & Kang T.W. (2008c). ZnO filled opal arrays: Photo- and cathodoluminescence studies. Solid State Communications, 145 (11-12), pp. 577-581.

Kurbanov S.; Panin G. & Kang T.M. (2009a). Spatially resolved investigations of the emission around 3.31 eV (A-line) from ZnO nanocrystals. Appl. Phys. Lett. 95, pp. 211902-1- 211902-3.

Kurbanov S.; Panin G.N.; Kim T.W. & Kang T.W. (2009b). Strong violet luminescence from ZnO nanocrystals grown by the low-temperature chemical solution deposition. Journal of Luminescence 129, pp. 1099–1104.

Lany S. & Zunger A. (2005). Anion vacancies as a source of persistent photoconductivity in II-VI and chalcopyrite semiconductors. Phys. Rev. B 72, pp. 035215(13).

Lee Sang Wuk; Cho Hak Dong; Panin Gennady & Kang Tae Won (2011). Vertical ZnO nanorod/Si contact light-emitting diode. Appl. Phys. Lett. 98, pp. 093110-1-093110-3

Leiter F. H.; Alves H. R.; Hofstaetter A.; Hofmann D. M. & Meyer B. K. (2001). The Oxygen Vacancy as the Origin of a Green Emission in Undoped ZnO. Phys. Status Solidi B 226, pp. R4-R5.

Levitsky I.A., Liang J. & Xu J.M. (2002). Highly ordered arrays of organic–inorganic nanophotonic composites. Appl. Phys. Lett. 81, pp. 1696-1698.

Liu B & Zeng H C (2004) Room Temperature Solution Synthesis of Monodispersed Single-Crystalline ZnO Nanorods and Derived Hierarchical Nanostructures. Langmuir 20, pp. 4196-4204.

Liu M., Kitai A. H. & Mascher P. (1992). Point defects and luminescence centres in zinc oxide and zinc oxide doped with manganese. J. Lumin. 54, pp. 35-42.

Look, C. (2001) Recent advances in ZnO materials and devices. Mater. Sci. Eng., B 80, pp. 383-387.

Look D. C., Reynolds D. C., Litton C. W., Jones R. L., Eason D. B. & Gantwell G. (2002). Characterization of homoepitaxial p-type ZnO grown by molecular beam epitaxy. Appl. Phys. Lett. 81, pp. 1830-1832.

Lyapina O. A., Baranov A. N., Panin G. N., Knotko A. V. & Kononenko O. V. (2008). Synthesis of ZnO Nanotetrapods. Inorganic Materials, Vol. 44, No. 8, pp. 845–851.

Makino T., Tamura K., Chai C.H., Segawa Y., Kawasaki M., Ohtomo A. & Koinuma H. (2002). Radiative recombination of electron–hole pairs spatially separated due to quantum-confined Stark and Franz–Keldish effects in $ZnO/Mg_{0.27}Zn_{0.73}O$ quantum wells. Appl. Phys. Lett. 81, pp. 2355-2357.

Mass J., Avella M., Jiménez J., Rodríguez A., Rodríguez T., Callahan M., Bliss D. & Wang B. (2008). Cathodoluminescence study of ZnO wafers cut from hydrothermal crystals. J. Cryst. Growth 310, pp. 1000-1005.

Massines F., Tiemblo P., Teyssedre G. & Laurent C. (1997),On the nature of the luminescence emitted by a polypropylene film after interaction with a cold plasma at low temperature. J. Appl. Phys. 81, pp. 937-943.

Matsumoto T., Kato H., Miyamoto K., Sano M. & Zhukov M. (2002). Correlation between grain size and optical properties in zinc oxide thin films. Appl. Phys. Lett. 81, pp. 1231-1233.

Meulenkamp E. A. (1998). Synthesis and Growth of ZnO Nanoparticles. J. Phys. Chem. B 102, pp. 5566-5572.

Meulenkamp E. A. (1998). Size Dependence of the Dissolution of ZnO Nanoparticles. J. Phys. Chem. B 102, pp. 7764-7769.

Meyer B. K., Alves H., Hofmann D. M., Kriegseis W., Forster D., Bertram F., Christen J., Hoffmann A., Straßburg M., Dworzak M., Haboeck U. & Rodina A. V. (2004). Bound exciton and donor–acceptor pair recombinations in ZnO. Phys. Status Solidi B 241, pp. 231-260.

Micheletto Ruggero Yoshimatsu Nobuki, Yokokawa Masatoshi, An Taekyung, Lee Haiwon & Okazaki Satoshi (2001). Optical study of a polymeric LED with a nano-sized electrode realized by a modified SNOM setup. Opt. Commun. 196, pp. 47-53.

Minne S.C., Manalis S.R. & Quate C.F. (1995). Parallel atomic force microscopy using cantilevers with integrated piezoresistive sensors and integrated piezoelectric actuators. Appl. Phys. Lett. 67, pp. 3918-3920.

Mitra A., Thareja R. K., Ganesan V., Gupa A., Sahoo P. K. & Kulkarni V. N., (2001). Synthesis and characterization of ZnO thin films for UV laser. Appl. Surf. Sci. 174, pp. 232-239.

Mordkovich V. Z., Hayashi H., Haemori M., Fukumura T. & Kawasaki M. (2003). Discovery and Optimization of New ZnO-Based Phosphors Using a Combinatorial Method. Adv. Funct. Mater. 13, pp. 519-524.

Motyka M., Sek G., Kudrawiec R., Misiewicz J., Li L. H. & Fiore A. (2006). On the modulation mechanisms in photoreflectance of an ensemble of self-assembled InAs/GaAs quantum dots. J. Appl. Phys. 100, pp. 073502(5).

Norberg N.S.& Gamelin D.R. (2005). Influence of Surface Modification on the Luminescence of Colloidal ZnO Nanocrystals. J. Phys. Chem. B 109, pp. 20810-20816.

Ohtomo A., Kawasaki M., Koida T., Masubuchi K., Koinuma H., Sakurai Y., Yoshida Y., Yasuda T. & Segawa Y. (1998). $Mg_xZn_{1-x}O$ as a II–VI widegap semiconductor alloy. Appl. Phys. Lett. 72, pp. 2466-2468.

Ohtomo A., Shiroki R., Ohkubo I., Koinuma H. & Kawasaki M. (1999). Thermal stability of supersaturated $Mg_xZn_{1-x}O$ alloy films and $Mg_xZn_{1-x}O$/ZnO heterointerfaces. Appl. Phys. Lett. 75, pp. 4088-4090.

Ohtomo A., Tamura K., Kawasaki M., Makino T., Segawa Y., Tang Z.K., Wong G.K.L., Matsumoto Y., & Koinuma H. (2000) Room-temperature stimulated emission of excitons in ZnO/(Mg,Zn)O Superlattices. Appl. Phys. Lett. 77, pp. 2204-2206.

Özgür Ü.; Alivov Ya. I.; Liu C.; Teke A.; Reshchikov M. A.; Doğan S.; Avrutin V.; Cho S.-J. & Morkoç H. (2005). A comprehensive review of ZnO materials and devices. J. Appl. Phys. 98, pp. 041301(103).

Panin G. N.; Kang T.W. & Jang M. S. (2003a). Spatially Resolved Study of Luminescent and Electrical Properties of ZnO/GaAs Structures. Journal of the Korean Physical Society, Vol. 42, pp. S357-S360.

Panin G.N.; Kang T.W.; Kim T.W.; Park S.H.; Si S.M.; Ryu Y.S. & Jeon H.C. (2003b). Semiconductor quantum dots created by postgrowth treatment. Physica E 17, pp. 484 -488

Panin G.N., Kang T.W. & Lee H. (2004a). Light emission from the polythiophene derivative/ITO structure under electron beam excitation. Physica E 21, pp. 1074 - 1078.

Panin Gennady N., Kang Tae Won & Lee Haiwon (2004b). Electron beam induced light emission from the polythiophene derivative/ITO structure. phys. stat. sol. (b) 241, No. 12, pp. 2862–2865

Panin G. N., Baranov A. N., Oh Young-Jei & Kang Tae Won (2004c). Luminescence from ZnO/MgO nanoparticle structures prepared by solution techniques. Curr. Appl. Phys. 4, pp. 647-650

Panin G. N.; Park Y. S.; Kang T. W.; Kim T. W.; Wang K. L. & Bao M. (2004d). Self-Assembled GaN Quantum Dots in GaN/AlxGa1-xN Structures Grown by PAMBE. Journal of the Korean Physical Society, Vol. 45, pp. S840-S843.

Panin G. N., Kang T.W., Aleshin A. N., Baranov A. N., Oh Y.-J. & Khotina I. A. (2005a). Electric field switching between blue-green and red cathodoluminescence in poly.4,4'- diphenylene diphenylvinylene. mixed with ZnO:Mg nanoparticles. Appl. Phys. Lett. 86, pp. 113114(3).

Panin G.N., Baranov A.N., Oh Y.-J., Kang T.W. & Kim T.W. (2005b). Effect of thermal annealing on the structural and the optical properties of ZnO/MgO nanostructures. Journal of Crystal Growth 279, pp. 494–500

Panin G. N.; Park Y. S.; Kang T. W.; Kim T. W.; Wang K. L. & Bao M. (2005c). Microstructural and optical properties of self-organized GaN quantum-dot assemblies. J. Appl. Phys., 97, pp. 043527.

Panin Gennady. N.; Baranov Andrey. N.; Kononenko Oleg. V.; Dubonos Sergey. V. & Tae Won Kang (2007a). Resistance Switching Induced by an Electric Field in ZnO:Li, Fe Nanowires. AIP Conf. Proc. 893, pp. 743 -744.

Panin, G.N., Baranov, A.N., Kang, T.W., Kononenko, O.V., Dubonos, S.V., Min S.K. & Kim H.J. (2007b). Electrical and magnetic properties of doped ZnO nanowires. MRS Symp. Proc, 957, pp. 49-54.

Panin G.N.; Baranov A.N.; Kang T.W.; Min S.K. & Kim H.J. (2007c). Spatially-resolved study of magnetic properties of Mn-doped ZnO quantum wires. Journal of the Korean Physical Society 50 (6), pp. 1711-1715.

Panin G. N.; Kim H. J.; Kim S. Y.; Jung J. H.; Kim T. W.; Jeon H. C.; Kang T. W. & Kim M. D. (2007d). Formation and Optical Properties of ZnSe Self-assembled Quantum Dots in Cl-doped ZnSe Thin Films Grown on GaAs (100) Substrates. Solid State Phenomena, 124-126, pp. 567.

Panin Gennady N., Baranov Andrey N., Khotina Irina A. & Kang Tae W. (2008). Luminescent Properties of ZnO/MgO Nanocrystal/Polymer Composite structure. Journal of the Korean Physical Society, Vol. 53, No. 5, pp. 2943-2946.

Pekar S I 1958 Sov. Phys.—JETP 6, pp. 785

Pesika S., Stebe K. J. & Searson P. C. (2003). Determination of the Particle Size Distribution of Quantum Nanocrystals from Absorbance Spectra. Adv. Mater. 15, pp. 1289-1291.

Raghavan S., Hajra J.P., Iyengar G. N. K. & Abraham K.P. (1991). Terminal solid solubilities at 900–1000°C in the magnesium oxide-zinc oxide system measured using a magnesium fluoride solid-electrolyte galvanic cell. Thermochim. Acta 189, pp. 151-158.

Ramakrishna G., G. & Ghosh H.N. (2003). Effect of Particle Size on the Reactivity of Quantum Size ZnO Nanoparticles and Charge-Transfer Dynamics with Adsorbed Catechols. Langmuir 19, pp. 3006-3012.

Reynolds D. C. & Collins T. C. (1969). Excited Terminal States of a Bound Exciton-Donor Complex in ZnO. Phys. Rev. 185, pp. 1099-1103.

Reynolds D. C., Look D. C. & Jogai B. (2001). Fine structure on the green band in ZnO. J. Appl. Phys. 89, pp. 6189-6191.

Riehl N. & Ortman O. Z. (1952) Elektrochem. 60, pp. 149 [in German].

Roessler D.M. & Walker W.C. (1967). Electronic Spectrum and Ultraviolet Optical Properties of Crystalline MgO. Phys. Rev. 159, pp. 733-738.

Rowland G. L., Hosea T. J. C., Malik S., Childs D. & Murray R. (1998). A photomodulated reflectance study of InAs/GaAs self-assembled quantum dots. Appl. Phys. Lett. 73, pp. 3268-3270.

Ryu M. K.; Lee S.; Jang M. S.; Panin G. N. & Kang T. W. Kang (2002). Postgrowth annealing effect on structural and optical properties of ZnO films grown on GaAs substrates by the radio frequency magnetron sputtering technique. J. Appl. Phys., Vol. 92, No. 1, pp. 154-158.

Sakohara S., Ishida M. & Anderson M. A. (1998). Visible Luminescence and Surface Properties of Nanosized ZnO Colloids Prepared by Hydrolyzing Zinc Acetate. J. Phys. Chem. B 102, pp. 10169-10175.

Schmeller A., Hansen W., and Kotthaus J. P. Trinkle G. & Weimann G. (1994) Franz-Keldysh effect in a two-dimensional system. Appl. Phys. Lett. 64 (3), pp. 330-332.

Shalish I., Temkin H., H. & Narayanamurti V. (2004).Size-dependent surface luminescence in ZnO nanowires. Phys. Rev. B 69, pp. 245401(4).

Shannon R.D. (1976). Acta Crystallogr., Sect. A: Cryst. Phys., Diffr., Theor. Gen. Crystallogr. 32, pp. 751.

Sharma A.K., Narayan J., Muth J.F., Teng C.W., Jin C., Kvit A., Kolbas R.M. & Holland O.W. (1999). Optical and structural properties of epitaxial $Mg_xZn_{1-x}O$ alloys. Appl. Phys. Lett. 75, pp. 3327-3329.

Spanhel L. & Anderson M. A. (1991). Semiconductor clusters in the sol-gel process: quantized aggregation, gelation, and crystal growth in concentrated zinc oxide colloids. J. Am. Chem. Soc. 113, pp. 2826-2833.

Shih H. Y.; Chen Y. T.; Huang N. H.; Wei C. M. & Chen Y. F. (2011)Tunable photoluminescence and photoconductivity in ZnO one-dimensional nanostructures with a second below-gap beam. J. Appl. Phys. 109, 103523-1-5.

Sun H.D., Makino T., Tuan N.T., Segawa Y., Tang Z.K., Wong G.K.L., Kawasaki M., Ohtomo A., Tamura K. Koinuma H. (2000). Stimulated emission induced by exciton–exciton scattering in ZnO/ZnMgO multiquantum wells up to room temperature. Appl. Phys. Lett. 77, pp. 4250-4252.

Studenikin S. A., Golego N. & Cocivera M. (1998a). Optical and electrical properties of undoped ZnO films grown by spray pyrolysis of zinc nitrate solution. J. Appl. Phys. 83, pp. 2104-2111.

Studenikin S. A., Golego N. & Cocivera M. (1998b). Fabrication of green and orange photoluminescent, undoped ZnO films using spray pyrolysis. J. Appl. Phys. 84, pp. 2287-2295.

Studenikin S. A., Golego N. & Cocivera M. (1998c). Density of band-gap traps in polycrystalline films from photoconductivity transients using an improved Laplace transform method. J. Appl. Phys. 84, pp. 5001-5004.

Studenikin S. A., Cocivera M., Kellner W. & Pascher H. (2000). Band-edge photoluminescence in polycrystalline ZnO films at 1.7 K. J. Lumin. 91, pp. 223-232.

Studenikin S. A. & M. Cocivera (2002). Time-resolved luminescence and photoconductivity of polycrystalline ZnO films. J. Appl. Phys. 91, pp. 5060-5065.

Takagaki Y., Wiebicke E., Riedel A., Ramsteiner M., Kostial H., Hey R. & Ploog K. H. (2003). Hybrid optical modulator based on surface acoustic waves fabricated using imprint lithography and the epitaxial lift-off technique. Semicond. Sci. Technol. 18, pp. 807-811.

Teke A., Özgür Ü., Doğan S., Gu X. & Morkoç H. (2004). Excitonic fine structure and recombination dynamics in single-crystalline ZnO.Phys. Rev. B 70, pp. 195207(10).

Tokumoto M. S., Pulcinelli S. H., Santilli C. V. & Valérie Briois (2003). Catalysis and Temperature Dependence on the Formation of ZnO Nanoparticles and of Zinc Acetate Derivatives Prepared by the Sol–Gel Route. J. Phys. Chem. B 107, pp. 568-574.

Tomzig E. & Helbig R. (1976). Band-edge emission in ZnO. J. Lumin. 14, pp. 403-415.

Tong Y.H., Liu Y.C., Lu S.X., Dong L., Chen S.J. & Xiao Z.Y. (2004). The Optical Properties of ZnO Nanoparticles Capped with Polyvinyl Butyral. J. Sol- Gel Sci. Technol. 30, pp. 157-161.

Van de Walle C. G. (2001). Defect analysis and engineering in ZnO. Physica B (Amsterdam) 308-310, pp. 899-903.

Van der Horst J.W., Bobbert P.A. & Michels M.A.J. (2002). Electronic and optical excitations in crystalline conjugated polymers. Phys. Rev. B 66, pp. 035206(7).

VanDijken A., Meulenkamp E. A., Vanmaekelbergh D. & Meijerink A. (2000a). The luminescence of nanocrystalline ZnO particles: the mechanism of the ultraviolet and visible emission. J. Lumin. 87-89, pp. 454—456.

Van Dijken A., Meulenkamp E.A., Vanmaekelbergh D. & Meijerink A. (2000b). The Kinetics of the Radiative and Nonradiative Processes in Nanocrystalline ZnO Particles upon Photoexcitation. J. Phys. Chem. B 104, pp. 1715-1723.

Van Dijken A., Makkinje J. & Meijerink A. (2001). The influence of particle size on the luminescence quantum efficiency of nanocrystalline ZnO particles. J. Lumin 92, pp. 323-328.

Vanheusden K.; Seager C. H.; Warren W. L.; Tallant D. R. & Voigt J. A. (1996a). Correlation between photoluminescence and oxygen vacancies in ZnO phosphors. Appl. Phys. Lett. 68, pp. 403-405.

Vanheusden K.; Warren W. L.; Seager C. H.; Tallant D. R. & Voigt J. A. (1996b). Mechanisms behind green photoluminescence in ZnO phosphor powders. J. Appl. Phys. 79, pp. 7983-7990.

Vanheusden K., Seager C. H., Warren W. L., Tallant D. R., Caruso J., Hampden-Smith M. J. & Kodas T. T. (1997). Green photoluminescence efficiency and free-carrier density in ZnO phosphor powders prepared by spray pyrolysis. J. Lumin. 75, pp. 11-16.

Vogel R., Meredith P., Harvey M.D. & Rubinsztein-Dunlop H. (2004). Absorption and fluorescence spectroscopy of rhodamine 6G in titanium dioxide nanocomposites. Spectrochim. Acta Part A 60, pp. 245-249.

Wang X, Ding Y, Summers C J & Wang Z L (2004). Large-Scale Synthesis of Six-Nanometer-Wide ZnO Nanobelts. J. Phys. Chem. B 108, pp. 8773-8777.

Wang H., Xie Ch. & Zeng D. (2005) Controlled growth of ZnO by adding H_2O. J. Cryst. Growth 277, pp. 372-377.

Welter S., Brunner K., Hofstraat J. W. & De Cola L. (2003). Electroluminescent device with reversible switching between red and green emission. Nature, 421, pp. 54-57.

Wong. E. M., Hoertz P. G., Liang C. J., Shi Bai-Ming, Meyer Gerald J. & Searson Peter C. (2001). Influence of Organic Capping Ligands on the Growth Kinetics of ZnO Nanoparticles. Langmuir 17, pp. 8362-8367.

Woo H.S., Czerw R., Webster S. Carroll D. L., Ballato J., Strevens A. E., O'Brien D. & Blau W. J. (2000). Hole blocking in carbon nanotube–polymer composite organic light-emitting diodes based on poly (m-phenylene vinylene-co-2, 5-dioctoxy-p-phenylene vinylene). Appl. Phys. Lett. 77, pp. 1393-1395.

Wood D. L. & Rabinovich E.M. Study of Alkoxide Silica Gels by Infrared Spectroscopy. Appl. Spectrosc. 43, pp. 263-267.

Wu X.L., Siu G.G., Fu C.L. & Ong H.C. (2001). Photoluminescence and cathodoluminescence studies of stoichiometric and oxygen-deficient ZnO films. Appl. Phys. Lett. 78, pp. 2285-2287.

Xia Y, Yang P, Sun Y, Wu Y, Mayers B, Gates B, Yin Y, Kim F & Yan H (2003). One-Dimensional Nanostructures: Synthesis, Characterization, and Applications. Adv. Mater 15, pp. 353-389.

Yang P, Yan H, Mao S, Russo R, Johnson J, Saykally R, Morris N, Pham J, He R & Choi H. J. (2002). Controlled Growth of ZnO Nanowires and Their Optical Properties. Adv. Funct. Mater. 12, pp. 323-331.

Yang C.L., Wang J.N., Ge W.K., Guo L., Yang S.H. & Shen D.Z. (2001). Enhanced ultraviolet emission and optical properties in polyvinyl pyrrolidone surface modified ZnO quantum dots. J. Appl. Phys. 90, pp. 4489-4493.

Yang X. D., Xu Z. Y., Sun Z., Sun B. Q., Ding L., Wang F. Z. & Ye Z. Z. (2006). Recombination property of nitrogen-acceptor-bound states in ZnO. J. Appl. Phys. 99, pp. 046101(3).

Yao B D, Chan Y F & Wang N (2002). Formation of ZnO nanostructures by a simple way of thermal evaporation. Appl. Phys. Lett. 81, pp.757-759.

Yin M, Gu Y, Kuskovsky I L, Andelman T, Zhu Y, Neumark G F & O'Brien S (2004). Zinc Oxide Quantum Rods. J. Am. Chem. Soc. 126, pp. 6206-6207.

Ying J.Y., Benziger J.B. & Navrotsky A. (1993). Structural Evolution of Alkoxide Silica Gels to Glass: Effect of Catalyst pH. J. Am. Ceram. Soc. 76, pp. 2571-2582.

Zhang B. P., Binh N. T., Segawa Y., Wakatsuki K. & Usami N. (2003). Optical properties of ZnO rods formed by metalorganic chemical vapor deposition. Appl. Phys. Lett. 83, pp. 1635(3).

Zhang S. B., Wei S.-H. & Zunger A. (2001). Intrinsic n-type versus p-type doping asymmetry and the defect physics of ZnO. Phys. Rev. B 63, pp. 075205(7).

Zhang S. K., Santos P. V. & Hey R. (2001). Photoluminescence modulation by high-frequency lateral electric fields in quantum wells. Appl. Phys. Lett. 78, pp. 1559-1561.

Zhang X H, Xie S Y, Jiang Z Y, Xie Z X, Huang R B, Zheng L S, Kang J Y & Sekiguchi T. (2003). Microwave plasma growth and high spatial resolution cathodoluminescent spectrum of tetrapod ZnO nanostructures. J. Solid State Chem. Chem. 173, pp. 109-113.

Zhao Q.X., Klason P., Willander M., Zhong H.M., Lu W. & Yang J.H. (2005). Deep-level emissions influenced by O and Zn implantations in ZnO. Appl. Phys. Lett. 87, pp. 211912(3).

Zimmermann S.; Govorov A. O.; Hansen W.; Kotthaus J. P.; Bichler M. & Wegscheider W. (1997). Lateral superlattices as voltage-controlled traps for excitons, Phys. Rev. B 56, pp. 13414–13421

Zimmermann S.; Schedelbeck G.; Govorov A. O.; Wixforth A.; Kotthaus J. P.; Bichler M.; Wegscheider W.; & Abstreiter G. (1998). Spatially resolved exciton trapping in a voltage-controlled lateral Superlattice. Appl. Phys. Lett., Vol. 73, No. 2, pp. 154-156.

Zimmermann S.; Wixforth A.; Kotthaus J. P.; Wegscheider W.; & Bichler M. (1999). A Semiconductor-Based Photonic Memory Cell. Science 283, pp. 1292-1295.

7

Cathodoluminescence from Amorphous and Nanocrystalline Nitride Thin Films Doped with Rare Earth and Transition Metals

Muhammad Maqbool[1], Wojciech M. Jadwisienczak[2] and
Martin E. Kordesch[2]
[1]Ball State University
[2]Ohio University
USA

1. Introduction

Rare earth (RE) ion luminescence has long been used in laser and optical fiber communications technology. Bulk RE doped oxides were widely used in color phosphors for Cathode Ray Tubes. The wide band gap (WBG) semiconductors and insulators have been used for visible emission at 300 K from RE ions since the reports first by Zanata (Zanatta and Nunes 1998) for Er in silicon nitride (photoluminescence) and then shortly thereafter by Steckl (Steckl and Birkhahn 1998) for Er in GaN. The III-nitrides were emerging as Light Emitting Diodes and semiconductors at the time, and it was reported that the luminescence intensity of the RE ions was improved by a wide band gap host. Silicon (band gap ≈ 1.1 eV) is not suited for most visible RE ion emission. Glasses and oxides were used for the infra red (IR), especially for the emission from Er^{3+} ion at ~ 1.5 µm. In 1998, RE ion incorporation into a crystalline host was often accomplished by ion implantation at low atomic concentrations, or by in situ doping (again with low atomic percentages) of the RE ions. It was believed that the quality of the host lattice was essential to the RE luminescence. However, Zanata and Nunes observed green room temperature luminescence from an Er-doped silicon nitride film deposited by reactive sputtering in nitrogen. Visible emission was observed with an estimated 10 at. % dopant concentration in an amorphous material. Both the amorphous and crystalline hosts discovered by the Zanata and Steckl groups set in motion the (enduring) pursuit of practical visible light emission devices using RE ions in wide band gap materials.

In 1999, Gurumurugan (Gurumurugan, Chen et al. 1999) sputter deposited amorphous AlN doped with Er (3.4 at.%) and observed the full range of Er emission lines in cathodoluminescence (CL) in amorphous AlN. There is no visible defect band in the large gap of AlN (~6 eV), in contrast to sputtered GaN which suffers from an intense, broad yellow emission from defects. The AlN band gap is almost identical with the crystalline AlN bandgap, and the bandgap was not reduced significantly by the RE dopant. The AlN film was transparent, amorphous, and could be deposited at 300 K. Gurumarugan et al. also showed that the amorphous AlN films could be heated to 1300 K without crystallization. The heating improved the RE emission intensity. Also, it was noted that in these films the intensity of some of the Er^{3+} transitions *decreased* with rising temperature, and some *increased* with rising temperature.

The crystalline III-nitrides are now extensively used for LED's, lasers and solid state lighting, high power electronic devices and other passive electronics uses related to the thermal conductivity of AlN. The light emission from direct transitions in GaN and In-Al-Ga-N alloys has displaced the III-nitride RE devices in lighting applications. The low intensity and long lifetime of the RE^{3+} transitions have also hindered the application of RE luminescence for on-chip communications. Thin film materials doped with RE's are used for phosphors, and specialty applications such as in situ thermometry.

2. Growth, doping and measurement techniques

2.1 Sputter deposition

Radio frequency (RF) sputtering is a physical vapour deposition process that uses RF excitation to cause gas molecules to collide with a target, removing the target material by the mechanical impact of the sputtering gas. The RF sputtering process is well known and used for many industrial applications. The details can be obtained from several textbooks. RF magnetron sputtering is a variation most often used for the efficient deposition of thin films.

In the sputtering process, a target of a metal or insulator might be made from solid material or pressed powders. The sputter gas used is often Argon (Ar). The gas pressure is adjusted so that a plasma of Ar ions is created by the RF field above the target. In magnetron sputtering, a system of magnets is used to confine the RF field and the ions so that the target is used efficiently. In this example, the sputter gas is inert, and physically collides with the target surface to eject a target atom that is then condensed onto the substrate. In this simple example, the kinetic energy of the Ar ion is used to eject the target atom. No reaction takes place, and ideally the condensed film is composed only of the target atoms. For AlN or GaN, sputterd with Ar, the condensed film is ideally a stoichiometric film of AlN or GaN. Sputtered films are often disordered, because the process does not generate or require the input of large amounts of heat such as thermal evaporation or chemical vapor deposition. The substrate can be cold. As a consequence, the condensed atoms do not have sufficient energy on the substrate to organize into a crystalline lattice.

In reactive sputtering, the target material is sputtered in a reactive gas. The RF plasma creates ions of the reactive gas, so that impact with the target results in a compound of the target and sputter gas formed by a chemical reaction which then condenses onto the substrate. For nitrogen and aluminium, for example, the nitrogen decomposes into N ions, and reacts with the Al metal target to form AlN. For simple compounds, where there are few alternative compositions and structures, the stoichiometric product is often the most likely compound to be deposited. The purity of the initial components determines the composition of the deposited film. In the previous example, reactive sputtering of Al in nitrogen, water vapor or oxygen in the sputter gas can produce aluminium oxide along with AlN.

A very significant aspect of sputter deposition is the option to grow unusual phases or compositions that could not be grown in equilibrium processes at high temperature.

Because the plasma is "hot", reactions are fast and complete; rapid thermal quenching by condensation onto a (relatively) cold substrate preserves the composition achieved in the plasma. Phase segregation can be avoided as long as the films are not heated to the point where the different phases are free to form. Alternatively, sputtering onto a heated substrate can produce polycrystalline films and even epitaxy.

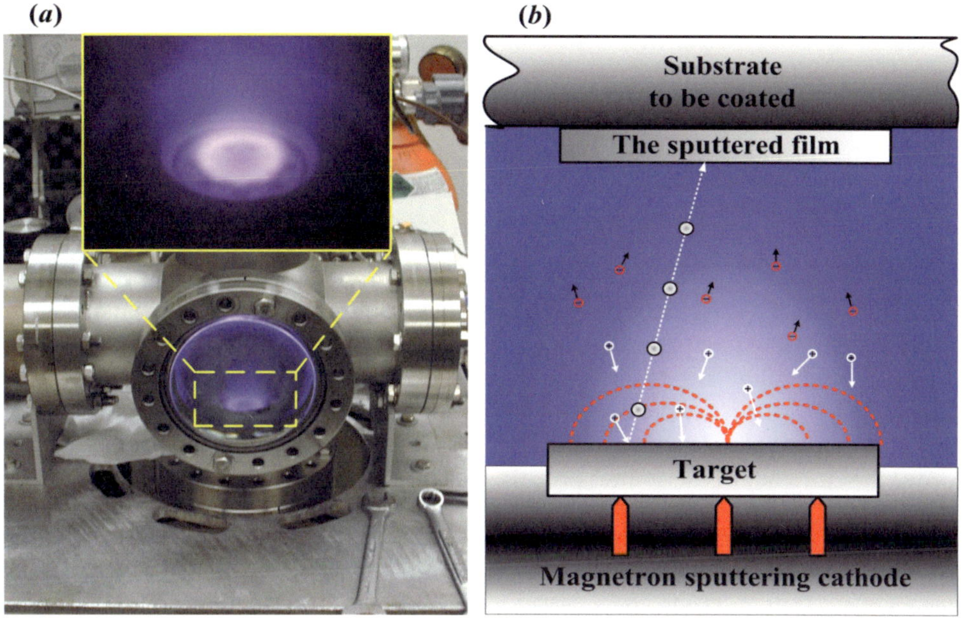

Fig. 2.1. RF magnetron sputtering system. a) actual system; b) schematic diagram. From (Ebdah 2011), used with permission.

Practical considerations can encourage the use of a particular target. Al sputtered in nitrogen is an ideal system. Only AlN forms from the reaction. The Al itself is inexpensive. Targets can even be made from industrial grade Al plate or bars (however, these materials usually contain small amounts of Cr, and chromium oxide CL is observed), making the use of multiple targets simple and convenient. In Kordesch's laboratory at Ohio University, multiple targets were fabricated from 2 inch diameter (50 mm) Al bars or from targets purchased from commercial vendors. Small plugs of the RE or transition metals were pressed into the aluminium targets. In this way, multiple dopants or multiple dopant percentages could be obtained with very simple methods. Zanatta used plates or chips of metal on the silicon target. For materials that do not melt, or alloy with the sputter target, this method could also be used.

Gallium was sputtered in our laboratory at Ohio University from a pressed powder target and also from a liquid metal target. In this case, the powder target was more difficult to use with regard to the dopant material, because there is no good mechanical contact between the powder and the metal insert. For the liquid target, some droplet ejection occurs during sputtering. The existence of metal droplets on the thin film surface is detrimental to device performance, especially if contact or insulating layers are deposited over the luminescent layer.

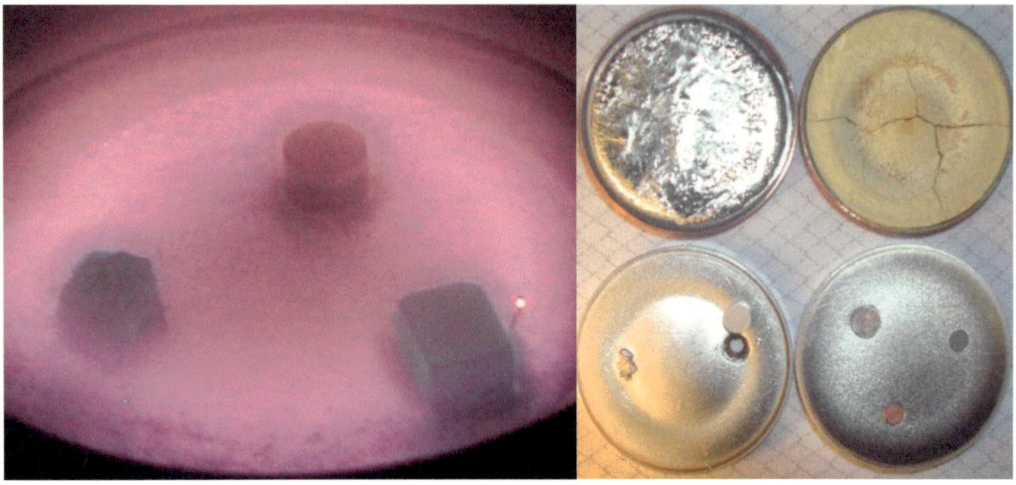

Fig. 2.2. Left: Target with (clockwise from top) Sm, Er, and Tb pieces during sputtering. Right: clockwise from top: Liquid metal Ga target in a copper cup, GaN powder target in a copper cup, aluminium target with Mn , Tb and Cu plugs, aluminium target with RE plug removed, and Tm dendrites pressed into the target. The circular depression in the targets at about ½ the radius is the "racetrack", where the plasma is confined by the magnets to improve the efficiency of the sputter process. All targets are 50 mm diameter.

In practice, the targets used were mostly 50 mm diameter, 6 mm thick disks of aluminium. The RE or transition metal plugs were obtained commercially as rods of 3 - 12 mm diameter. Plugs were cut from the rods and pressed into holes in the aluminium target. Manganese, for example, could only be purchased as a powder, so the powder was pressed into the 6 mm diameter hole. Some RE metals were obtained as dendrites or split fragments. These could be pressed into the softer aluminium target by force, even if voids were left around the RE material. Typical power levels were from 100-200 Watts RF (13.56 MHz) power at 2-10 mTorr pressures of nitrogen. This is the equivalent of 5-10 Watts/cm^2. The substrates could be heated from the back by using a quartz lamp with a parabolic reflector focussed onto the back of the substrate, or cooled by clamping the substrates to a copper cold finger cooled with dry ice. We have not used liquid nitrogen to cool the samples. First, the lower temperature will condense water onto the sample, and second, there are stresses that come from warming a film grown at low temperature to 300 K, just as there are stresses in films cooled from a high temperature deposition.

Aluminum, gallium, beryllium, scandium, boron, silicon, silicon nitride, aluminium oxide, zinc, hafnium, titanium and alloy targets were used. Dopant RE metals such as Er, Tb, Gd, Tm, Sm, Pr and Ho were used, as well as Cr, Cu, Mn, Mg, Si, C, Ti, Sn, Ag. The sputter target is always below the sample/substrate to avoid particulates formed in the sputter process from falling onto the substrate (see Fig. 2.1).

An important consideration for RE doping in the nitrides is the stability of the host material. Many RE nitrides decompose when in contact with water or humid air, which causes a reaction resulting in ammonia and a RE oxide. AlN, BN, GaN, SiN and ScN are not soluble in water. In our experiments, beryllium nitride was also stable in humid air. There is considerable controversy over the role of oxygen in promoting the luminescence of RE ions in the nitrides. It is not known if the RE nitrides in a nitride host decompose or remain bonded to the N atom.

Finally, it should be pointed out that no electronic doping is observed in the amorphous nitride films. Partially, this is due to the fact that most of the amorphous nitride films are insulators, with the exception of n-type GaN. Electronic doping with Mg or Si was not successful, because the dopants react to form nitrides in the plasma, making the dopants commonly used in crystalline GaN for example, ineffective in amorphous nitrides.

2.2 Cathodoluminescence measurements

2.2.1 CL at room temperature and above

For CL measurements at 300 K and above, a simple system build in a six-way stainless steel cross with copper sealed flanges was used. The system was pumped with a 450 L/sec turbomolecular pump. A CRT electron gun with deflection plates was adapted to a multi-pin electrical feedthrough. The electron gun was powered with a laboratory-built power supply. The electron beam voltage could be varied up to 2.8 kV. Focus controls allowed the beam to be focussed onto the sample and to observe the spot on the sample through a vacuum viewport for optimization. The CL spot on the sample was then focussed onto the entrance slits of a monochromater using a lens.

The sample was illuminated with the electron beam from one side, the viewport was at 90 degrees from the electron gun. The sample was placed at 45 degrees to both arms of the cross. An aperture was placed in front of the sample between the sample and electron gun to block the visible light spot from the filament. The deflectors on the electron gun make it possible to steer the electron beam through the aperture while the direct light from the hot filament is blocked by the aperture.

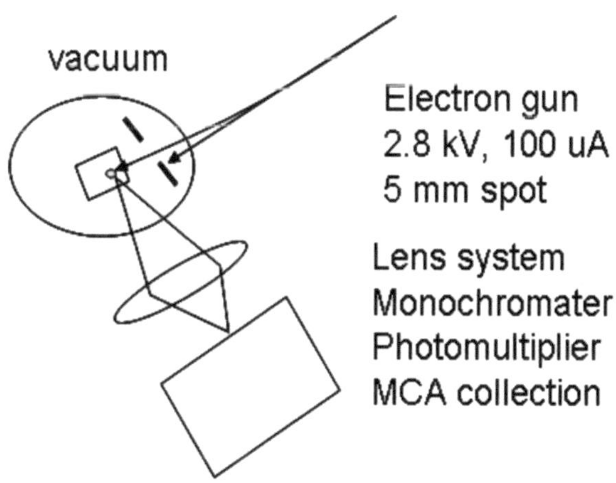

Fig. 2.3. Experimental setup for CL.

A commercial Luminoscope was also used for some CL measurements. The Luminoscope is meant to be used for geological specimens, and is equipped with a vacuum chamber that replaces the stage of a light-optical microscope. A window allows microscopic observation of the sample surface during electron illumination. A gas discharge electron source is used to generate an electron beam of up to 30 kV. The Luminoscope can be used to take color micrographs of the sample surface. A drawback to spectroscopic measurements in the Luminoscope is the necessity to remove the gas discharge spectrum (He) from the CL spectra.

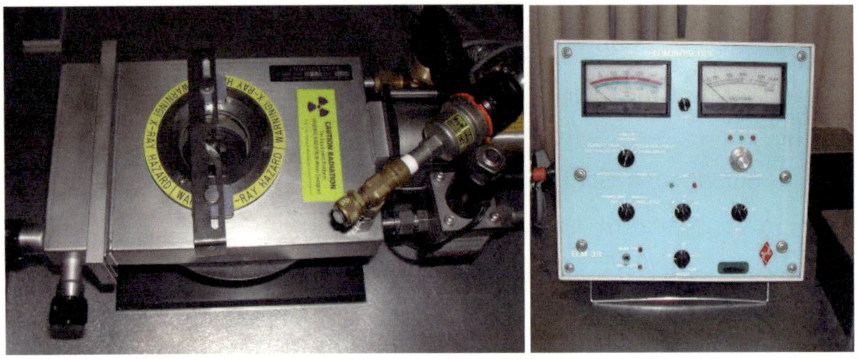

Fig. 2.4. Left: Lumiscope sample holder. Right: Lumiscope control and electronics.

2.2.2 CL at low temperature

Low temperature CL spectroscopic analysis was measured in our laboratory at cryogenic temperatures and excitation conditions using experimental setup shown in Fig.2.5. Typically samples were mounted on the cold finger of a closed-cycled helium refrigerator operating down to 6K. The CL was generated by the Staib Instruments, Inc. Electron Gun EK-20-R equipped with a beam blanker (repetition rate: DC to 1 MHz) electron gun system being in common vacuum (of 5×10^{-7} Torr) with the cryostat. The electron beam was incident upon the sample at a 45° angle from an electron gun. The CL depth of the excitation could be easily varied by varying the electron acceleration voltage between 500eV up to 20 keV.

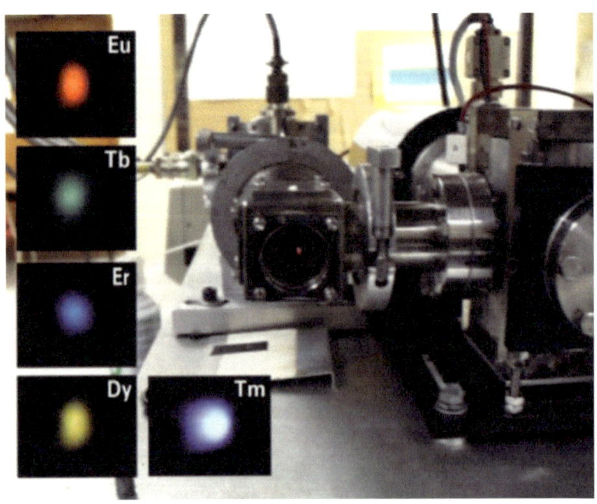

Fig. 2.5. Front view of a low temperature cathodoluminescence experimental setup operating between 6K- 330K and electron acceleration voltage up to 20 kV. Insets show color emissions from different RE-doped III-nitrides.

2.2.3 CL in the electron microscope

Cathodoluminescence attachments can be obtained for both scanning electron microscopes (SEM) and transmission electron microscopes (TEM). In the simplest SEM versions, a parabolic mirror is mechanically inserted into the specimen chamber above the sample, with a hole in the mirror for the passage of the electron beam. The light from the sample CL is reflected onto a photomultiplier tube and used to form a scanned image based on the CL yield. In advanced systems, and monochromater is added to the optical system so that spectroscopically resolved images can be generated. Similar systems are used in TEM, but without a CL image. Only spectroscopy is possible in the TEM. CL based images can be obtained in scanning TEM, STEM.

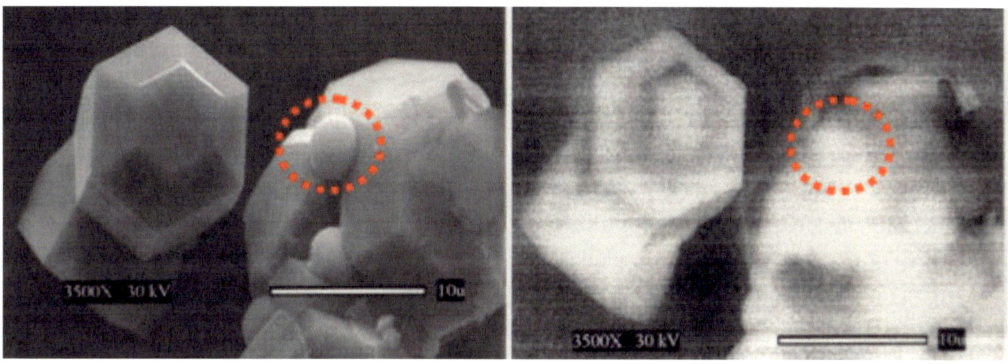

Fig. 2.6. Left: GaN crystallites in the SEM. A bit of a-GaN is shown in the stippled circle. Right: The same area, showing CL from the a-GaN and c-GaN. This image was one of the first experimental observations of a-GaN. (Hassan 1998).

3. Cathodoluminescence of rare earth doped amorphous nitrides

In the last two decades it has been shown that the incorporation of RE elements as dopant atoms into III-nitride semiconductors such as GaN, AlN and their alloys, both crystalline and amorphous, leads to a temperature-stable luminescence whose wavelength is nearly independent of the specific semiconductor host (O'Donnell 2010). Luminescence from these materials having amorphous morphology grown on variety of substrates, and doped by implantation or during the growth with Ce (Aldabergenova, Osvet et al. 2002) , Pr (Maqbool, Ahmad et al. 2007; Maqbool, Richardson et al. 2007; Maqbool and Ahmad 2009), Sm (Zanatta, Ribeiro et al. 2001; Weingartner, Erlenbach et al. 2006; Maqbool and Ali 2009), Eu (Aldabergenova, Osvet et al. 2002), (Weingartner, Erlenbach et al. 2006), (Caldwell, Van Patten et al. 2001), Gd (Maqbool, Ahmad et al. 2007),(Maqbool, Kordesch et al. 2009), Tb (Aldabergenova, Osvet et al. 2002), (Weingartner, Erlenbach et al. 2006),(Richardson, Van Patten et al. 2002), (Jadwisienczak, Lozykowski et al. 2000), Dy (Weingartner, Erlenbach et al. 2006), Ho (Maqbool, Ali et al.; Aldabergenova, Frank et al. 2006; Maqbool, Kordesch et al. 2009), Er (Gurumurugan, Chen et al. 1999; Chen, Gurumurugan et al. 2000; Dimitrova, Van Patten et al. 2000; Dimitrova, Van Patten et al. 2001; Zanatta, Ribeiro et al. 2005), Tm (Maqbool, Kordesch et al. 2009) and Yb (Weingartner, Erlenbach et al. 2006) has been reported so far. In particular, RE doped GaN-nitrides based electroluminescent devices (ELDs) have been shown to have a versatile approach for the fabrication of variety of electrically driven optical light sources with narrow line-width emissions from the ultraviolet to the infrared (Steckl 1999). Thus optoelectronic devices utilizing $4f^n$ transitions appearing in the nitrides forbidden band gap "window" are practically viable with these semiconductors (O'Donnell 2010). Recently it was demonstrated that the low voltage-operation of current-injected red emission from a crystalline *p*-type/Eu-doped/*n*-type GaN epilayers light emission diode operating at room temperature (Nishikawa, Kawasaki et al. 2009; Kasai, Nishikawa et al. 2010; Nishikawa, Furukawa 2010; Dierolf 2011) together with demonstrations of stimulated emission from Eu^{3+} doped GaN and AlGaN layers on a Si substrate proved in principal that optoelectronic devices covering the UV, visible and IR regions might be fabricated (Park and Steckl 2004; Park and Steckl 2005; Park and Steckl 2006).

These new results suggest a novel way to realize III-nitride semiconductors-based red emitting, current driven light emitting devices, as well as other primary colors and their mixture, monolithic devices. In the past the feasibility of using the RE-doped amorphous III-nitrides (a-III-nitrides) for light emitting applications was also demonstrated (Dimitrova, Van Patten et al. 2000; Dimitrova, Van Patten et al. 2001; Richardson, Van Patten et al. 2002; Maqbool, Kordesch et al. 2009), (Kim, Shepherd et al. 2003; Kim and Holloway 2004). However, to make these devices commercially viable, the internal quantum efficiency of the active RE-doped layers has to be significantly improved. The future success of the RE-doped III-nitrides optoelectronics, both crystalline and amorphous based will most probably depend on engineering of multilayer structures. In these devices an enhancement of RE^{3+} ion luminescence intensity can be achieved through *e.g.* (a) modification of the RE^{3+} center environment and (b) localization of carriers in the vicinity of emitting RE^{3+} ion center. The former one can be achieved through engineering stress/strain parameters during the growth process, whereas the last one will results from carriers confinement in the quantum structures.

In general, the majority available research papers focus on RE-doped crystalline III-nitride semiconductors with less emphasis on a-III-nitrides used as hosts for optically, electrically and/or magnetically active RE dopants. The amorphous semiconductors including a-III-nitrides as hosts for RE doping have many of the desirable qualities of the crystalline materials; however they also offer unique features rooted in the nature of the amorphous matrix (Adachi 1999; Singh 2003; Street 2010). The most important are that a-III-nitrides can be achieved at higher growth rates and lower temperatures, substrate selection and resulting lattice mismatch is not a significant obstacle here, they do not easily recrystallize when subjected to thermal processing and that they can adapt the RE ions in concentrations far beyond those given by the solubility limit found in their crystalline counterparts. Furthermore, it is known that a-III-nitrides show a natural tendency to develop a state free gap [(Chen and Drabold 2002; Drabold 2010). Also a-III-nitrides, unlike other amorphous systems or glasses which look locally very similar to the crystal, apparently have local environments very different from the main crystalline morphology what make them potentially useful as hosts for optically and electrically active RE dopants. It was demonstrated that the RE ions luminescence intensity, when intended for optoelectronic applications, can be increased by proper structural tailoring of a-III-nitride matrices. At the same time the RE^{3+} ion excitation and de-excitation processes are more dependent on the local RE ion environment than in crystalline semiconductors due to the lack of long range ordering in amorphous host (Zanatta, Ribeiro et al. 2006). The *f-f* transitions are forbidden, but they are partially allowed when the wave functions are mixed with those of opposing parity. This always occurs on localized ions in non-central symmetrical lattices. The forbidden nature of these transitions results in long radiative lifetimes when the selection rules are relaxed. Placing trivalent rare earth ions in amorphous solids can still result in characteristic emission from these intra-$4f$ transitions despite the lack of a crystalline host structure (O'Donnell 2010). In both cases RE luminescence depends very little on the nature of the host and the ambient temperature. The $4f$ orbitals of RE ions incorporated into semiconductors, including a-III-nitrides, are so deeply buried within the electronic shell that the energy levels of the $4f^n$ configuration are only slightly perturbed compared to free ion energy levels indicating large independence from the chemical surrounding and thus the choice of the host material (Wybourne and Smentek 2007). RE ions when doped to a-III-nitrides are known to exhibit sharp and well-defined luminescence which arises mostly from internal transitions between $4f$ levels within RE^{3+} ion. Because of the shield provided by the outer $5s^2$ and $5p^6$ shells, the wavelength involved in these intra-$4f$ transitions is less temperature sensitive and rather weakly depends on the a-III-nitride host. On the other hand the mechanism behind the activation of RE are known to be sensitive to a local host details such as atomic structure and optical band gap which defines which RE transitions can be excited.

3.1 Rare earth doping

Doping of a-III-nitrides is generally achieved *in situ* ; however it can be also done by RE ions implantation (Aldabergenova, Frank et al. 2006). In the former case doping typically takes place during sputtering process in which RE-doped a-III-nitride is formed from Ga, Al and RE atoms removed from metal targets by momentum transfer from an RF plasma. The sputtering growth of a-III-nitrides is explored much less than other technologically important amorphous semiconductors e.g a-Si doped with RE impurities (Iacona, Franzo et al. 2009). Ion implantation has been used extensively to *ex situ* dope III-nitride crystalline semiconductors with variety of rare earth ions. Excellent review concerning this issue is available (O'Donnell 2010). There are a limited number of papers on RE ions implantation to a-III-nitrides and resulting luminescence spectra (Jadwisienczak, Lozykowski et al. 2000). Ion implantation is a well established technique, which provides good control of the concentration profile in the sample. However, because of the unavoidable damage associated with the implantation process, typically a high temperature annealing step is imperative. This is due to the fact that when heavy atoms like RE metals are implanted to an amorphous semiconductor they can introduce both a chemical change in the substrate, in that they can introduce a different element than the host or induce a nuclear transmutation, and a structural change, in that the morphology of the host can be damaged or even destroyed by the energetic collision cascades induced by implanted heavy RE ions (Kucheyev, Williams et al. 2004).

It is known that the thermal anneal of the a-AlN films affects not only the luminescent properties of the films, but also their optical band gap and chemical characteristics. As a result of thermal treatments of RE-doped a-III-nitrides at increasing temperatures the optical-absorption edge of amorphous films may change indicating the Al and N bonding environment modification (Zanatta, Ribeiro et al. 2005). Based on the available experimental results it is clear that RE-doped a-III-nitrides subjected to isochronal thermal annealing treatments indeed most likely underwent partial removal of both structural and chemical disorder resulting in modification of RE ions local environments as it is shown in Figs. 3.1 and 3.2 for Er-doped a-AlN films (Gurumurugan, Chen et al. 1999; Jadwisienczak, Lozykowski et al. 2000; Weingartner, Erlenbach et al. 2006).

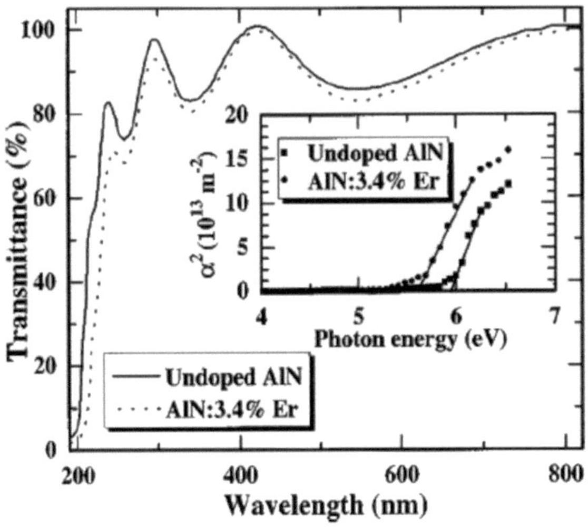

Fig. 3.1. Optical transmission spectra of undoped and Er-doped a-AlN thin films. Inset shows the photon energy dependence of the square of the absorption coefficient for undoped and Er-doped a-AlN thin films. After Ref. (Gurumurugan, Chen et al. 1999)

It is documented in the literature that thermally annealed amorphous semiconductors experience some structural reordering (not necessarily crystallization) (Gurumurugan, Chen et al. 1999). Figure 3.2 shows evolution of cathodoluminescence (CL) spectra of Tb-doped a-AlN films subjected to thermal annealing in nitrogen ambient up to 1100 °C.

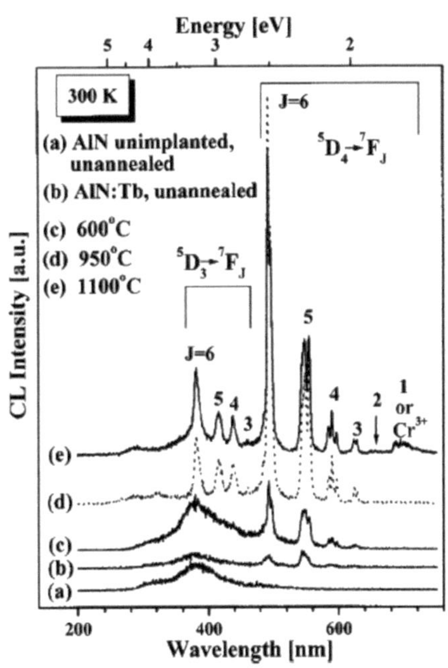

Fig. 3.2. Evolution of CL spectra in Tb-doped a-AlN films as a function of annealing temperature. After Ref. (Gurumurugan, Chen et al. 1999).

It is believed that when this takes place the dangling or broken bonds and/or the atomic rearrangements of the amorphous material may have great influence on the electronic states of amorphous host. The available reports show that the conducted x-ray diffraction analysis of as-grown or thermally treated RE-doped a-III-nitride thin films has not revealed any microscopic morphology changes in studied materials (see Fig. 3.3).

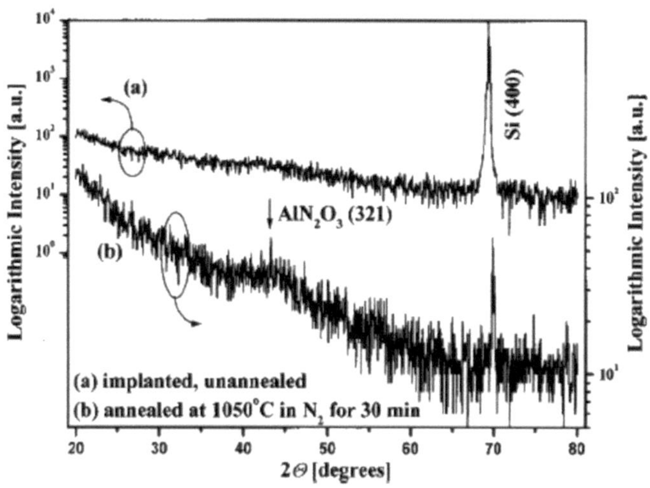

Fig. 3.3. XRD spectra of a-AlN thin films doped with Tb^{3+} ions. (a) unannealed and (b) annealed for 30 min at 1050 °C in nitrogen. Ref. (Jadwisienczak, Lozykowski et al. 2000).

As reported by Zanatta et al. (Zanatta, Khan et al. 2007) a-SiN doped with Sm and Tb exhibit significant enhancements of 4*f*-shell radiative recombinations upon thermal annealing due to a decrease in the density of deep and/or tail defects (suppression of the nonradiative processes). A similar observation was made for a-AlN:Ho system (Aldabergenova, Frank et al. 2006). Furthermore, in that study, the growth of small crystallites in initially mostly a-AlN:Ho host was observed after annealing above 900 °C. See Fig 3.4.

In order to avoid the implantation induced amorphous material morphological changes and defects due to ion implantation, simultaneous co-deposition of different RE metals during RF sputtering growth was demonstrated (Maqbool, Kordesch et al. 2009). This approach is an extension of a typical a-III-nitrides RF sputtering growth technique where RE ions concentration in resulting film is controlled by sputtering metals to target surface ratio. It is possible using this procedure to control the RE concentration in the sample over a wide range. Furthermore, this technique gave an opportunity to derive mixed RE-doped a-III-nitride systems where more complex interaction schemes defining energy transfer processes can be studied. However, there exist only limited papers reporting on this issue up to date (Maqbool, Kordesch et al. 2009). The single RE ion and multiple RE ions doping to a-III-nitrides should have typical doping effects in these hosts in the sense of controlling the electronic properties of the material and determining the Fermi level position. In contrast to unhydrogenated a-Si and hydrogenated a-Si:H where it was proposed that several different REs turned out to act as acceptors or donors (Tessler 1999) no similar postulate was considered in the case of RE-doped a-III-nitrides to date. It is uncertain at present if other than trivalent RE ions in a-III-nitrides can exist; however such a scenario is feasible especially if one considers a plethora of possible RE ions configurations in amorphous matrix as well as the fact that RE ions are strong oxygen and other impurities getters. Systematic RE ion doping studies with controlled co-dopant concentrations are needed to validate this consideration in the future.

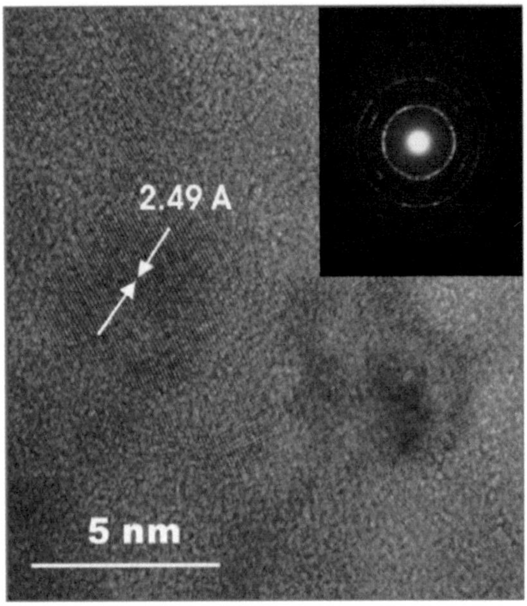

Fig. 3.4. High resolution micrograph showing small AlN crystallites embedded in as grown a-AlN film. The insert shows the pertaining diffraction pattern. After (Aldabergenova, Frank et al. 2006).

3.1.1 Rare earth excitation

The excitation processes of RE ions in a-III-nitrides, similarly to crystalline hosts, can be generally divided into two categories: direct and indirect excitation processes (Jadwisienczak, Lozykowski et al. 2000; Lozykowski, Jadwisienczak et al. 2000). The direct excitation process occurs in selective excitation of $4f^n$ electrons by photons (photoluminescence, PL) selective excitation or in cathodoluminescence (CL) and electroluminescence (EL) by collision with hot electrons. The indirect excitation process occurs via transfer of energy to the $4f^n$ electron system from electron–hole pairs generated by photons with higher energy than the band gap (PL excited above band gap), injected in forward bias *p-n* junctions, or generated by hot carriers in CL and EL. An excitation mechanism in CL and EL involves direct impact excitation of RE^{3+} ions by hot electrons, as well as an energy transfer from the generated electron–hole pairs or by impact excitation (or ionization) involving impurity states outside the $4f$ shell, with subsequent energy transfer to this shell. The most important, from an applications point of view, is the excitation of the RE ions by energy transfer processes from electron-hole (e-h) pairs or excitons. This process most probably involves the rare earth ion isovalent traps (Lozykowski, Jadwisienczak et al. 2000). Since there is no charge involved, the isoelectronic center forms the bound states by short range central cell potential. After an isoelectronic trap has captured an electron or a hole, the isoelectronic trap is negatively or positively charged, and by Coulomb interaction it will capture a carrier of the opposite charge creating a bound exciton. There are three possible mechanisms of energy transfer (Lozykowski 1993). The first is the energy transfer process from excitons bound to structured isoelectronic centers to the core electrons. This takes place as a result of the electrostatic perturbation between the core electrons of the RE structured impurity and the exciton, effective-mass-like particles. The second mechanism is the transfer of energy to the core electrons involving the structured isoelectronic trap occupied by electron (hole) and free holes (electrons) in the valence (conduction) band. The third mechanism is the transfer through an inelastic scattering process in which the energy of a free exciton near a RE structured trap is given to the localized core excited states. If the initial and final states are not resonant, the energy mismatch must be distributed in some way, e.g., by phonon emission or absorption. If the atomic core excitations are strongly coupled to the host phonons, the energy transfer probability is likely to be higher (Lozykowski 1993). Strong phonon coupling may also be desirable in ensuring that relaxation down the ladder of the core excited state occurs quickly, thus preventing back transfer. However, for efficient radiative recombination, the phonon coupling should not be strong, in order to prevent core de-excitation by nonradiative multiphonon process. It is natural to assume that when isoelectronic impurity atoms, in this case RE ions, are incorporated in a-III-nitride host, they modify local vibrational properties in a definite way. Thus, there should be observed distinct vibrational frequencies associated with localized motion of the RE impurity in addition to the host lattice vibrations. In III-nitrides the threshold energy for electron damage in amorphous films is considerably smaller contrary to crystalline compounds (Zanatta, Ribeiro et al. 2005). Moreover, energetic electron irradiation may create defects that are not possible with photons with energies in the visible energy range. These additional defects, mainly broken bonds and atomic displacements, act as nonradiative centers that considerably reduce the luminescence efficiency of a-III-nitride semiconductor. On the other hand, RE ions are high efficiency recombination centers that effectively compete with other non-radiative processes taking place in a semiconductor host. As a result, when irradiating the RE-doped a-III-nitrides with up to a few keV energy electrons, most of the electron-hole pairs recombine preferentially through the RE ions. Luminescence after photon excitation, on the other hand, behaves in a different manner and is very strongly excitation wavelengths dependent (Gurumurugan, Chen et al. 1999; Jadwisienczak, Lozykowski et al. 2000; Zanatta, Ribeiro et al. 2005).

In photoluminescence each absorbed photon with energy higher than the band gap produces a single electron-hole pair while in cathodoluminescence, high energy single electrons generate a huge number of "hot" e–h pairs, reducing the energy from tens of keV to zero (Jadwisienczak, Lozykowski et al. 2000). In general, the optical excitation process demands that the *4f* electrons are excited absorbing energy. In a-III-nitride materials with large band gap like a-GaN and a-AlN this can occur with a direct absorption to one of the transitions of upper energy of the RE^{3+} ion degenerated with the conduction band or via transfer of energy from a defect related Auger resonant process between the RE^{3+} ion states and the dipole formed by the *s*-like conduction band states and *p*-like dangling bond states. The PL of RE^{3+} ion in a-III-nitrides is obtained by an indirect excitation process via transfer of energy to the $4f^n$ electron system from e-h pairs. In electron beam excitation (CL), however, the RE^{3+} ions are excited by direct impact with hot electrons, as well as by energy transfer processes from the generated e–h pairs or by impact excitation (or ionization) involving other impurities (or complex defects) with subsequent energy transfer to the RE 4*f*-shell electrons. Generally the excitation by energetic electrons produces emission via all possible luminescence mechanisms available in a semiconductor. Another factor which may play a role in excitation and emission processes is the charged nature of excitation: uncharged photons in PL versus negatively charged electrons in CL. The excitation depth in PL and CL are also different due to the strong absorption of the excitation photons in a-III-nitrides within a few tens of nanometers layer; whereas the electron penetration depth is electron acceleration energy dependent and can reach up to a few hundreds nanometers (Jadwisienczak, Lozykowski et al. 2000). Figure 3.5 and Fig. 3.6 show examples of CL spectra of the rare earth (Tm, Tb, Dy, Sm, Eu, Er and Yb) doped a-AlN layers [(Gurumurugan, Chen et al. 1999; Jadwisienczak, Lozykowski et al. 2000; Weingartner, Erlenbach et al. 2006). The transitions corresponding to the strongest emission lines are indicated by their energy level assignments. It was reported in these studies that all RE-doped a-AlN layers show (even untreated) pronounced RE^{3+} ion optical spectra at room temperature.

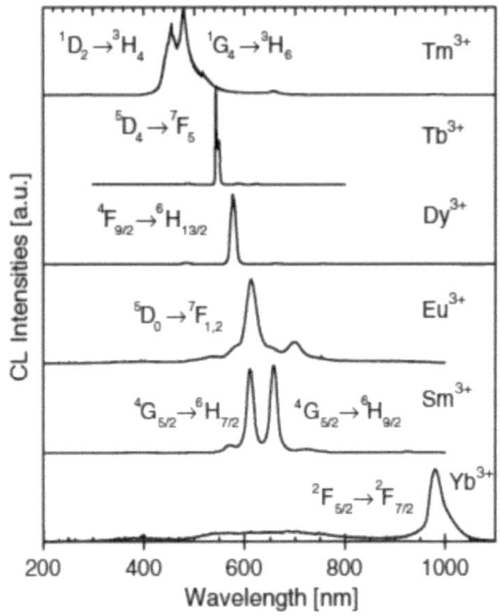

Fig. 3.5. Survey of CL spectra of RE-doped a-AlN thin films. After Ref. (Weingartner, Erlenbach et al. 2006).

The RE ion excitation process responsible for observed EL from RE-doped a-III-nitrides is very similar to described above CL. Thus CL and EL spectra, in general, shows similar spectra features. It was reported that in cases of the Er and Tb-doped a-AlN alternating-current thin-film electroluminescent devices the EL and CL spectra are very similar (Dimitrova, Van Patten et al. 2000; Dimitrova, Van Patten et al. 2001; Richardson, Van Patten et al. 2002).

Typical transient analysis of luminescence observed from RE-doped III-nitride crystalline semiconductors stimulated by different means provided detailed insight into energy migration and excitation processes observed in these technologically important materials. There are a very limited number of papers reporting on luminescence (CL, EL) kinetics of RE-doped a-III-nitride up to date (Richardson, Van Patten et al. 2002). It is known that the nonradiative decays of excited RE ions in semiconductors, including RE-doped a-III-nitrides, are phonon dependent (Jadwisienczak, Lozykowski et al. 2000). Furthermore, at sufficiently high concentrations of RE ions, nonradiative decay can occur via ion pair cross-relaxation or energy migration and finally transfer to quenching centers (Jadwisienczak, Lozykowski et al. 2000). For low RE concentration, where ion–ion pair relaxation or energy migration and transfer to quenching centers has low probability, excited RE electronic states gave up energy radiatively by purely electronic or phonon assisted transitions or nonradiatively by the emission of a single phonon or multiphonon. Nonradiative relaxation between RE^{3+} ion *J*-manifolds in solid state host, including reported CL kinetics of Tb-doped a-AlN , typically requires the participation of several phonons and occurs at slow rates, while relaxation between crystal field levels of a *J*-manifold by single or two phonon processes is much faster. It is known that the rate of multiphonon emission is strongly dependent upon the number of phonons required to conserve energy, and hence on the size of the energy gap to the next lowest energy level. The detailed investigations of nonradiative decay of RE ions in a-III-nitrides have not been done yet; however it is expected that that multiphonon transitions involving the emission of a maximum of a few phonons will effectively compete with radiative transitions. Furthermore, it is expected that the temperature dependence of the decay rates will show that relaxation occurs mainly by high energy phonons. It is worth noting that energy separation of *J*-crystal field levels shows only small changes with the host (Jadwisienczak, Lozykowski et al. 2000; Jadwisienczak, Lozykowski et al. 2000). Differences in the rate of nonradiative decay from a particular RE energy level arise from the phonon energy spectrum of the host material and the strength of the RE ion–lattice coupling. Local vibrational modes related to RE ions, their complexes, or others impurities present in a-III-nitride host will also play an important role in the above-discussed process; however these assumptions will have to be critically confirmed in the future.

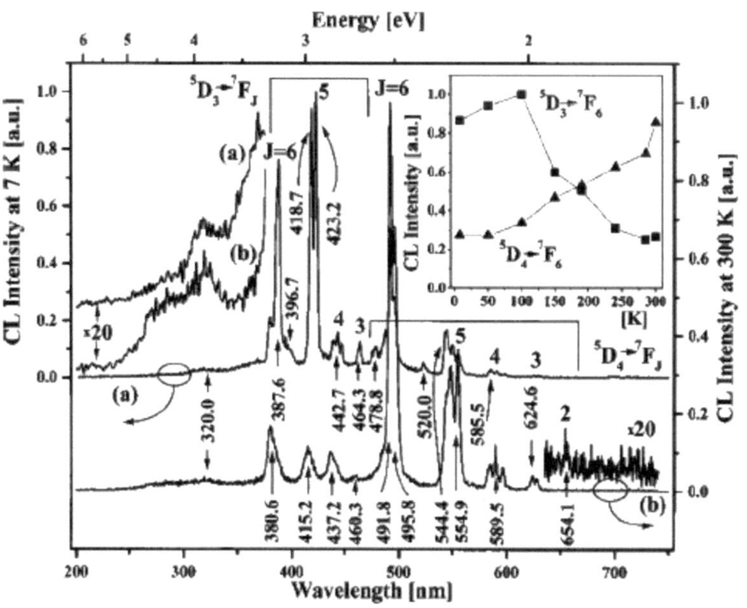

Fig. 3.6. CL spectra of a-AlN:Tb recorded at cryogenic and room temperature. The inserts show the total integral intensity of the dominant 4*f*-shell transition lines as a function of temperature. After Ref. [(Jadwisienczak, Lozykowski et al. 2000)].

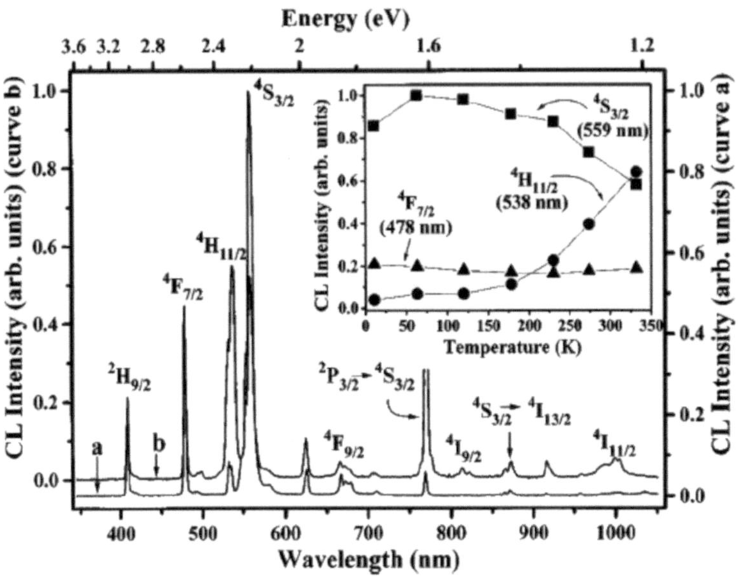

Fig. 3.7. CL spectra of a-AlN:Er recorded at cryogenic and room temperature. The inserts show the total integral intensity of the dominant 4*f*-shell transition lines as a function of temperature. After Ref. [(Gurumurugan, Chen et al. 1999)].

The local environment of the RE^{3+} ions, and consequent luminescence intensity, in a-III-nitrides can be greatly influenced by the presence of nitrogen atoms, intrinsic impurities, larger clusters and thermal treatments stimulating morphological changes at micro-scale. This in turn affects the optical activity of RE^{3+} ion by enhancing or quenching luminescence originating from 4*f*-shalls transitions. The vulnerability of amorphous semiconductors to defects generation in CL and EL is rather straight forward considering the threshold energy for electron damage in amorphous networks being much smaller (1 keV) comparing to crystalline semiconductors (100 keV) (Zanatta, Ribeiro et al. 2005). This fact creates an interesting research opportunity for RE-doped a-III-nitrides in the search for more effective energy transfer process between RE ions and amorphous host. It is known that the III-nitride quantum structures with dimensions close to the excitonic Bohr radius exhibit electronic and optical properties affected by the confinement of electrons in one, two, or three dimensions (Ihn 2010). The principal consequences of quantum confinement are an increase in the band gap energy and increased probability of radiative transitions. Confinement of carriers in real space causes their wavefunctions to spread out in momentum space, increasing the probability of radiative processes due to greater wavefunction overlap. Because of the modification of their band structure, doping such confined systems with RE^{3+} ions can also help to overcome some of the nonradiative de-excitation problems associated with RE-doped III-nitride. Quantum confinement affects also carrier lifetimes and their degree of localization in real space. This in turn modifies the Auger back-transfer processes that limit luminescence efficiency and increases the interaction probability between the confined carriers and the RE^{3+} ions. In general, the RE^{3+} ion radiative quantum efficiency strongly depends on the carrier mediated energy transfer processes, which have to compete with nonradiative recombination channels abundant in a-III-nitrides. It is known that RE^{3+} ions induce significant local site distortion due to their large radii as well as gettering effect (O'Donnell 2010). It was theoretically shown that the Coulomb excitation of 4*f* electrons near interface of heterostructures, where certain degree of disorder is expected, is more effective than a similar excitation in the bulk semiconductors (Zegrya and Masterov 1995; Zegrya and Masterov 1996; Zegrya and Masterov 1998; Masterov and Gerchikov 1999). Also, as discussed above, the thermal annealing process of RE-doped a-III-nitride changes grain boundary, where RE ions most probably reside, resulting in change of RE ion local environments and their optical activities due to relaxed 4*f*-shell electron-lattice coupling. In that sense one may assume that RE^{3+} ions located at the surface of the a-III-nitride nano-grains may act as grain boundary activators. It has now been well established that semiconductor nano-clusters can act as efficient luminescence sensitizers for RE ions in wide band gap solid states (Iacona, Franzo et al. 2009). Moreover, it appears that it is not necessary for the semiconductor inclusions to be crystalline - amorphous nano-clusters are at least as efficient as nano-crystals (Kenyon and Lucarz). The customary observation is that absorption of photons by semiconductor nano-clusters results in excitation of RE ions via an efficient transfer mechanism to unexcited RE ions. Such a process is significant for two reasons: firstly, the effective absorption cross-section of the RE ion is increased by several orders of magnitude (Kenyon, Lucarz; Kik 2003), and secondly, it becomes possible to excite the RE ions via the broad-band absorption of the semiconductor nano-clusters (Kenyon, Chryssou et al. 2002; Kik 2003; Iacona, Franzo et al. 2009). Moreover, the prospect of engaging amorphous nano-clusters to activate RE ions in a-III-nitrides removes the prerequisite to control tightly the crystallinity and size distribution of nano-cluster sensitizers thus greatly simplifying material processing. More studies are necessary to clear this point. It is our belief that the quenching mechanism should not depend on the form of sample preparation rather than on the final RE ion local symmetry inducing site degeneracy required for observing peculiar intra 4*f*-shell transitions.

3.2 Amorphous nitrides: Examples

Most of the metal nitrides are $M^{3+}N^{3-}$ compounds, so that no special considerations are necessary for RE emission. There is still some controversy about the site of the RE ion in crystalline nitrides. In an amorphous solid, bonding to the N as a (3+) ion is the most obvious choice. However, it is the case that sharp lines are observed from RE ions in the amorphous hosts, and in some cases the analogue of "crystal field splitting". There are probably several types of sites for the RE ion, but not enough to broaden the emission peaks excessively.

Because the emission lines in RE ions are well known, in most cases the experimental task is to see which transitions are observed in each host. Post growth heat treatment of the as-gown nitrides often improves the intensity of the RE emission in the nitrides. Because there is no "annealing" in the sense of crystal structure or long range lattice improvement, the heating process can cause the removal or transport of materials unrelated to the nitride structure. There could be removal of hydrogen, for example, or diffusion of oxygen. The reaction of water trapped in the film due to background gasses in the vacuum or in the heat treatment gas during post growth thermal treatment would include the removal of N atoms as ammonia, affecting the basic M-N bonds. Some rearrangement of the short range nitride structure is possible.

Even though there can be much larger RE concentrations in sputter deposited nitride films compared to ion implanted films, there is no dramatic increase in intensity with higher RE concentrations. It must be assumed that a large number of the RE ions are not optically active in the as-grown amorphous films. Heat treatment or oxygen additions increase the number of active RE ions. Co-doping with other (3+) ions sometimes also improves the emission intensity of RE ion luminescence. One of the possibilities is that the (3+) impurity occupies the inactive sites and displaces the RE ions onto active sites. Optical interactions are also possible.

3.2.1 Amorphous silicon nitride

The bulk of studies on amorphous silicon nitride doped with RE ions are due to Zanata and co-workers. Silicon is multivalent, so that a-SiN may well result in a RE^{3+} ion without any further need for charge compensation.

3.2.1.1 Erbium, Holmium, Dysprosium Samarium and Praseodymium

Zanatta and co-workers studied Pr, Sm, Dy, Ho, and Er in amorphous SiN films.

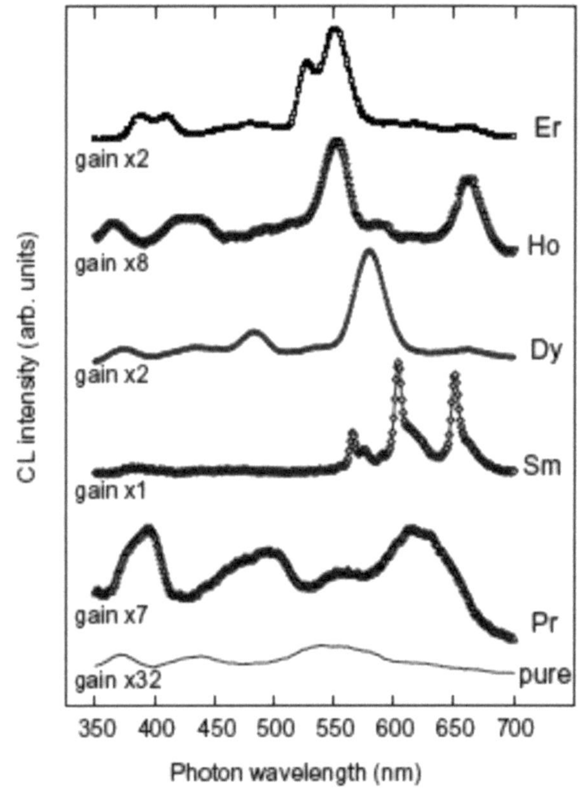

Fig. 3.8. Cathodoluminescence spectra of RE^{3+}-doped a-SiN films deposited by co-sputtering. The spectra have been acquired at room temperature by using 15 keV electrons and a current density of 10 nA. All spectra were normalized and vertically shifted for comparison. From (Zanatta, Ribeiro et al. 2004).

The CL spectra of the present RE-doped a-SiN films are shown in Fig. 3.8. In addition to the levels close to 490 nm these new transitions correspond to Er^{3+} ions ($^4G_{11/2}$ to $^4I_{15/2}$ at 390 nm, $^2H_{9/2}$ to $^4I_{15/2}$ at 410 nm); Ho^{3+} ions ($^5G_5$ to $^5I_5$ at 430 nm, $^3H_5$ to $^5I_5$ at 360 nm); Sm^{3+} ($^4K_{11/2}$ to $^6H_{5/2}$ at 380 nm); and Pr^{3+} ions ($^1S_0$ to $^1D_2$ at 370 nm, $^1S_0$ to $^1I_6$ or $^3P_1$ at 395 nm). The variations observed in the luminescence intensity probably occur because of differences in the RE content; Er (0.5 at.%), Ho (0.3 at.%), Dy (0.3 at.%), Sm (0.5 at.%), and Pr (0.2 at.%). (Zanatta, Ribeiro et al. 2004)

3.2.1.2 Samarium and terbium

Most of the luminescence features present in the 400–850 nm wavelength range (Fig. 3.9) correspond to optical transitions due to the Sm^{3+} and Tb^{3+} ions (Dieke 1968): Sm1 at ~565 nm ($^4G_{5/2} \rightarrow {}^6H_{5/2}$), Sm2 at ~605 nm ($^4G_{5/2} \rightarrow {}^6H_{7/2}$), Sm3 at ~650 nm ($^4G_{5/2} \rightarrow {}^6H_{9/2}$), Tb1 at ~485 nm ($^5D_4 \rightarrow {}^7F_6$), Tb2 at ~545 nm ($^5D_4 \rightarrow {}^7F_5$), Tb3 at ~590 nm ($^5D_4 \rightarrow {}^7F_4$), and Tb4 at ~625 nm ($^5D_4 \rightarrow {}^7F_3$). In addition to these, we also observe the Sm4 light emission at ~725 nm ($^4G_{5/2} \rightarrow {}^6H_{11/2}$), which is associated with the superposition of the $^4G_{5/2} \rightarrow {}^6H_{11/2}$ transition due to Sm^{3+} ions and the $^5D_0 \rightarrow {}^7F_0$, $^5D_0 \rightarrow {}^7F_1$, and $^5D_0 \rightarrow {}^7F_2$ transitions of Sm^{2+} ions. The infrared contribution Sm5 at ~810 nm corresponds to the $^5D_0 \rightarrow {}^7F_4$ transition and is exclusively due to Sm^{2+} ions (Dieke 1968).

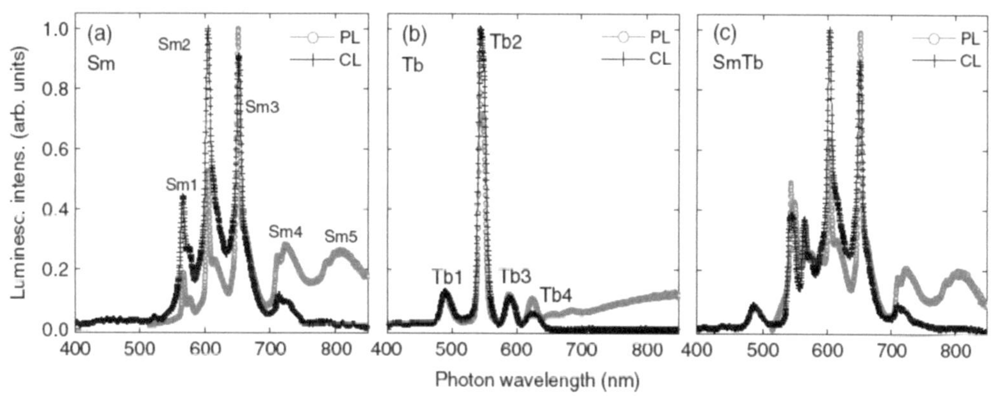

Fig. 3.9. Smarium and Terbium CL in amorphous silicon nitride. From (Zanatta, Khan et al. 2007).

3.2.2 Amorphous aluminum nitride

3.2.2.1 Erbium and terbium

The first RE spectrum in amorphous Aluminum Nitride (a-AlN) was that of Gurumurugan et al. (Gurumurugan, Chen et al. 1999). The spectra are shown in Figures and 3.7 (Jadwisienczak, Lozykowski et al. 2000).

3.2.2.2 Europium

Europium doped a-AlN was studied by Caldwell et al. (Caldwell, Van Patten et al. 2001).

The most significant result of this study was that thermal activation of the RE ion in a-AlN, which was found generally to increase the luminescence yield by a factor of up to 100, could be matched or exceeded by the addition of oxygen during the growth of the film by sputtering. The luminescence yield increased over 600 times with the addition of oxygen to the sputter gas.

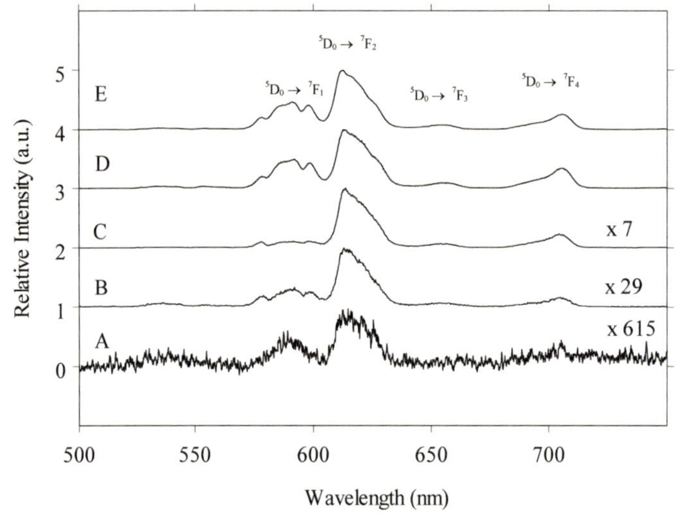

Fig. 3.10. CL spectra of a-AlN:Eu processed as follows: a) no oxygen, and no heat treatment, b) no oxygen, heated to 923 K. c) grown with 1.6% oxygen in nitrogen, d) grown in 3.8% oxygen in nitrogen, e) grown in 20% oxygen, with the balance nitrogen. From (Caldwell, Van Patten et al. 2001).

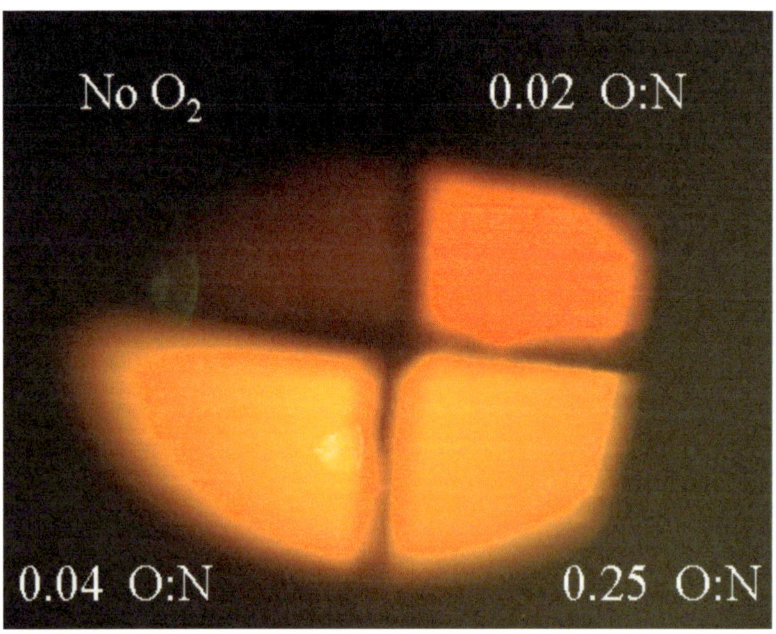

Fig. 3.11. CL from some of the types of a-AlN:Eu films described in Fig. 3.10.

3.2.2.3 Holmium and gadolinium

Data for Ho and Gd in a-AlN are given by Maqbool (Maqbool 2005).

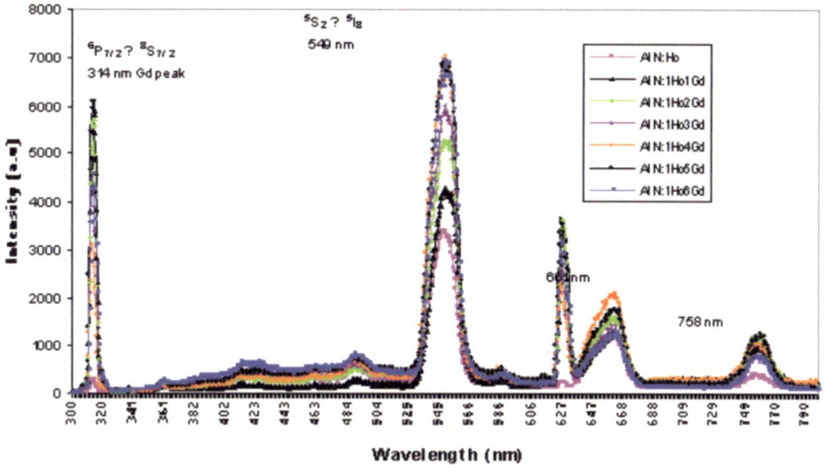

Fig. 3.12. CL from codoped Gd and Ho a-AlN films.

Material	Transition Number	Transition	Wavelength (nm)	Relative Intensity
AlN:Ho	1	$^5G_5 \rightarrow {}^5I_8$	362	0.355
	2	$^5G_4 \rightarrow {}^5I_8$	394	0.530
	3	$^5F_1 \rightarrow {}^5I_8$	461	0.300
	4	$^5S_2 \rightarrow {}^5I_8$	549	1.000
	5	$^5F_3 \rightarrow {}^5I_8$	659	0.277
	6	$^5S_2 \rightarrow {}^5I_7$	758	
AlN:Gd	1	$^6P_{7/2} \rightarrow {}^8S_{7/2}$	314	

Table 3.1. Summary of Ho^{3+} and Gd^{+3} ions emission from a-AlN:Ho and a-AlN:Gd.

Ho relative concentration	Gd relative Concentration	Relative Intensity of 549 nm peak	% increase in intensity
1	0	1	0 (reference)
1	1	1.275	27.5
1	2	1.564	56.4
1	3	1.747	74.7
1	4	2.074	107.4
1	5	2.047	104.7
1	6	2.054	105.4

Table 3.2. Effect of Gd concentration on the luminescence of Ho $^5S_2 \rightarrow {}^5I_8$ transition at 549 nm.

3.2.2.4 Praseodymium

Data for Pr in a-AlN are given by Maqbool(Maqbool 2005).

Material	Transition Number	Transition Assignment	Wavelength (nm)	Intensity(a.u)
AlN:Pr	1	$^1S_0 \rightarrow {}^1D_2$	335	162
	2	$^1S_0 \rightarrow {}^1I_6$	385	438
	3	$^3P_2 \rightarrow {}^3H_4$	439	362
	4	$^3P_0 \rightarrow {}^3H_4$	488	1002
	5	$^3P_0 \rightarrow {}^3H_4$	504	1024
	6	$^3P_1 \rightarrow {}^3H_5$	526	4577
	7	$^3P_0 \rightarrow {}^3H_5$	573	767
	8	$^3P_0 \rightarrow {}^3H_6$	618	643
	9	$^3P_0 \rightarrow {}^3F_2$	652	1227
	10	$^3P_0 \rightarrow {}^3F_3$	710	410
	11	$^3P_0 \rightarrow {}^3F_4$	738	520

Table 3.3. Summary of Pr^{3+} ion emissions from Pr-doped a-AlN.

Pr Transition	Wavelength (nm)	Gd Relative Concentration	Relative Intensity	% Increase in Intensity
$^3P_0 \rightarrow {}^3H_4$	490	0	1	0 (reference)
		1	1.58	58
		4	4.37	337
$^3P_0 \rightarrow {}^3H_6$	618	0	1	0
		1	1.73	73
		4	4.14	314
$^3P_0 \rightarrow {}^3F_2$	649	0	1	0
		1	1.66	66
		4	3.4	240

Table 3.4. Effect of Gd concentration on the luminescence of Pr^{3+} ion transitions.

3.2.2.5 Thulium

Data for Tm in a-AlN are given by Maqbool (Maqbool 2005).

Material	Transition Number	Transition Assignment	Wavelength (nm)	Relative Intensity(a.u)
AlN:Tm	1	$D_2 \rightarrow {}^3H_6$	371	0.126
	2	${}^1D_2 \rightarrow {}^3F_4$	467	1.000
	3	${}^1D_2 \rightarrow {}^3H_6$	480	0.648
	4	${}^1D_2 \rightarrow {}^3H_5$	528	0.092
	5	${}^1G_4 \rightarrow {}^3F_4$	650	0.126
	6	${}^1D_2 \rightarrow {}^3H_4$	685	0.061
	7	${}^3H_4 \rightarrow {}^3H_6$	802	0.563
	8	${}^3H_4 \rightarrow {}^3H_6$	808	0.539

Table 3.5. Summary of Tm^{3+} ion emissions from Tm-doped AlN.

3.2.2.6 Samarium

Data for Sm in a-AlN are given by Maqbool (Maqbool 2005).

Material	Transition Number	Transition Assignment	Wavelength (nm)	Relative Intensity(a.u)
AlN:Tm	1	${}^4G_{5/2} \rightarrow {}^6H_{5/2}$	564	0.425
	2	${}^4G_{5/2} \rightarrow {}^6H_{7/2}$	600	1.000
	3	${}^4G_{5/2} \rightarrow {}^6I\,I_{9/2}$	648	0.686
	4	${}^4G_{5/2} \rightarrow {}^6H_{11/2}$	711	0.312
	5	Gd impurity	314	0.276
	6	Cr impurity	692	0.220

Table 3.6. Summary of Sm^{3+} ion emissions from Sm-doped a-AlN.

3.2.3 Amorphous gallium nitride

Amorphous GaN usually has a large defect luminescence in the yellow region of the visible spectrum. In Fig. 3.13, an a-GaN:Er thin film is shown in CL. This film was heated to 950 °C in nitrogen for 30 minutes. The Er^{3+} ion transitions are clearly visible on the yellow defect emission band. The defect band problem is much less in crytsalline GaN, which was extensively investigated by several groups, most notably Steckl and his coworkers.

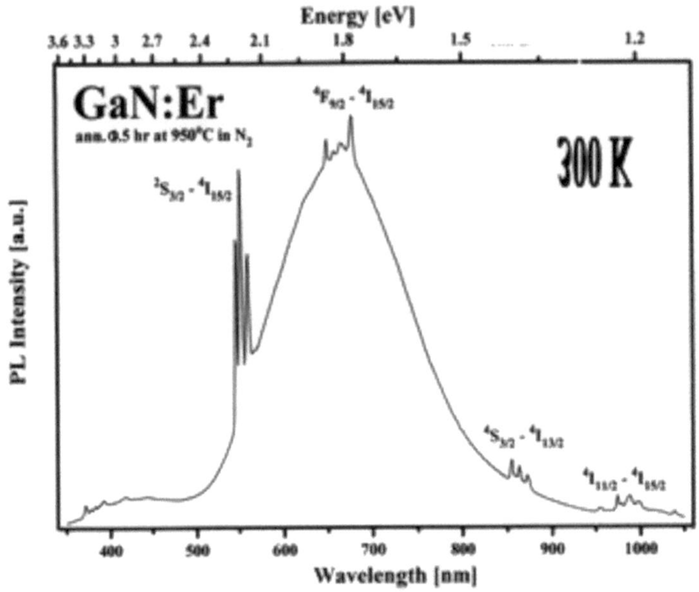

Fig. 3.13. CL from a-GaN: Er.

Maqbool (Maqbool, Richardson et al. 2005) investigated Pr^{3+} ion luminescence in a-GaN. The data are given in Table 3.7 .

Material	Transition Number	Transition Assignment	Wavelength (nm)	Relative Intensity(a.u)
	3	$^3P_2 \rightarrow {}^3H_4$	418	559
	4	$^3P_0 \rightarrow {}^3H_4$	493	5687
	6	$^3P_1 \rightarrow {}^3H_5$	532	1895
	8	$^3P_0 \rightarrow {}^3H_6$	621	4015
	9	$^3P_0 \rightarrow {}^3F_2$	650	4227
	10	$^3P_0 \rightarrow {}^3F_3$	713	1508
	11	$^3P_0 \rightarrow {}^3F_4$	736	1627

Table 3.7. GaN:Pr

3.2.4 Other amorphous nitrides

3.2.4.1 BN

Sputtered boron nitride was shown to be a useful host for RE ions by Maqbool et al. (Maqbool, Richardson et al. 2005) and Allen et al. (Allen S. 2000).

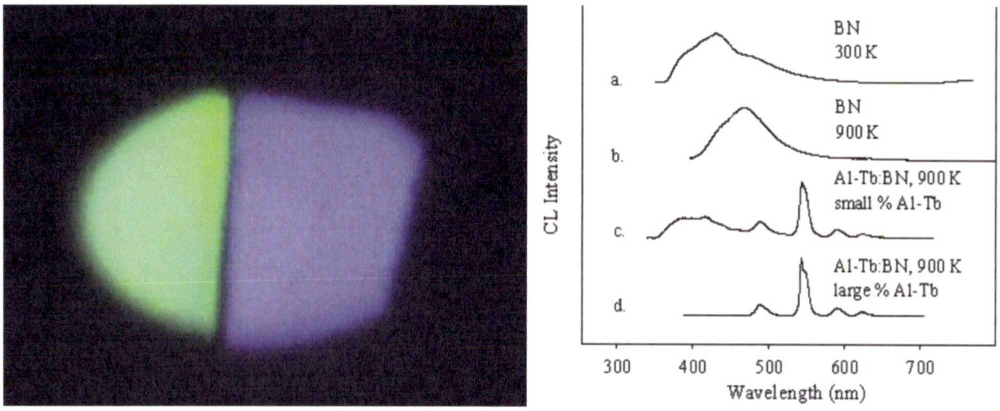

Fig. 3.14. Left: CL from sputter deposited BN:Tb and pure BN. Right: CL spectra of BN and BN:Tb, as deposited and heated to 900 K. At large Tb concentrations, the blue BN luminescence is not visible.

Maqbool et al. (Maqbool, Richardson et al. 2005) investigated the luminescence of Pr in BN.

Data are given below in Table 3.8 (Maqbool 2005)

Material	Transition Number	Transition Assignment	Wavelength (nm)	Relative Intensity(a.u)
	3	BN transition	415	372
	4	$^3P_0 \rightarrow {}^3H_4$	492	973
	6	$^3P_1 \rightarrow {}^3H_5$	544	752
	8	$^3P_0 \rightarrow {}^3H_6$	628	842
	9	$^3P_0 \rightarrow {}^3F_2$	651	748
	10	$^3P_0 \rightarrow {}^3F_3$	713	257
	11	$^3P_0 \rightarrow {}^3F_4$	736	289

Table 3.8. BN:Pr

3.2.4.2 BeN

Amorphous Beryllium Nitride, produced by reactive sputtering of Be in nitrogen, was studied with samarium (Zanatta, Richardson et al. 2007).

PL and CL excitation was achieved with 532 nm photons and 10 keV electrons, respectively. Light emission is due to the Sm^{3+} ions and correspond to the following transitions (see inset on the left in Fig. 3.15): $^4G_{5/2} \rightarrow {}^6H_{5/2}$ (A), $^4G_{5/2} \rightarrow {}^6H_{7/2}$ (B), $^4G_{5/2} \rightarrow {}^6H_{9/2}$ (C), and $^4G_{5/2} \rightarrow {}^6H_{11/2}$ (D). A photograph illustrating the cathodoluminescence image of the Sm-doped a-BeN sample is presented in the upper-right corner of Fig. 3.15.

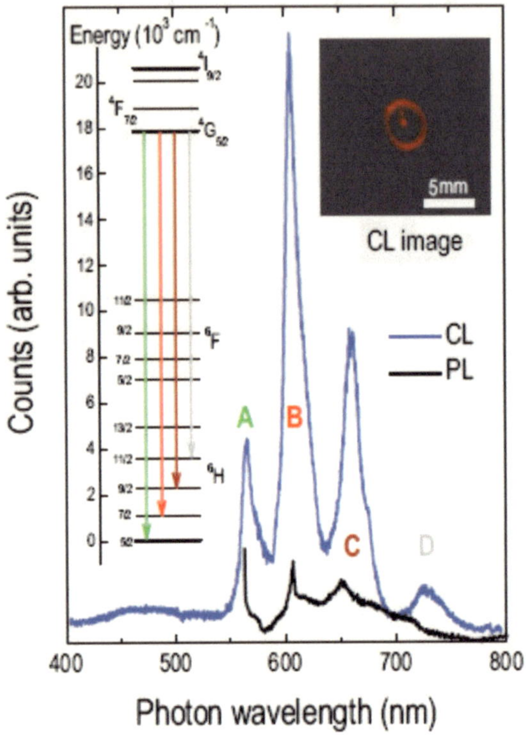

Fig. 3.15. Room-temperature luminescence spectra of a Sm-doped a-BeN film. From(Zanatta, Richardson et al. 2007)

4. Transition metal cathodoluminescence

Elements in the periodic table that make a bridge between the first two groups and last six groups of the main elements are defined as transition metals (TM). This bridging role has given these elements the name *transition* metals. These metals consist of many elements, however, we shall limit our discussion to a few of them, mainly $_{22}Ti$, $_{24}Cr$, $_{39}Y$, and $_{74}W$. These metals are specially characterized by their common feature: partial occupancy of the d-shells. This partially filled *d* shell configuration assigns special properties to these elements, including their strong suitability for use as optical and luminescent materials. To study their optical and luminescent properties it will be useful to understand some of their basic characteristics. Housecroft (Housecroft et al 2007) and Mackay (Mackay 1996) reported some of the basic characteristics of transition metals. The discussion of these characteristics in section 4.1 follows Housecroft and Mackay.

4.1 Characteristics of transition metals

4.1.1 Density and metallic radii

The transition elements are much denser than the s-block elements and show a gradual increase in density from scandium to copper. This trend in density can be explained by the small and irregular decrease in metallic radii coupled with the relative increase in atomic mass.

4.1.2 Melting and boiling points

The melting points and the molar enthalpies of fusion of the transition metals are both high in comparison to main group elements. This arises from strong metallic bonding in transition metals which occurs due to delocalization of electrons facilitated by the availability of both d and s electrons.

4.1.3 Ionization energies

In moving across the series of metals from scandium to zinc a small change in the values of the first and second ionization energies is observed. This is due to the build-up of electrons in the immediately underlying d-sub-shells that efficiently shields the 4s electrons from the nucleus and minimizing the increase in effective nuclear charge from element to element. The increases in third and fourth ionization energy values are more rapid. However, the trends in these values show the usual discontinuity half way along the series. The reason is that the five d electrons are all unpaired, in singly occupied orbitals. When the sixth and subsequent electrons enter, the electrons have to share the already occupied orbitals resulting in inter-electron repulsions, which would require less energy to remove an electron. Hence, the third ionization energy curve for the last five elements is identical in shape to the curve for the first five elements, but displaced upwards by 580 kJ mol^{-1}.

4.1.4 Electronic configuration

The electronic configuration of the atoms of the first row transition elements are basically the same. It can be seen in Table 4.1 that there is a gradual filling of the 3d orbitals across the series starting from scandium. This filling is, however, not regular, since at chromium and copper the population of 3d orbitals increase by the acquisition of an electron from the 4s shell. This illustrates an important generalization about orbital energies of the first row transition series. At chromium, both the 3d and 4s orbitals are occupied, but neither is completely filled in preference to the other. This suggests that the energies of the 3d and 4s orbitals are relatively close for atoms in this row.

In the case of copper, the 3d level is full, but only one electron occupies the 4s orbital. This suggests that in copper the 3d orbital energy is lower than the 4s orbital. Thus the 3d orbital energy has passed from higher to lower as we move across the period from potassium to zinc. However, the whole question of preference of an atom to adopt a particular electronic configuration is not determined by orbital energy alone. In chromium it can be shown that the 4s orbital energy is still below the 3d which suggests a configuration $[Ar]3d^4 4s^2$. However due to the effect of electronic repulsion between the outer electrons the actual configuration becomes $[Ar]3d^5 4s^1$ where all the electrons in the outer orbitals are unpaired. Table 1 gives some of the physical properties and free atom electronic configuration of transition metals.

4.1.5 Oxidation states

Oxidation states of transition metals are very important to study the spectroscopy and luminescence from these metal ions. The partially filled d-shell electrons play important role in the oxidation states of transition metals. To fully understand the phenomena of oxidation states of transition metals, we have to understand how the unpaired d-orbital electrons bond. There are five orbitals in a d subshell manifold. As the number of unpaired valence electrons increases, the d-orbital increases, the highest oxidation state increases. This is because unpaired valence electrons are unstable and eager to bond with other chemical species. This means that the oxidation states would be the highest in the very middle of the transition metal periods due to the presence of the highest number of unpaired valence electrons. To determine the oxidation state, unpaired d-orbital electrons are added to the 2s-orbital electrons since the 3d-orbital is located before the 4s-orbital in the periodic table. For example: Scandium has one unpaired electron in the d-orbital. It is added to the 2 electrons of the s-orbital and therefore the oxidation state is +3. So that would mathematically look like: 1s electron + 1s electron + 1d electron = 3 total electrons = oxidation state of +3.

The formula for determining oxidation states would be (with the exception of copper and chromium):

$$\begin{aligned}&\textit{Highest Oxidation State for a Transition metal} = \\ &= \textit{Number of Unpaired } d-\textit{electrons} + \textit{Two } s-\textit{orbital electrons}\end{aligned}$$

Element	**Group**	**Density (g/cm³)**	**M.P. (°C)**	**B.P. (°C)**	**Radius (pm)**	**Free atom configuration**	**ionization energy (kJ mol⁻¹)**	**Oxidation state**
Sc	3	2.99	1541	2831	164	[Ar] $3d^1 4s^2$	631	+3
Ti	4	4.50	1660	3287	147	[Ar]$3d^2 4s^2$	658	+2, +3, +4
V	5	5.96	1890	3380	135	[Ar]$3d^3 4s^2$	650	+2, +3, +4, +5
Cr	6	7.20	1857	2670	129	[Ar]$3d^5 4s^1$	653	+2, +3, +6
Mn	7	7.20	1244	1962	137	[Ar]$3d^5 4s^2$	717	+2, +3, +4, +6, +7
Fe	8	7.86	1535	2750	126	[Ar]$3d^6 4s^2$	759	+2, +3
Co	9	8.90	1495	2870	125	[Ar]$3d^7 4s^2$	758	+2, +3
Ni	10	8.90	1455	2730	125	[Ar]$3d^8 4s^2$	737	+2
Cu	11	8.92	1083	2567	128	[Ar]$3d^{10} 4s^1$	746	+2
Zn	12	7.14	420	907	137	[Ar]$3d^{10} 4s^2$	906	+2

Table 4.1. Physical properties and free atom electronic configuration of transition metals.

Scandium is one of the two elements in the first transition metal period which has only one oxidation state (zinc is the other, with an oxidation state of +2). All the other elements have at least two different oxidation states. Manganese, which is in the middle of the period, has the highest number of oxidation states, and indeed the highest oxidation state in the whole period since it has five unpaired electrons.

It was mentioned previously that both copper and chromium do not follow the general formula for transition metal oxidation states. This is because copper has 9 d-electrons, which would produce 4 paired d-electrons and 1 unpaired d-electron. Since copper is just 1 electron short of having a completely full d-orbital, it steals an electron from the s-orbital, allowing it to have 10 d-electrons. Likewise, chromium has 4 d-electrons, only 1 short of having a half-filled d-orbital, so it steals an electron from the s-orbital, allowing chromium to have 5 d-electrons.

4.2 Luminescence from pure TM doped in nitride semiconductors

Luminescence and spectroscopic properties of materials play an important role in optical device fabrication, display technologies and the laser industry; transition metals are not exceptions. The optical properties of transition metal ion solids have been studied for many years. Hidalgo (Hidalgo, Mendez et al. 1998) and Muller (Muller, Zhou et al. 2009) studied that partially occupied d-shells of these elements play important role in the luminescence and spectroscopy of these elements. Thurbide (Thurbide and Aue 2002), Gedam (Gedam, Dhoble et al. 2007), Grinberg (Grinberg, Barzowska et al. 2001) and Lapraz (Lapraz, Iacconi et al. 1991) reported that in many host materials, these partially occupied d-shells give rise to several important technological applications and in particular, production of high resistivity materials and photonic devices through transition-metal doping has been widely used. Maqbool (Maqbool, Wilson et al. 2010) and Martin (Martin, Spalding et al. 2001) showed that thermal activation, the oxygen effect and co-dopants are good tools to obtain high efficiency and improved luminescence from these metal ions.

Recent progress toward nitride-based light-emitting diodes and electroluminescent devices (ELDs) has been made using crystalline and amorphous nitride semiconductors doped with a variety of transition metals: Richardson (Richardson, Van Patten et al. 2002), Maqbool (Maqbool, Main et al. 2010; Maqbool, Wilson et al. 2010) and Caldwell (Caldwell, Martin et al. 2001). The amorphous III-nitride semiconductors are equally important as their crystalline counterpart because the amorphous material can be grown at room temperature with little stress due to lattice mismatch. They may also be more suitable for waveguides and cylindrical and spherical laser cavities because of the elimination of grain boundaries at low temperature growth. Cathodoluminescence, spectroscopy and the effects of various factors on the luminescence of a few TMs are given below.

4.2.1 Chromium doped in amorphous aluminum nitride

Thin films of a-AlN:Cr were prepared and deposited on Si (100) substrate by the method of plasma magnetron sputtering at low temperature as described earlier. The x-ray diffraction (XRD) analysis confirmed that the deposited films were amorphous.

Figure 4.1 shows the XRD analysis of the a-AlN:Cr films deposited on flat Si(100) substrate. Only one peak can be observed in the film at 69.1° which corresponds to Si(100). No other peak is present in the figure, indicating that the films deposited on flat silicon substrates are amorphous. Thermal activation of the films at 1200 K has not changed the structure of the films.

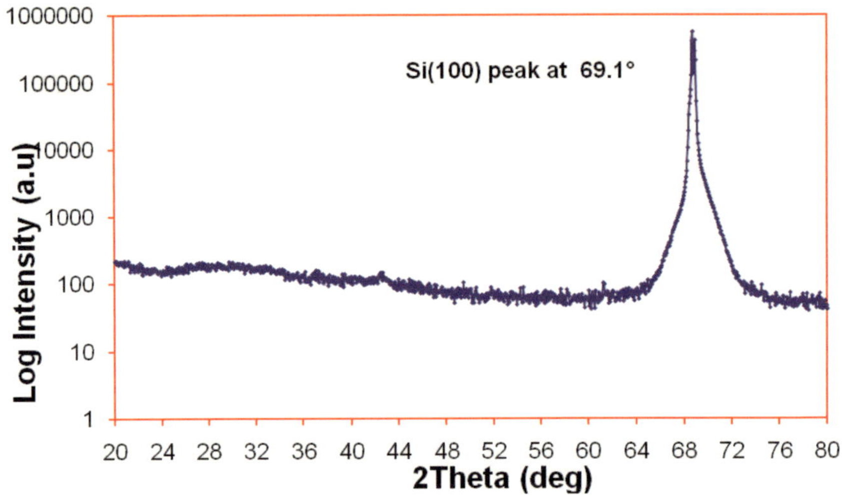

Fig. 4.1. XRD analysis of the a-AlN:Cr films deposited on flat Si (100) substrate.

Cathodoluminescence (CL) of the Cr-doped a-AlN was obtained to study the suitability of this TM for various applications in optical and display technology. Figure 4.2 shows the CL spectrum of the as-deposited and thermally activated a-AlN:Cr in 600 nm to 750 nm range.

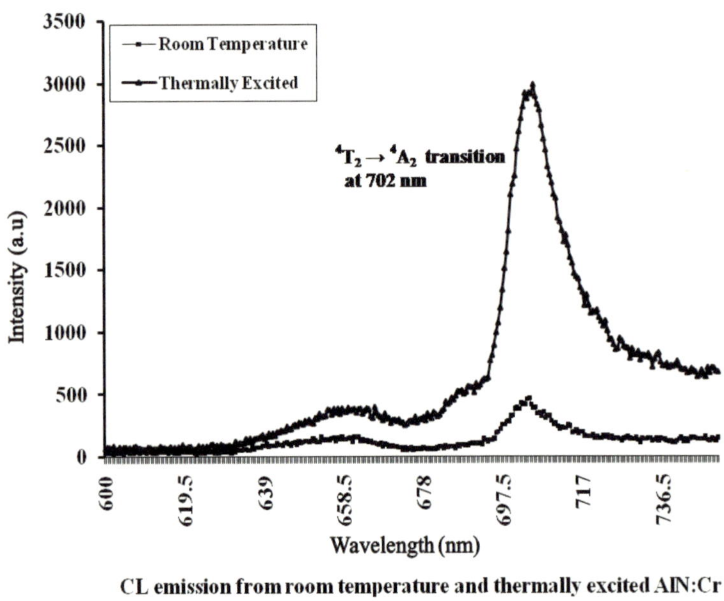

Fig. 4.2. CL spectra of room temperature and thermally activated a-AlN:Cr films.

A strong emission has occurred at 702 nm indicated by a sharp peak in Fig. 4.2. This peak corresponds to $^4T_2 \rightarrow {}^4A_2$ transition. It is clear from the figure that thermal activation enhances the luminescence intensity six times. A small peak at 661 nm is also observed. This peak corresponds to Ho^{+3} ion, indicating the presence of holmium impurity in the films.

Figure 4.2 reports that the intensity of the emission is not only strong but an enhancement in the luminescence is possible by various means like thermal activation. The strong intensity enabled us to see the red emission in CL apparatus directly with naked eye when the excitation current is reasonably high. Six times increase in the emitted light intensity makes it possible that a-AlN:Cr can serve as a potential candidate for a laser production at 702 nm and other optical devices applications. Moreover the significant increase in the intensities of luminescence from Cr^{+3} ions by thermal activation can be explained on the basis of luminescence from the triply ionized chromium ions. Luminescence occurs from Cr^{+3} ions and not from Cr^{+2} or Cr^{+1}. During the film deposition it is most likely that some of Al^{+3} of AlN may be replaced by Cr^{+3} but there are also chances for imperfections and defects giving rise to Cr^{+2} and Cr^{+1}during film growth. These ions do not contribute to luminescence. The smaller the number of these ions, more will be Cr^{+3} ions and hence luminescence will be higher. When these films are activated thermally at a higher temperature then most of Cr^{+2} and Cr^{+1} impurities ionize and convert to Cr^{+3} ions giving a path to enhanced luminescence. Moreover when the films are transferred to the furnace and thermally activated after removed from the deposition chamber, they are exposed to air. Thus oxidation of the surface of the film cannot be ignored. Oxygen enhances the luminescence of TM ions giving rise to the enhanced luminescence after thermal activation of the films. Chen (Chen, Chen et al. 2000), Little and Kordesch (Little and Kordesch 2001) and Suyver (Suyver et al. 2005) have reported such results in other materials as well.

The results show that amorphous AlN:Cr is a promising candidate for its use in optical and photonic devices and communication tools. The strong red-IR emission makes this material a potential candidate for making laser cavities, quantum dots and other wave-guided applications. Due to the high penetration ability of near infrared light in human tissues, it can also be used in biomedical applications.

4.2.2 Tungsten doped in amorphous aluminum nitride

Another important member of the TM family is Tungsten (W). Tugsten has the highest melting point and lowest vapor pressure of all metals. It has a very high tensile strength. Our investigations revealed that along with other physical properties tungsten can also be used for visible light emission applications. The cathodoluminescence of tungsten shows that it gives a very broad emission under cathode ray excitation. The emission spectrum is so broad that it covers the entire visible range of the electromagnetic spectrum from 350 nm up to 700 nm. However the dominant portion of the spectrum comes in blue region with a peak at 491 nm. Another peak in blue is also observed at 429 nm. Because of the huge portion of the spectrum and its peaks lying in blue the films also looked blue in appearance when directly exposed to the electron beam in CL. This broad CL emission from tungsten is shown in Fig. 4.3.

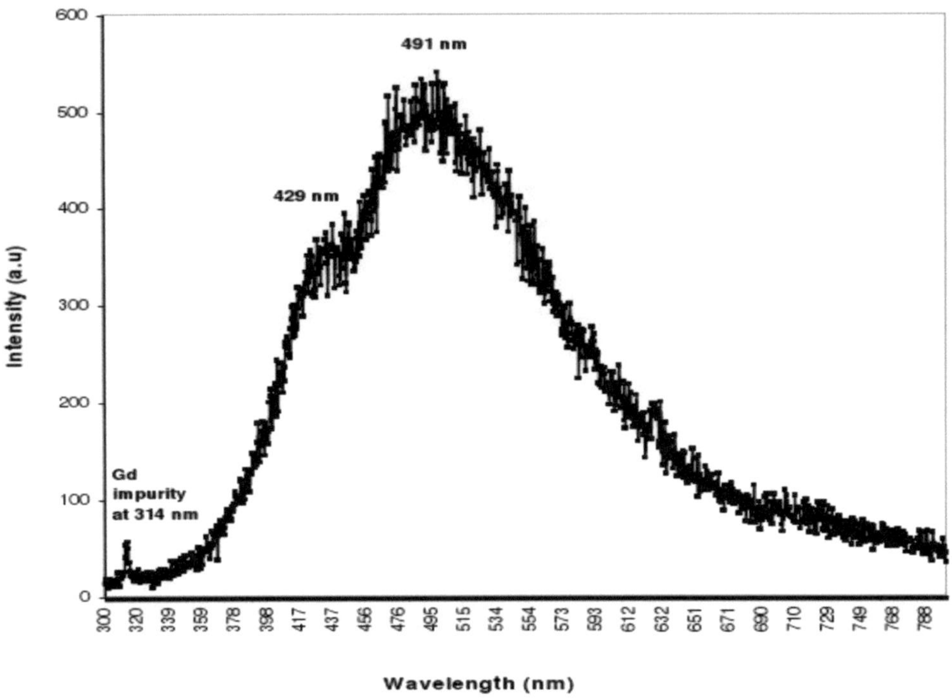

Fig. 4.3. CL spectrum of amorphous AlN:W

4.2.3 Interaction of co-doped Gd^{+3} with AlN:W

Due to its broad emission W^{+3} is also able to interact with other ions when co-doped with it. For this purpose it was also co-doped with holmium and gadolinium separately to test for

luminescence enhancement in the green emission from holmium ions or luminescence enhancement in the AlN:W by gadolinium. It was observed that no change has occurred in Ho luminescence by the addition of W. However the luminescence from Gd has enhanced W luminescence. Figure 4.4 shows how the addition of Gd^{+3} affects the light emission from W^{+3}.

Figure 4.4 shows luminescence from amorphous AlN films co-doped with 1 piece of W and 1-3 pieces of Gd. It is clear from the figure that the addition of Gd has a dominant effect on the luminescence from W. Luminescence from AlN:1W2Gd is 3 times the luminescence from AlN:1W1Gd and luminescence from AlN:1W3Gd is almost 6 times the luminescence from AlN:1W1Gd. This fact shows a huge increase in intensities in the W luminescence by Gd.

Figure 4.4 also reveals some other information. It can be observed that the main peak in blue shifts with the relative increase in Gd concentration. With one piece of Gd the main peak of W appears at 506 nm. With two pieces of Gd added to AlN:W the peak gets bigger but shifts to 534 nm. However with the addition of another piece of Gd the peak gets bigger but shifts back to 523 nm. A possible explanation for this shift in the peaks may be the activation of new radiative energy levels in W by the Gd. Further, the intensity of Gd peak is reduced with higher concentration of Gd. Table 4.2 is giving the increase in the W intensity with the concentration of Gd and also the intensity of Gd peak with the concentration of Gd.

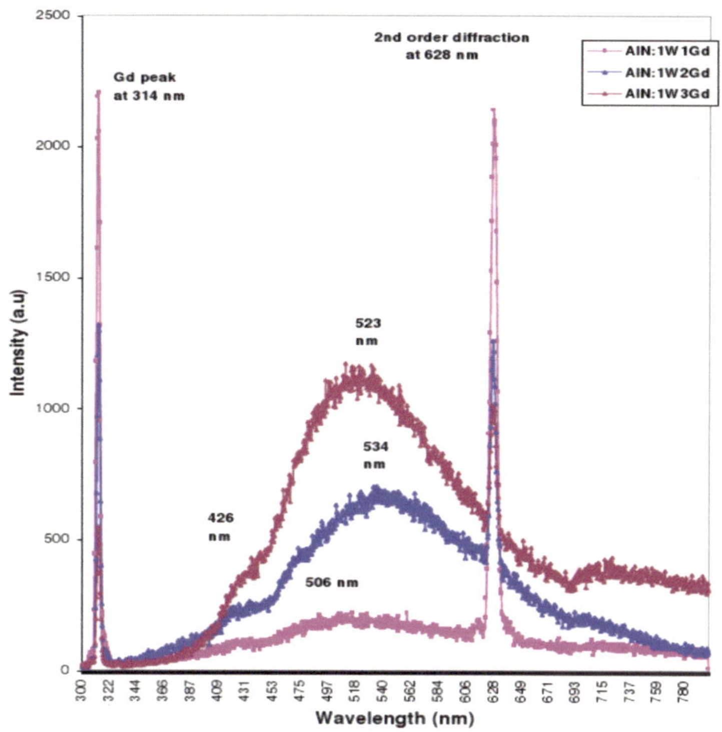

Fig. 4.4. The effect of Gd concentration on the luminescence of a-AlN:W films.

4.2.4 Interaction of co-doped Ho^{+3} with AlN:W

Tungsten is added to Holmium (Ho) in order to observe any enhancement in the green luminescence from Ho. Figure 4.5 shows the effect of varying W concentration on Ho. A comparison between amorphous AlN:1Ho1W, AlN:1Ho3W and thermally activated AlN:1Ho1W is given in this figure. It is clear from the figure that there is no effect of increasing W concentration on the Ho luminescence. The W concentration is increased 3 times which made the W peak broad but no enhancement in Ho. Further, if we compare this figure with the pure W spectrum then a shift can be seen in the W emission wavelength just like that happened due to Gd. And we see from the figure that in the presence of Ho, tungsten emits a single blue light with a wavelength of 461 nm rather than emissions at 429 nm and 491 nm in pure tungsten. A possible explanation may be the holmium has affected the energy level distribution in tungsten. Further it can also be given in explanation that this shift of wavelength did not occur in rare-earth elements discussed in previous chapters but the transition element W has suffered a lot from this shift. Since rare-earth elements are favorable for their internal f-f transitions. These f-levels lying inside the other shells and hence any external change barely affect these transitions. However it is not true for transition metals and they can be significantly affected by other dopant impurities. Moreover this figure also gives us the effect of thermal annealing on the luminescence of W and Ho. It can be easily deduced from the figure that thermal annealing has activated Ho luminescence more than W. There is also existence of a ruby impurity, which is considerably enhanced in luminescence by thermal activation.

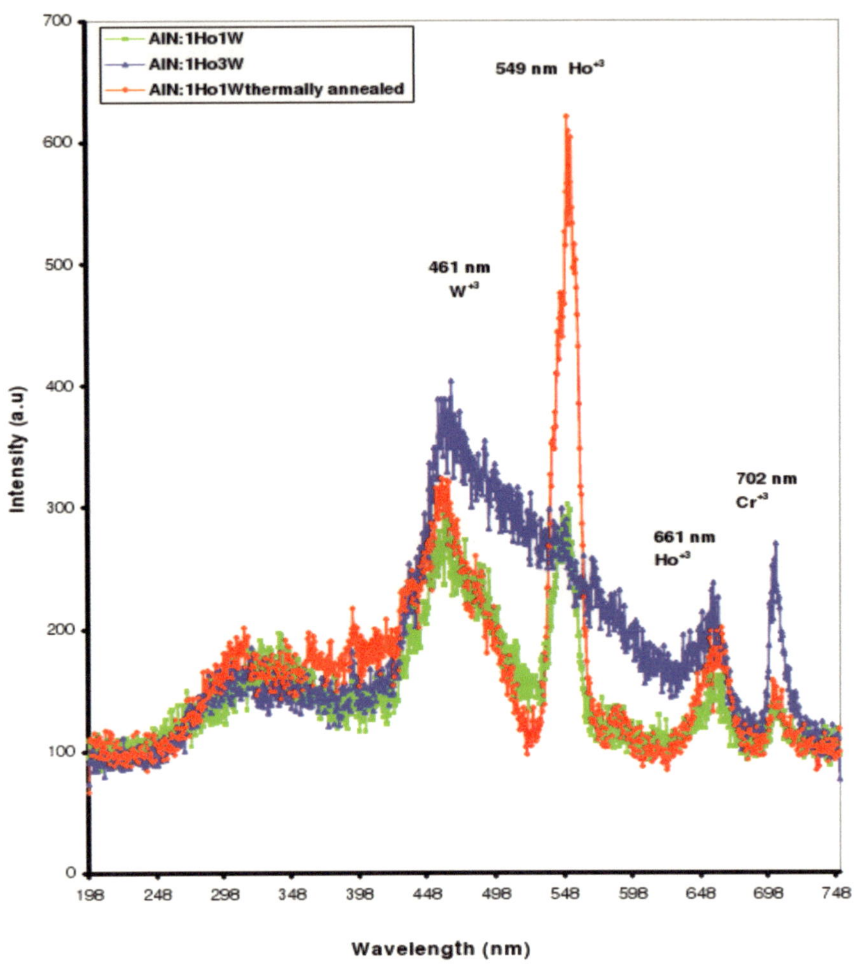

Fig. 4.5. The effect of thermal activation and W concentration on Ho luminescence.

Gd Concentration	Intensity of Gd luminescence (a.u)	Intensity of W luminescence (a.u)	Percent increase in luminescence of W
1	2207	208	Reference
2	1318	705	339
3	538	1159	557

Table 4.2. The effect of Gd^{+3} concentration on the W and Gd emission.

4.2.5 Cathodoluminescence and thermal activation of AlN doped with Yttrium (Y)

Cathodoluminescence of cold deposited Yttrium doped AlN films were characterized for CL in the same way described earlier for other materials. Figure 4.6 represents the luminescence from AlN:Y films when exposed to cathode rays in CL assembly. This figure confirms emission from amorphous AlN:Y films in UV , blue and in bluish green regions. There is also a peak in IR which may correspond to Yttrium or possibly a ruby impurity. The wavelengths correspond to these emissions are 360 nm, 421 nm, 518 nm and 705 nm respectively. The emission 705 nm could be from AlN:Y or from Cr^{+3} The bluish green and blue emissions are the dominant in intensity and that is why the films appear blue to naked eye when exposed to electron beam in CL. However we already know from the previous materials investigations that thermal annealing tremendously enhance the luminescence from Cr^{+3} and hence we can analyze this peak obtaining the CL after performing thermal annealing.

Figure 4.7 gives a comparison of CL spectra from amorphous AlN:Y before and after thermal activation. Films were thermally activated at 900 °C for one hour and then characterized for CL. From the figure it is clear that thermal activation has almost no effect on the Yttrium luminescence at 360 nm, 421 nm and 518 nm. However the peak at 705 nm is strongly enhanced by thermal activation. This enhancement in the peak at 705 nm is more than three and half times the luminescence from the same peak in the film, which is not thermally activated. Further, this peak was less than half the intensity of peak at 518 nm before thermal activation but after activation it is about one and half time more intense than the peak at 518 nm. One can guess at this stage that the peak at 705 is most probably due to chromium oxide (Ruby) rather than emission from yttrium ions itself.

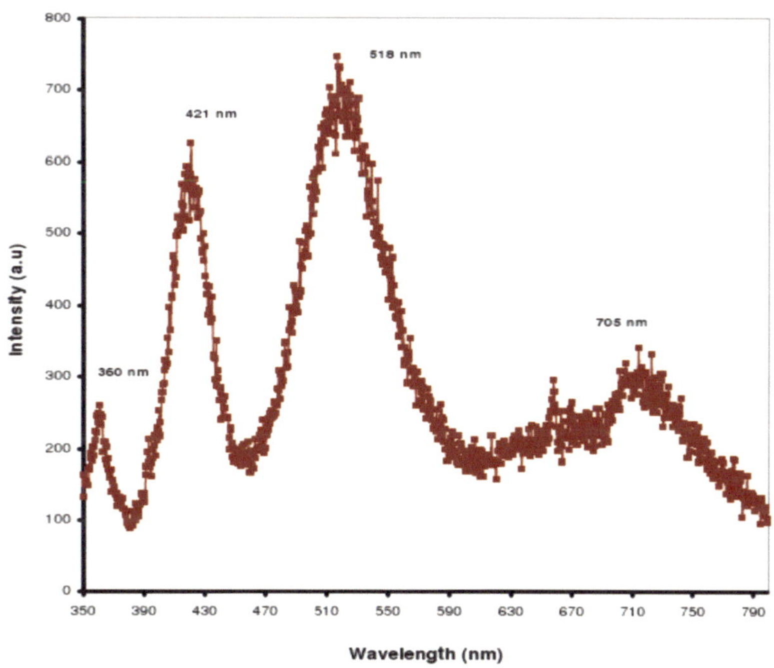

Fig. 4.6. CL spectrum of amorphous AlN:Y

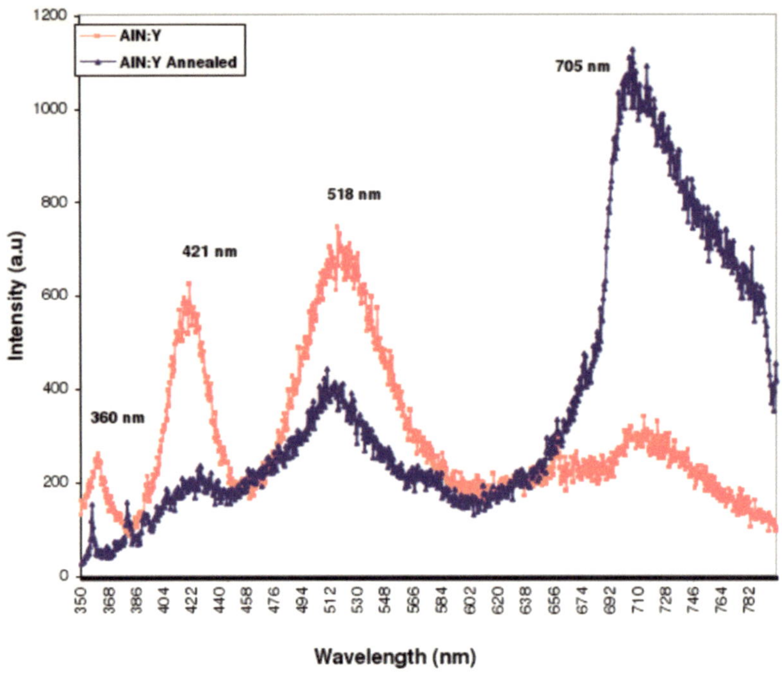

Fig. 4.7. Comparison of the CL spectra of thermally activated and as deposited amorphous AlN:Y films.

4.2.6 Titanium doped amorphous aluminum nitride

Titanium is one of the important members of the TM family. It is lightweight, strong, corrosion resistant metal and the ninth most plentiful amongst all the elements in the earth's crust. This metal, when doped in nitride semiconductors, is proved to be a very good source to make laser cavities and optical devices. The cathodoluminescence of Ti^{+3} doped in amorphous AlN shows a broad emission over a wide range from 650 nm up to 900 nm with a peak around 760 nm. Figure 4.8 shows the CL spectrum of a-AlN:Ti. The broad emission is clear from the spectrum.

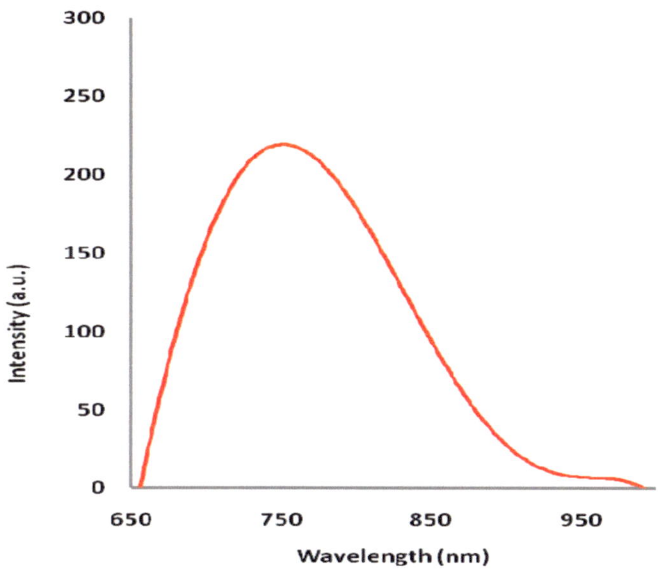

Fig. 4.8. Cathodoluminescence spectrum of amorphous AlN:Ti.

5. Applications and special geometries

5.1 Amorphous AlN:Tm (Ti) on optical fiber in a cylindrical and ring geometry

The broad emission from a-AlN:Ti, as discussed in the previous section, makes it hard to use it directly for any particular wavelength application. However it has a great potential for narrowing the peak by a resonance emission to produce a laser out of Ti^{+3}. The resonance emission to produce a laser will definitely need a cavity. Rather than making a traditional cavity, however, we made a cavity by depositing a-AlN:Ti, around optical fiber and obtained a microlaser with emission wavelength of 780.5 nm. The fiber acts as a cavity when the light emitted from the deposited film circulates around the fiber. Only those light waves will enhance each other whose wavelength is an integral multiple of their path around the fiber.

Figure 5.1 represent thin film deposition and light propagation around optical fiber, where 5.1(a) shows longitudinal view of the fiber containing thin film around it. The dark red region around the fiber is the a-AlN:Ti, film deposited uniformly around the fiber. Figure 5.1(b) is a cross sectional view of the fiber with the film, showing light propagation in whispering gallery mode (WGM), The diameter 's' of this section is 12 μm. The dark red region around the fiber is the a-AlN:Ti film deposited on the fiber. The thickness 'd' of this film is 4 micron. This makes the total diameter of the fiber and the film around it to be 20 micron ($D = 2d + s = 20$ μm). The white lines in the film represent the propagation of light in the film. The pattern of light propagation is restricted to the a-AlN:Ti film only, without touching the fiber itself. Such arrangement is known as whispering gallery. Figure 5.1(c) shows pump laser coupled to fiber and microring emission. Optical fiber with a-AlN:Ti film was held such that it's longitudinal axis is perpendicular to the pump laser beam direction. The pump laser hits the thin films around the fiber directly, exciting a-AlN:Ti for infrared emission, that generates whispering gallery modes in the fiber. The laser produced in the fiber cavity comes out as a microring emission from the fiber.

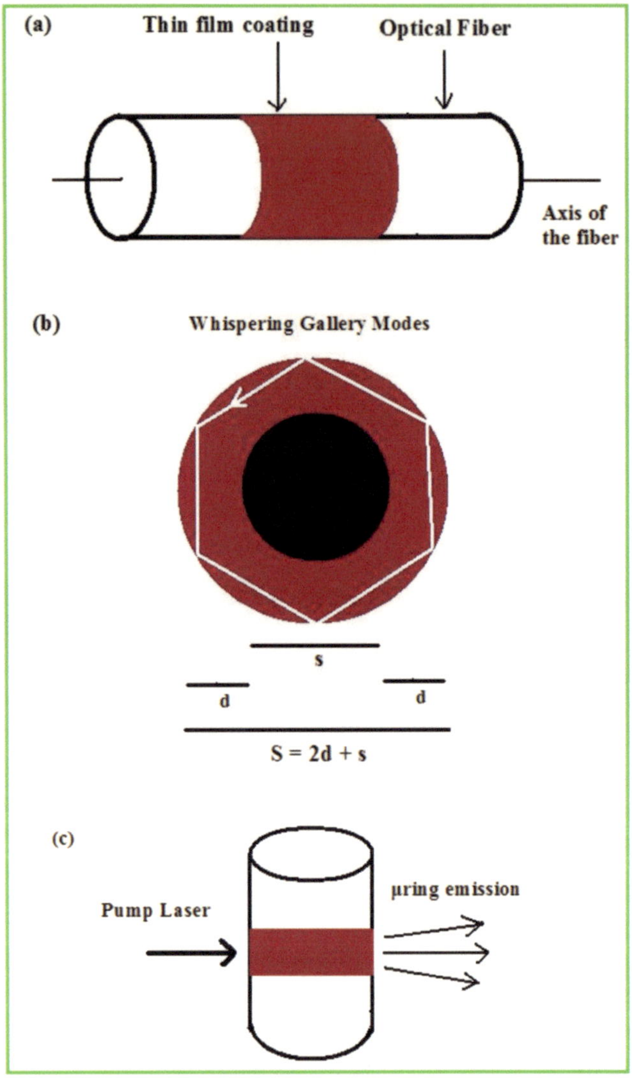

Fig. 5.1. Whispering gallery mode microlaser cavity formation around optical fiber.

Figure 5.2 shows the band narrowing and laser action in a-AlN:Ti films deposited around the optical fiber excited by 532 nm Nd:YAG laser. The power of the excitation laser is varied between 7.5 mW and 30 mW. The figure shows two emission spectra. The broad spectrum showing just the fluorescence emission from the deposited films was obtained when 15 mW of power from the Nd:YAG laser is used for excitation. This broad emission spectrum is about 20 nm wide and no lasing action is observed in this spectrum under 15 mW excitation power. The sharp and very intense emission with narrow bandwidth shows lasing action in the film. The laser is produced in the whispering gallery mode at 780.5 nm. We also observed the secondary modes in this spectrum, confirming the resonances in the lasing action. The primary mode lasing and the secondary mode emission peak show gain of ~ 20 at an input power around 30 mW.

Figure 5.3 gives increase in the observed near infrared laser in a-AlN:Ti with the increasing power of Nd:YAG pump laser. The curve shows that no laser action can be achieved in a-AlN:Ti if the power P of excitation laser is less than a threshold value P_0 . For $P > P_0$ an almost linear increase in the titanium microlaser emission was observed with increasing excitation laser power. The threshold power P_0 was found to be about 23.5 mW. For $P < P_0$ we just observed a fluorescence emission from the fiber with an emission width of 35 nm.

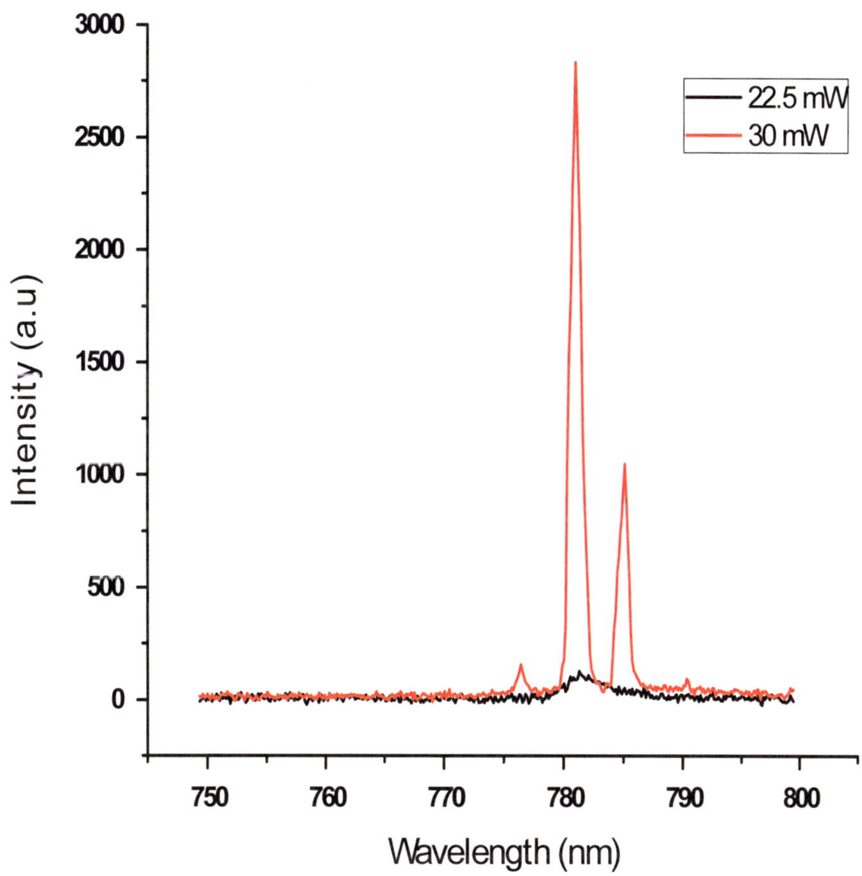

Fig. 5.2. Laser action at 780.5 nm in a-AlN:Ti thin films around optical fiber. The Equally spaced peaks at 776.4 nm and 785.5 nm are the secondary modes of laser. Mode spacing is 4.6 nm.

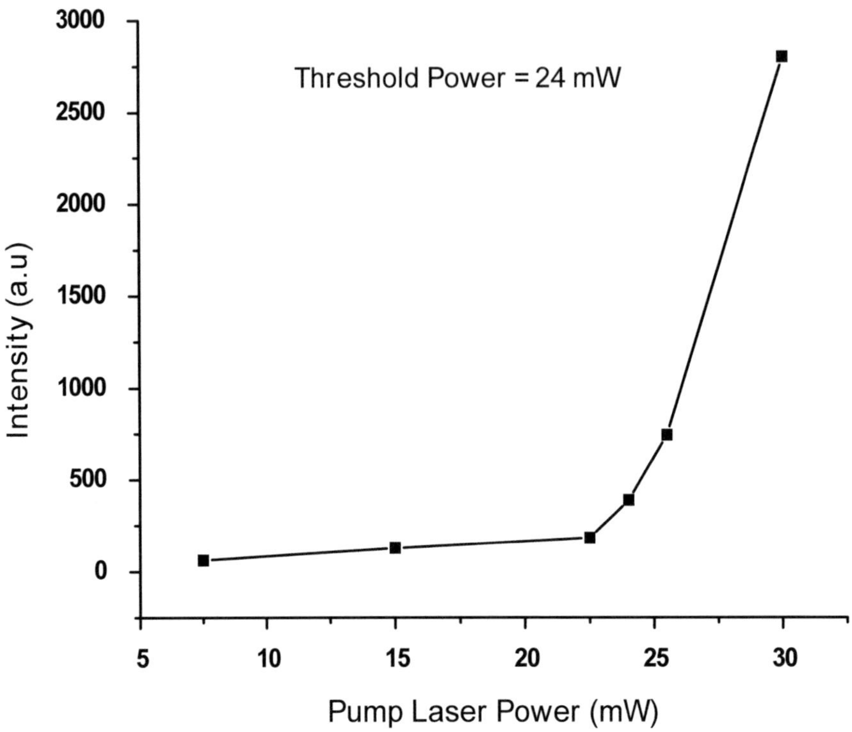

Fig. 5.3. Increase in the a-AlN:Ti laser intensity with the pump laser power. A threshold power of 24 mW can be observed.

In micro-ring cavities, the thin a-AlN:Ti film deposited around the optical fiber supports whispering gallery modes. Frolov (Frolov, Fujii et al. 1998) (Frolov, Vardeny et al. 1998) and Maqbool and Kordesch (Maqbool, Main et al. 2010) figured out that the angular momentum mode number (n) for a WGM is given by equation (1), where D is the diameter of the optical fiber with thin film deposited on it, λ_n is the wavelength of the WGM and m is the index of refraction for the a-AlN film.

$$n = \pi D m / \lambda_n \qquad (1)$$

Khoshman and Kordesch (Khoshman and Kordesch 2005) worked on amorphous AlN and found that the refractive index in the near infrared region (780.5 nm) is 1.95. This value of the index of refraction of amorphous AlN is obtained using the films deposited in the same deposition system and the same conditions that we used for our work. Thus, m = 1.95 is the most appropriate value to use in our work. Our results satisfy the given equation for the integers n = 157. Because the film around the optical fiber is thick enough to satisfy the WGM condition ($d/D \geq 0.2$), we did not observe waveguide modes. The a-AlN:Ti doped optical fiber was placed in vertical direction so that the ring laser formation occur in a horizontal axis. The resonance wavelength λ_n should also support the mode separation $\Delta\lambda$ equation, given by;

$$\Delta\lambda = \lambda_n - \lambda_{n+1} = \lambda^2 / \pi D m \qquad (2)$$

Using n = 156, the outer diameter D = 20 micron, and m = 1.95, equation (2) gives $\Delta\lambda$ = 4.9 nm. From figure 5.2 we find that our experimental results give $\Delta\lambda$ = 4.6, which is in quite good agreement with the theoretical calculations.

The a-AlN:Ti laser, we produced, is very important from biomedical applications point of view. Along with its uses in optics and photonics this laser can also be very beneficial in health sciences applications, particularly for diagnostic and therapeutic purposes. Researchers have reported that near infrared light with a wavelength between 700 nm and 900 nm has minimum absorption and the greatest penetration in body tissues (Noriyuki, Ohdan et al. 1997; Cerussi, Shah et al. 2006). Our laser produced at 780.5 nm is in this range and hence, can be used for diagnosis of deep tissues abnormality and tumors, and laser surgery of deep body tissues due to its high penetration ability in the human body.

5.2 Textiles

The deposition of amorphous AlN onto textiles was examined by Kordesch and Richardson (Kordesch and Richardson 2003). The non-woven textiles (clean room suit material) is made by thermally binding polymer threads. The threads are bound by many pads so that no loose threads are lost. The material is sensitive to heat, so that it is a good test for the successful deposition of a RE - doped Al coated textile. In this case the dopant was Tb.

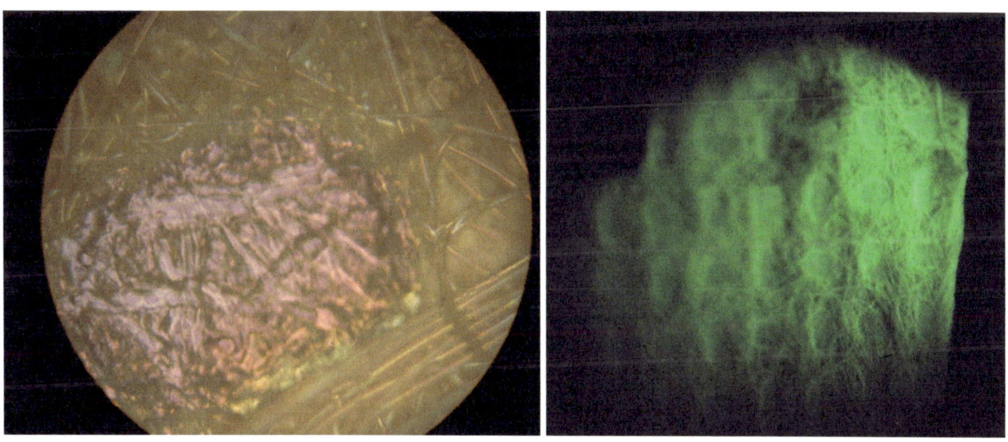

Right: Cathodoluminescence from the a-AlN:Ti coated textile. The voltage and current density used was 2.8 kV and 0.15uA/cm^2. Field of View 30mm.

Fig. 5.4. Left: a-AlN:Ti film on a thermally bonded textile pad. Field of view is 3mm.

5.3 Thermometery

A novel use of a-AlN:Ti thin films has been developed by Richardson and co-workers (Carlson, Khan et al. 2011; Wang, Carlson et al. 2011).

While the example given in Figure 5.5 is necessarily photoluminescence because the intention of the study is to determine the heating due to a gold particle in a liquid or biological environment, Er based thermometry could be used in CL and in other environments. Several other RE based thermometry systems are possible(Alden, Omrane et al.; Lai, Feng et al.).

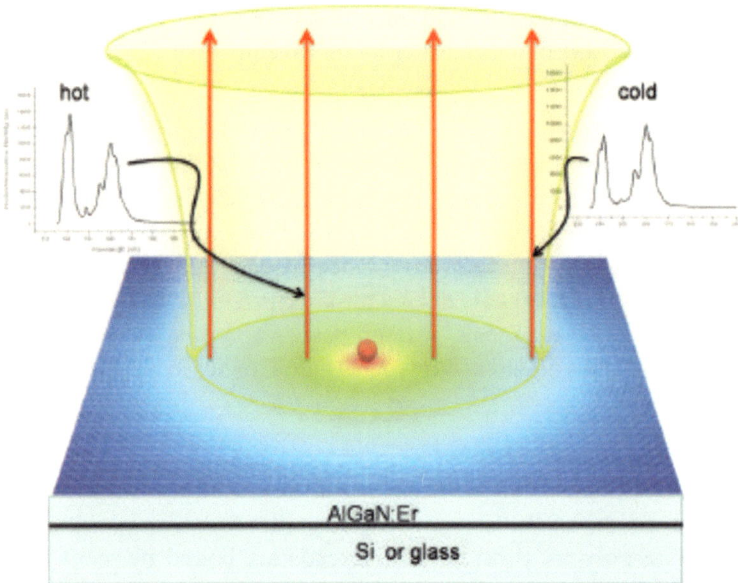

Fig. 5.5. Schematic of the Er thermometer. The Er:AlGaN film is deposited onto a substrate. The gold particle on the film surface is excited with laser light through the objective of a microscope. The photoluminescence spectrum of the Er peaks at 540 and 565 nm are used to determine the temperature of the film surrounding the substrate. From (Carlson, Khan et al. 2011), with permission.

6. Acknowledgments

We would like to thank all of the former members of our research teams who have worked on Rare Earth luminescence in the amorphous nitrides. In particular, we thank Professors Henryk J. Lozykowski, A. Ricardo Zanatta, and Hugh H. Richardson. This work was funded by several agencies, including grants from the Ballistic Missile Defence Organization and the Defence Advanced Research Projects Agency administered by the Office of Naval Research: N00014-96-1782 entitled "Growth, Doping and Contacts from Wide Band Gap Semiconductors" and grant N00014-99-1-0975 entitled "Band-Gap Engineering of the Amorphous In-Ga-Al Nitride Semiconductor Alloys for Luminescent Devices from the Ultraviolet to the Infrared."

7. References

Adachi, S. (1999). *Optical Constants of Crystalline and Semiconductors:Numerical Data and Graphical Information*, Springer.

Aldabergenova, S. B., G. Frank, et al. (2006). "Structure changes of AlN: Ho films with annealing and enhancement of the Ho3+ emission." *Journal Of Non-Crystalline Solids* 352(9-20): 1290-1293.

Aldabergenova, S. B., A. Osvet, et al. (2002). "Blue, green and red emission from Ce3+, Tb3+ and Eu3+ ions in amorphous GaN and AlN thin films." *Journal Of Non-Crystalline Solids* 299: 709-713.

Alden, M., A. Omrane, et al. "Thermographic phosphors for thermometry: A survey of combustion applications." *Progress In Energy And Combustion Science* 37(4): 422-461.

Allen S., R., H.H. and Kordesch. M.E. (2000). Luminescence of Transition and Lanthanide Metal Ions in Sputtered BN Thin Films, Ohio Univerity.

Caldwell, M. L., A. L. Martin, et al. (2001). "Emission properties of an amorphous AlN: Cr3+ thin-film phosphor." *Applied Physics Letters* 78(9): 1246-1248.

Caldwell, M. L., P. G. Van Patten, et al. (2001). "Visible luminescent activation of amorphous AlN: Eu thin-film phosphors with oxygen." *Mrs Internet Journal Of Nitride Semiconductor Research* 6(13): 1-8.

Carlson, M. T., A. Khan, et al. (2011). "Local Temperature Determination of Optically Excited Nanoparticles and Nanodots." *Nano Letters* 11(3): 1061-1069.

Cerussi, A., N. Shah, et al. (2006). "In vivo absorption, scattering, and physiologic properties of 58 malignant breast tumors determined by broadband diffuse optical spectroscopy." *Journal Of Biomedical Optics* 11(4).

Chen, H., K. Y. Chen, et al. (2000). "Band gap engineering in amorphous AlxGa1-xN: Experiment and ab initio calculations." *Applied Physics Letters* 77(8): 1117-1119.

Chen, H., K. Gurumurugan, et al. (2000). "Visible and infrared emission from GaN: Er thin films grown by sputtering." *Mrs Internet Journal Of Nitride Semiconductor Research* 5: art. no.-W3.16.

Chen, K. Y. and D. A. Drabold (2002). "First principles molecular dynamics study of amorphous AlxGa1-xN alloys." *Journal Of Applied Physics* 91(12): 9743-9751.

Dieke, G. H. (1968). *Spectra and Energy Levels of Rare Earth Ion in Crystals*. New York, Wiley.

Dierolf, V., et al., eds. (2011). *Rare-Earth Doping of Advanced Materials for Photonic Applications*.

Dimitrova, V. I., P. G. Van Patten, et al. (2001). "Photo-, cathodo-, and electroluminescence studies of sputter deposited AlN: Er thin films." *Applied Surface Science* 175: 480-483.

Dimitrova, V. I., P. G. Van Patten, et al. (2000). "Visible emission from electroluminescent devices using an amorphous AlN: Er3+ thin-film phosphor." *Applied Physics Letters* 77(4): 478-479.

Drabold, D. A. a. E., S., eds. (2010). *Theory of Defects in Semiconductors*, Springer.

Ebdah, M. (2011). Engineering the Optical, Structural, Electrial and Magnetic Properties of Oxides and Nitrides of In-Ga-Zn Thin Films Using Nanotechnology. *Physics and Astronomy*. Athens, Ohio University. PhD: 155.

Frolov, S. V., A. Fujii, et al. (1998). "Cylindrical microlasers and light emitting devices from conducting polymers." *Applied Physics Letters* 72(22): 2811-2813.

Frolov, S. V., Z. V. Vardeny, et al. (1998). "Plastic microring lasers on fibers and wires." *Applied Physics Letters* 72(15): 1802-1804.

Gedam, S. C., S. J. Dhoble, et al. (2007). "Dy3+ and Mn2+ emission in KMgSO4Cl phosphor." *Journal Of Luminescence* 124(1): 120-126.

Grinberg, M., J. Barzowska, et al. (2001). "Inhomogeneous broadening of the dominant Cr3+ sites in LiTaO3 system." *Journal Of Luminescence* 94: 85-90.

Gurumurugan, K., H. Chen, et al. (1999). "Visible cathodoluminescence of Er-doped amorphous AlN thin films." *Applied Physics Letters* 74(20): 3008-3010.

Hassan, Z. (1998). Growth, Characterizationation and Fabrication of GaN-based Device Structures. *Physics and Astronomy*. Athens, Ohio University. PhD: 1-158.

Hidalgo, P., B. Mendez, et al. (1998). "Luminescence properties of transition-metal-doped GaSb." *Physical Review B* 57(11): 6479-6484.

Housecroft, C., and Sharpe, A.G. (2007) Inorganic Chemistry (3rd Edition), Upper Saddle River, NJ, Prentic Hall.

Iacona, F., G. Franzo, et al. (2009). "Er-based materials for Si microphotonics." *Optical Materials* 31(9): 1269-1274.

Ihn, T. (2010). *Semiconductor Nanostructures: Quantum states and electronic transport*, Oxford University Press.

Jadwisienczak, W. M., H. J. Lozykowski, et al. (2000). Visible emission from AlN doped with Eu, Tb, and Er ions. *2001 Ieee International Symposium On Compound Semiconductors*: 489-494.

Jadwisienczak, W. M., H. J. Lozykowski, et al. (2000). "Luminescence of Tb ions implanted into amorphous AlN thin films grown by sputtering." *Applied Physics Letters* 76(23): 3376-3378.

Kasai, H., A. Nishikawa, et al. (2010). "Improved Eu Luminescence Properties in Eu-Doped GaN Grown on GaN Substrates by Organometallic Vapor Phase Epitaxy." *Japanese Journal Of Applied Physics* 49(4).

Kenyon A.J. and Lucarz, F. "A Critique of the Existing Model for Excitation Exchange Between Silicon Nanostructures and Erbium Ions in SIlica." from http://discovery.ucl.ac.uk/1306794.

Kenyon, A. J., C. E. Chryssou, et al. (2002). "Luminescence from erbium-doped silicon nanocrystals in silica: Excitation mechanisms." *Journal Of Applied Physics* 91(1): 367-374.

Khoshman, J. M. and M. E. Kordesch (2005). Spectroscopic ellipsometry characterization of amorphous aluminum nitride and indium nitride thin films. *Physica Status Solidi C - Conferences and Critical Reviews, Vol 2, No 7*. 2: 2821-2827.

Kik, P. G. a. P., A. (2003). *Towards the First Silicon Laser*, Kluwer Academic Publishers.

Kim, J. H. and P. H. Holloway (2004). "Room-temperature photoluminescence and electroluminescence properties of sputter-grown gallium nitride doped with europium." *Journal Of Applied Physics* 95(9): 4787-4790.

Kim, J. H., N. Shepherd, et al. (2003). "Visible and near-infrared alternating-current electroluminescence from sputter-grown GaN thin films doped with Er." *Applied Physics Letters* 83(21): 4279-4281.

Kordesch, M. E. and H. H. Richardson (2003). Electroluminescent textiles using sputter-deposited amorphous nitride-Rare-Earth ion coatings. *Electronics On Unconventional Substrates-Electrotextiles And Giant-Area Flexible Circuits*. 736: 61-66.

Kucheyev, S. O., J. S. Williams, et al. (2004). "Dynamic annealing in III-nitrides under ion bombardment." *Journal Of Applied Physics* 95(6): 3048-3054.

Lai, B. Y., L. Feng, et al. "Optical transition and upconversion luminescence in Er(3+) doped and Er(3+)-Yb(3+) co-doped fluorophosphate glasses." *Optical Materials* 32(9): 1154-1160.

Lapraz, D., P. Iacconi, et al. (1991). "Thermostimulated Luminescence And Fluorescence Of Alpha-Al2o3-Cr-3+ Samples (Ruby) - Influence Of The Cr-3+ Concentration." *Physica Status Solidi A-Applied Research* 126(2): 521-531.

Little, M. E. and M. E. Kordesch (2001). "Band-gap engineering in sputter-deposited ScxGa1-xN." *Applied Physics Letters* 78(19): 2891-2892.

Lozykowski, H. J. (1993). "Kinetics Of Luminescence Of Isoelectronic Rare-Earth Ions In Iii-V Semiconductors." *Physical Review B* 48(24): 17758-17769.

Lozykowski, H. J., W. M. Jadwisienczak, et al. (2000). "Photoluminescence and cathodoluminescence of GaN doped with Pr." *Journal Of Applied Physics* 88(1): 210-222.

MacKay, K.M., et al., (1996) Introduction to Modern Inorganic Chemistry, Boca Raton, CRC Press.

Maqbool, M. (2005). Growth Characterization and Luminescence and Optical Properties of Rare Earth elements and Transition Metals doped in Wide Gap Nitride Semiconductors. *Physics and Astronomy*. Athens, Ohio University. PhD: 1-180.

Maqbool, M. and I. Ahmad (2009). "Ultraviolet spectroscopy of Pr+3 and its use in making ultraviolet filters." *Current Applied Physics* 9(1): 234-237.
Maqbool, M., I. Ahmad, et al. (2007). "Direct ultraviolet excitation of an amorphous AlN: praseodymium phosphor by codoped Gd3+ cathodoluminescence." *Applied Physics Letters* 91(19).
Maqbool, M., G. Ali, et al. "Nanocrystals formation and intense, green emission in thermally annealed AlN:Ho films for microlaser cavities and photonic applications." *Journal Of Applied Physics* 108(4).
Maqbool, M. and T. Ali (2009). "Intense Red Catho- and Photoluminescence from 200 nm Thick Samarium Doped Amorphous AlN Thin Films." *Nanoscale Research Letters* 4(7): 748-752.
Maqbool, M., M. E. Kordesch, et al. (2009). "Electron penetration depth in amorphous AlN exploiting the luminescence of AlN:Tm/AlN:Ho bilayers." *Current Applied Physics* 9(2): 417-421.
Maqbool, M., M. E. Kordesch, et al. (2009). "Enhanced cathodoluminescence from an amorphous AlN:holmium phosphor by co-doped Gd(+3) for optical devices applications." *Journal Of The Optical Society Of America B-Optical Physics* 26(5): 998-1001.
Maqbool, M., K. Main, et al. (2010). "Titanium-doped sputter-deposited AlN infrared whispering gallery mode microlaser on optical fibers." *Optics Letters* 35(21): 3637-3639.
Maqbool, M., H. H. Richardson, et al. (2005). Cathodoluminescence of praseodymium doped AlN, GaN and turbo static BN. *GaN, AlN, InN and Their Alloys*. 831: 417-421.
Maqbool, M., H. H. Richardson, et al. (2007). "Luminescence from praseodymium doped AlN thin films deposited by RF magnetron sputtering and the effect of material structure and thermal annealing on the luminescence." *Journal Of Materials Science* 42(14): 5657-5660.
Maqbool, M., E. Wilson, et al. (2010). "Luminescence from Cr(+3)-doped AlN films deposited on optical fiber and silicon substrates for use as waveguides and laser cavities." *Applied Optics* 49(4): 653-657.
Martin, A. L., C. M. Spalding, et al. (2001). "Visible emission from amorphous AlN thin-film phosphors with Cu, Mn, or Cr." *Journal Of Vacuum Science & Technology A-Vacuum Surfaces And Films* 19(4): 1894-1897.
Masterov, V. F. and L. G. Gerchikov (1999). "Mechanisms of excitation of the f-f emission in silicon codoped with erbium and oxygen." *Semiconductors* 33(6): 616-621.
Muller, S., M. J. Zhou, et al. (2009). "Intra-shell luminescence of transition-metal-implanted zinc oxide nanowires." *Nanotechnology* 20(13).
Nishikawa, A., N. Furukawa, et al. (2010). "Improved luminescence properties of Eu-doped GaN light-emitting diodes grown by atmospheric-pressure organometallic vapor phase epitaxy." *Applied Physics Letters* 97(5).
Nishikawa, A., T. Kawasaki, et al. (2009). "Room-Temperature Red Emission from a p-Type/Europium-Doped/n-Type Gallium Nitride Light-Emitting Diode under Current Injection." *Applied Physics Express* 2(7).
Noriyuki, T., H. Ohdan, et al. (1997). "Near-infrared spectroscopic method for assessing the tissue oxygenation state of living lung." *American Journal Of Respiratory And Critical Care Medicine* 156(5): 1656-1661.
O'Donnell, K. P., ed. (2010). *Rare-Earth Doped III-Nitrides for Optoelectronic and Spintronic Applications*, Springer.
Park, J. H. and A. J. Steckl (2004). "Laser action in Eu-doped GaN thin-film cavity at room temperature." *Applied Physics Letters* 85(20): 4588-4590.
Park, J. H. and A. J. Steckl (2005). "Demonstration of a visible laser on silicon using Eu-doped GaN thin films." *Journal Of Applied Physics* 98(5).

Park, J. H. and A. J. Steckl (2006). "Visible lasing from GaN: Eu optical cavities on sapphire substrates." *Optical Materials* 28(6-7): 859-863.

Richardson, H. H., P. G. Van Patten, et al. (2002). "Thin-film electroluminescent devices grown on plastic substrates using an amorphous AlN: Tb3+ phosphor." *Applied Physics Letters* 80(12): 2207-2209.

Singh, J. a. S., K., eds. (2003). *Advances in Amorphous Semiconductors, Advances in Condensed Matter Sciences*, CRC Press.

Steckl, A. J. and R. Birkhahn (1998). "Visible emission from Er-doped GaN grown by solid source molecular beam epitaxy." *Applied Physics Letters* 73(12): 1700-1702.

Steckl, A. J. a. Z., J.M (1999). "Optoelectronic Properties and Applications of Rare Earth." *MRS Bulletin* 24: 33-38.

Street, R. A., ed. (2010). *Technology and Applications of Amorphous Silicon*, Springer.

Tessler, L. R. (1999). "Erbium in a-Si: H." *Brazilian Journal Of Physics* 29(4): 616-622.

Thurbide, K. B. and W. A. Aue (2002). "Chemiluminescent emission spectra of lead, chromium, ruthenium, iron, manganese, rhenium, osmium and tungsten in the reactive flow detector." *Spectrochimica Acta Part B-Atomic Spectroscopy* 57(5): 843-852.

Wang, D., M. T. Carlson, et al. (2011). "Absorption Cross Section and Interfacial Thermal Conductance from an Individual Optically Excited Single-Walled Carbon Nanotube." *Acs Nano* 5(9): 7391-7396.

Weingartner, R., O. Erlenbach, et al. (2006). "Thermal activation, cathodo- and photoluminescence measurements of rare earth doped (Tm, Tb, Dy, Eu, Sm, Yb) amorphous/nanocrystalline AlN thin films prepared by reactive rf-sputtering." *Optical Materials* 28(6-7): 790-793.

Wybourne B.G. and Smentek, L. (2007). *Optical Spectroscopy of Lanthanides*, CRC Press.

Zanatta, A. R., A. Khan, et al. (2007). "Red-green-blue light emission and energy transfer processes in amorphous SiN films doped with Sm and Tb." *Journal Of Physics-Condensed Matter* 19(43).

Zanatta, A. R. and L. A. O. Nunes (1998). "Green photoluminescence from Er-containing amorphous SiN thin films." *Applied Physics Letters* 72(24): 3127-3129.

Zanatta, A. R., C. T. M. Ribeiro, et al. (2001). "Visible luminescence from a-SiN films doped with Er and Sm." *Applied Physics Letters* 79(4): 488-490.

Zanatta, A. R., C. T. M. Ribeiro, et al. (2004). "Photon and electron excitation of rare-earth-doped amorphous SiN films." *Journal Of Non-Crystalline Solids* 338: 473-476.

Zanatta, A. R., C. T. M. Ribeiro, et al. (2005). "Optoelectronic and structural characteristics of Er-doped amorphous AlN films." *Journal Of Applied Physics* 98(9).

Zanatta, A. R., C. T. M. Ribeiro, et al. (2006). "Thermally synthesized ruby microstructures and luminescence centers." *Journal Of Applied Physics* 100(11).

Zanatta, A. R., H. H. Richardson, et al. (2007). "Amorphous BeN as a new solid host for rare-earth-related luminescent materials." *Physica Status Solidi-Rapid Research Letters* 1(4): 153-155.

Zegrya, G. G. and V. F. Masterov (1995). "Mechanism Of The Intensification Of F-F Luminescence In Semiconductors." *Semiconductors* 29(10): 989-995.

Zegrya, G. G. and V. F. Masterov (1996). Two novel mechanisms of f-f-luminescence resonance excitation in semiconductors. *Tenth Feofilov Symposium On Spectroscopy Of Crystals Activated By Rare-Earth And Transitional-Metal Ions.* 2706: 235-240.

Zegrya, G. G. and V. F. Masterov (1998). "Mechanism of generation of f-f radiation in semiconductor heterostructures." *Applied Physics Letters* 73(23): 3444-3446.

Part 3

Application to Oxides and Minerals

8

Peculiarity of the Cathodoluminescence of Alpha- Alumina Prepared by Calcination of Gibbsite Powder or Generated by Oxidation of a Metallic FeCrAl Alloy

Djelloul Abdelkader and Boumaza Abdecharif
Laboratoire des Structures, Propriétés et Interactions Inter Atomiques (LASPI²A)
Centre Universitaire de Khenchela 40000
Algeria

1. Introduction

Most of metallic materials functioning at high temperature need to have oxidation resistance. This resistance can be achieved when the material develops, through oxidation, an oxide film which acts as a diffusion barrier while keeping a good adherence. In this respect, alpha alumina clearly acts as such. The oxides of aluminum have been the subject of many investigations because of their commercial importance and scientific interest. The thermal stability and optical properties of pure nanometer-sized alumina powder have received much attention because of their intrinsic interest and commercial value. Nanometer-sized alumina powders are widely applied today. One of its applications is in fluorescent lamps due to the absorption of ultraviolet light. In fact, it can also emit the light under excitation with a suitable wavelength. It is important to note that there are many works about alpha alumina using X-ray diffraction, but there is a need for a more detailed structural analysis. To achieve this more exhaustive structural characterization we have used the Rietveld refinement method and cathodoluminescence (CL) measurements.

CL spectroscopy is widely used as a contactless and relatively nondestructive method to provide microcharacterization of the optical and electronic properties of luminescent materials. Nevertheless, it is used comparatively rarely for the investigation of oxide semiconductor structures. The major advantage of CL spectroscopy in the case of such structures is that most of the anticipated products of oxidation are luminescent, and it is easy to get excitation across the bandgap of any dielectric with readily available electron beam voltages. The emission occurs for all the luminescent mechanisms present in the material.

Pure α-Al_2O_3 crystal is colorless and shows little absorption in the ultraviolet–visible (UV–Vis) range. But various impurities (Ti, Mn, Cr, and Fe) even a trace level causes apparent absorptions which are attributed to various emission centers (Jheeta et al., 2006).

The colors arise from very minor amounts of impurity (<1% of the Al^{3+} replaced by other cations) because the Al_2O_3 structure apparently does not tolerate substitutions. However, these trace substitutions can cause intense colors. Ruby is red because of its Cr^{3+} content. Yellow sapphire owes its color to Fe^{3+}. Blue sapphire derives its color from Fe^{2+}-Ti^{4+} and Fe^{2+}-Fe^{3+} intervalence charge transfer. Green sapphires contain a mixture of the blue and green colors.

Strong well-known ${}^2E \rightarrow {}^4A_2$ lines of Cr^{3+} (693 nm) with a long decay time characterize their luminescence spectra. Besides that, much weaker narrow lines are present, which are connected with Cr-pairs and more complicated complexes. The Mn^{4+} ion is isoelectronic with Cr^{3+}, i.e., both of them have the same electronic structure of the open shell ($3d^3$ configuration). Thus, the spectroscopic properties of α-Al_2O_3:Mn^{4+} are similar to those of ruby (α-Al_2O_3:Cr^{3+}). Octahedral Mn^{4+} ($3d^3$) would be expected to show the R-line fluorescence characteristic of isoelectronic Cr^{3+} and in approximately the same region. The dominate defects for the visible emission might be different for α-alumina powders formed by heating any of the hydrates of aluminium to a sufficiently high temperature.

In the presence of lattice defects, extra luminescence emissions can be observed in the ultra-violet (UV) region upon highly energetic excitation. The main intrinsic defects in the α-alumina crystals are oxygen vacancies in different charge states: a neutral vacancy, a vacancy capturing one electron (a F^+-center), and a vacancy capturing two electrons (a F-center) (Kislov et al., 2004; Michizono et al., 2007; Yu et al., 2002). The observed UV spectrum in α- alumina can be deconvoluted into two distinct sub-band components: an F^+-center band, located at around 3.8 eV, and a less intense F-center band, located at around 3.0 eV (Brewer et al., 1980; Boumaza & Djelloul, 2010; Boumaza et al., 2010). Depending on the defect introduction method one can create also F_2-centers, F_2^+-centers and F_2^{2+}-centers (double oxygen vacancy with four, three and two trapped electrons respectively). α-Al_2O_3 crystals with defects in the oxygen sublattice are actively studied as promising storage materials (Kortov & Milman, 1996). In this connection, it is interesting to study luminescence properties of the nanostructured aluminium oxide and compare them with analogous properties of crystalline samples.

In this chapter, we present X-ray diffraction (Rietveld analysis) and CL measurements of α-alumina powders formed by calcination of gibbsite or generated by oxidation of a metallic FeCrAl alloy. The peculiarity of the cathodoluminescence under comparable conditions of α- alumina is discussed.

2. Materials and experimental methods

Gibbsite powder, $Al_2(OH)_6$, from Prolabo (no 20 984.298) was used. The powder is made of platelet aggregates and was composed of 64.5–67% Al_2O_3 and max.: 0.01% Fe_2O_3, 0.02% SO_4, 0.002% heavy metals (as Pb), and 1.0% non precipitable by NH_4OH (as SO_4). The sample experienced an ignition loss of 33–34.5% at 1000 °C and had a purity grade of 99.7%. Its average particle size (20 μm) was due to the agglomeration of crystallites. The specific surface area of the original sample was 0.5 m^2/g.

The gibbsite platelets was calcined in ambient atmosphere (pO_2 = 0.21 atm) at 1573 K. The cycle was as follows: heating up to an isothermal temperature at 5 K/min, maintaining for 24h at the calcination temperature and fast cooling down to room temperature (air quench). The calcination temperature was maintained for 24 hours to obtain a well-crystallised product.

The present work was performed on ferritic ODS commercial FeCrAlY alloy PM2000 (20 wt% Cr, 5.8 wt% Al, 0.5 wt% Ti, 0.5 wt% Y_2O_3). This alloy is an alumina-forming alloy. The specimens, with dimensions 25×25×5 mm^3, are cut from a rolled plate. In the case of the PM2000, sample was oxidised in air at 1223 K for 72h. The thermogravimetric method (TG) and differential thermal analysis (DTA) data were recorded under a dry air flow with a heating rate of 10 K/min in a SETARAM TGDTA92-16.18 thermal analyser. TG measurements were corrected for temperature-dependent buoyancy by subtracting the data of a measurement carried out on an inert sample. The crystalline structure of the sample was investigated by XRD using a PANalytical X'Pert Pro MRD diffractometer configured as follows: Cu tube operating at 40 kV and 30 mA ($\lambda(K\alpha_1)$ = 0.15406 nm, $\lambda(K\alpha_2)$ = 0.15444 nm). The scan rate (2θ) was 1°/min at a step size of 0.025°. The data were processed to realize the conditions of the software program Fullprof Suite for the structure refinement.

The FTIR technique was used in the absorbance mode in the 200-4000 cm^{-1} range. For oxides all bands have characteristic frequencies between 200 and 1000 cm^{-1}. For the FTIR measurements, samples were prepared by grinding the oxide films scraped from the substrate (PM2000). After calcination, 10 to 100 µg of the powder was drawn, then grinded with 23 ±2 mg of CsBr in order to obtain, a pellet of 200-250 µm in thickness. After grinding, the powder was placed in a mould (5 mm diameter) and a cold isostatic pressure (CIP) of 150 MPa was applied for 5 min.

The FTIR spectra are obtained using a Perkin-Elmer spectrometer at resolution of 8 cm^{-1}. For each sample, 120 scans were used. The apparatus is equipped with a system allowing the reduction of the optical course in air in order to minimize the perturbations associated with ambient air (water vapour and CO_2). The uncertainty on the position of the various peaks is equal to ± 2 cm^{-1}.

The emitted light under electron beam excitation in a UHV system was analyzed through a quartz window with a Jobin Yvon CP 2000 spectrograph and a CCD detector. The wavelength range 200-1000 nm was investigated.

3. Results and discussions

3.1 Differential thermal analysis and thermal gravimetry analysis (DTA-TG) of gibbsite

Fig. 1 shows the typical TG–DTA curves of the gibbsite. TG and DTA curves are indicated as dotted and solid lines, respectively. Concerning the dehydration-dehydroxylation process of gibbsite, the dehydration appears to occur in two steps (around 598 and 803 K respectively) at higher temperature. The expected theoretical loss due to dehydration is 34.6%, the experimental loss is 34.3% a little lower. This difference 0.3% is a bit larger than the experimental uncertainty 0.1%, the starting gibbsite may be slightly dehydrated. The last step (803 to 1273 K) gives no thermal event but appears as a continuous mass loss (about 2%) which corresponds to the elimination of residual hydroxyls. The formation of α-alumina occurs between 1473 and 1533 K (MacKenzie et al., 1999). Finally, the structural transformations to well-crystallised α–alumina are described by nucleation (T<1470 K) and growth mechanisms (T>1470 K).

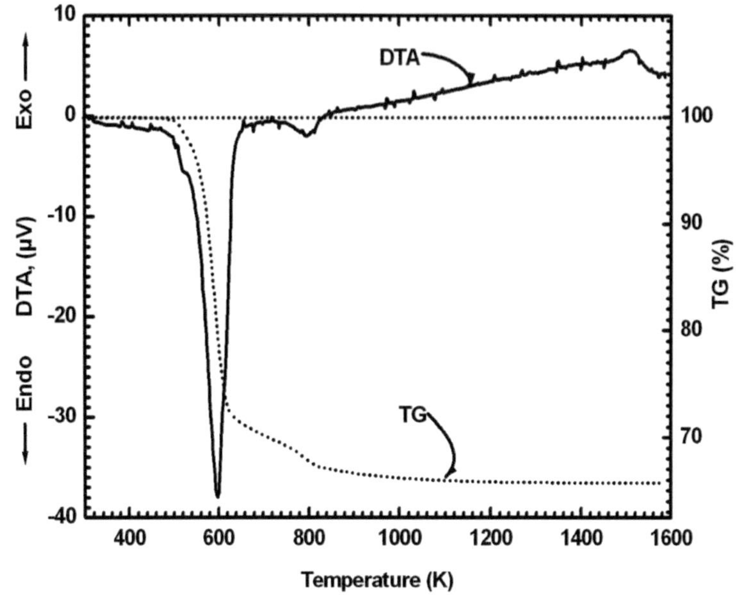

Fig. 1. TG and DTA curves of gibbsite.

3.2 Gibbsite

Gibbsite (γ-Al(OH)$_3$) has monoclinic symmetry (*a*=0.8684, *b*=0.5078, *c*=0.9736 nm, β=94.54°) with the space group P21/n, and the unit cell contains eight Al(OH)$_3$ units (Saalfeld & Wedde, 1974). Gibbsite is characterized by the stacking of two-layer units (AA or BB) of hydroxyl sheets with the sequence ABBAABBA... where hydroxyl sheets of the adjacent Gibbsite layers face the *c* direction (Kogure, 1999). In Fig. 2a, the XRD pattern obtained on the as received gibbsite powder shows a good agreement with the reference XRD pattern (33-0018 JCPDS file).

3.2.1 Phase transitions induced by heat treatment of gibbsite

When heating up fine-grained gibbsite, most OH groups are eliminated, and various forms of alumina are formed with the sequence: gibbsite $\rightarrow\chi$-$\rightarrow\kappa$-$\rightarrow\alpha$-Al_2O_3 when temperature increases. In order to study these phases, we performed XRD measurements on four samples prepared from gibbsite calcined for 24h at 773, 1073 (62h), 1173 and 1573K respectively.

According to Fig. 2b, at 773K ,the χ phase is expected. In spite of many investigations since the1950's (Brindley & Choe, 1961; Kogure, 1999; Saalfeld, 1960; Stumpf et al., 1950; Yu et al., 2002), the crystal structure of χ-alumina is still uncertain. Stumpth et al. (Stumpf et al., 1950) assumed a cubic (not spinel) unit cell of lattice parameter *a*=0.795 nm (04-0880 JCPDS file). On the other hand, two hexagonal structures have been suggested, either with the parameters *a*=0.556 nm and *c*=1.344 nm (Saalfeld, 1960) or with *a*=0.557 nm and *c*=0.864 nm (Brindley & Choe, 1961) (13-0373 JCPDS file). The two previous hexagonal unit cells may be described respectively as a stacking of 6 and 4 close-packed oxygen layers, of approximately the same thickness (0.224 and 0.216 nm) as the Al–OH layers in gibbsite (0.212 nm). More recently, Kogure (Kogure, 1999) proposed a hexagonal lattice with a=0.49 nm and an undefined c length indicating that χ-alumina structure can be regarded as random close packing of gibbsite-like layers.

For samples prepared at 1173 K, the κ phase is expected (see Fig. 2d). Contrary to the χ phase, the crystal structure of κ-alumina is well known (see for example Ref. (Ollivier et al., 1997) and references therein). κ-alumina is orthorhombic with the space group $pna2_1$ and results in ten independent atoms positions (four Al and six O). The experimental XRD pattern at 1173K is specific of a pure κ-alumina (Fig. 2d). Nevertheless, the presence of remnant χ-phase cannot be excluded as all the χ peaks also appear in κ structure. Note that the experimental XRD patterns show well crystallized phases, in contrast with the χ phase.

Fig. 3 give SEM images and XRD patterns of gibbsite powder (a) after calcination at (b)- 773 K; (c)- 1073 K; (d)- 1573 K, for 24 h.

When heating up at temperature above 1573K for 24h, gibbsite transforms into α-alumina, the stable structure. The XRD pattern of the 1573K sample (Fig. 6) shows that only α-Al_2O_3 is present when compared with 42-1468 JCPDS file. In this structure, tetrahedral Al^{3+} ions are no longer present and only AlO_6 octahedron remain.

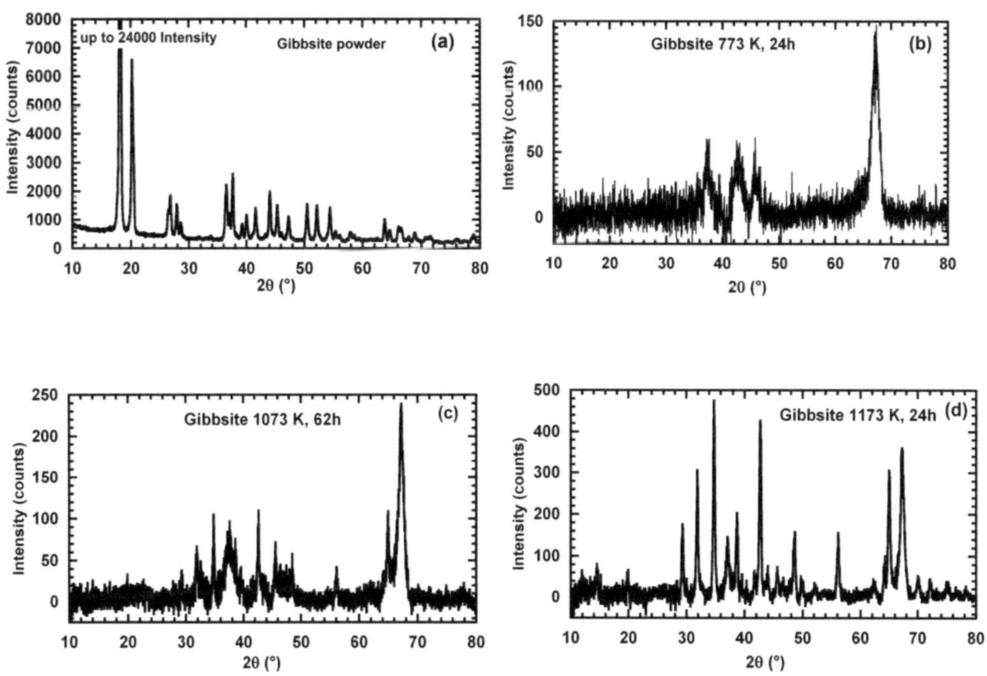

Fig. 2. XRD patterns: (a) Gibbsite (33-0018 JCPDS file), (b) alumina formed from gibbsite calcined for 24h at 773K (expected χ phase) (JCPDS 13-0373 file), (c) alumina formed from gibbsite calcined for 62h at 1073K, (d) alumina formed from gibbsite calcined for 24h at 1173K (expected κ phase) (JCPDS 88-0107 and 01-1305 file).

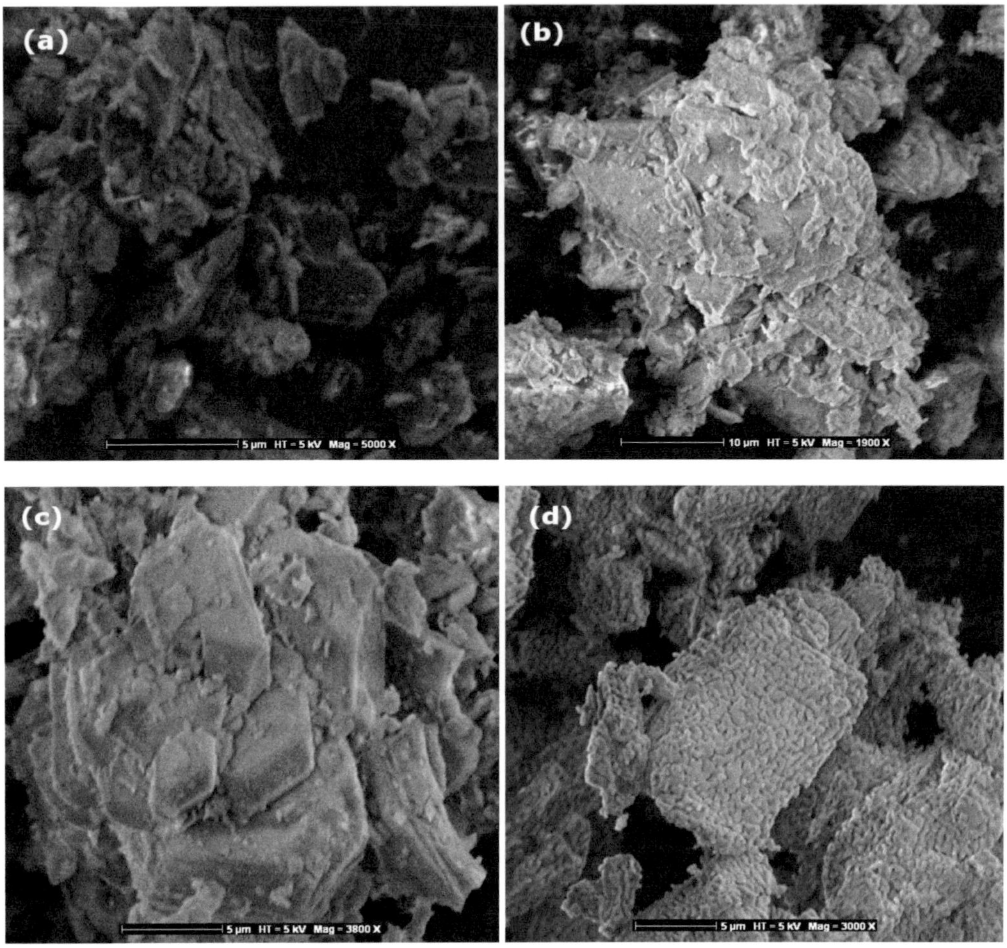

Fig. 3. SEM images: (a)- Gibbsite; (b)- 773 K; (c)- 1073 K; (d)- 1573 K.

3.2.2 General properties of alumina (Al_2O_3)

Aluminum oxide, commonly referred to as alumina, possesses strong ionic interatomic bonding giving rise to it's desirable material characteristics. It can exist in several crystalline phases which all revert to the most stable hexagonal alpha phase at elevated temperatures. This is the phase of particular interest for structural applications and the material available from Accuratus.

The exceptional properties of alumina (Al_2O_3), such as great hardness, high thermal and chemical stability, and high melting temperature, make it a very attractive material. The crystalline α-Al_2O_3 phase (corundum or sapphire) is the single stable modification of alumina. The crystalline α-Al_2O_3 has the band gap $E_g \approx 8.5$ eV and is widely used in optical devices. Sapphire doped with chrome (ruby) or titanium is applied as an active medium in laser systems. In microelectronics sapphire is used as a substrate for growing silicon and gallium nitride (GaN). Alumina is a highly radiation-resistant material and is used as a sensitive element in detectors when measuring the ionizing radiation parameters.

High purity alumina is usable in both oxidizing and reducing atmospheres to 1925°C. Weight loss in vacuum ranges from 10^{-7} to 10^{-6} g/cm^2.sec over a temperature range of 1700° to 2000°C. It resists attack by all gases except wet fluorine and is resistant to all common reagents except hydrofluoric acid and phosphoric acid. Elevated temperature attack occurs in the presence of alkali metal vapors particularly at lower purity levels.

The composition of the ceramic body can be changed to enhance particular desirable material characteristics. An example would be additions of chrome oxide or manganese oxide to improve hardness and change color. Other additions can be made to improve the ease and consistency of metal films fired to the ceramic for subsequent brazed and soldered assembly.

Mechanical, thermal and electrical properties of Al_2O_3 are summarized in table 1.

	Units of Measure	94% Al_2O_3	96% Al_2O_3	99.5% Al_2O_3
Mechanical				
Density	gm/cc	3.69	3.72	3.89
Porosity	%	0	0	0
Color	–	white	white	ivory
Flexural Strength	MPa	330	345	379
Elastic Modulus	GPa	300	300	375
Shear Modulus	GPa	124	124	152
Bulk Modulus	GPa	165	172	228
Poisson's Ratio	–	0.21	0.21	0.22
Compressive Strength	MPa	2100	2100	2600
Hardness	Kg/mm^2	1175	1100	1440
Fracture Toughness K_{IC}	MPa•m$^{1/2}$	3.5	3.5	4
Maximum Use Temperature	°C	1700	1700	1750
Thermal				
Thermal Conductivity	W/m•K	18	25	35
Coefficient of Thermal Expansion	10^{-6}/°C	8.1	8.2	8.4
Specific Heat	J/Kg•K	880	880	880
Electrical				
Dielectric Strength	ac-kV/mm	16.7	14.6	16.9
Dielectric Constant	At 25°C, 1 MHz	9.1	9.0	9.8
Dissipation Factor	At 25°C, 1 MHz	0.0007	0.0011	0.0002
Volume Resistivity	Ohm•cm	>10^{14}	>10^{14}	>10^{14}

Table 1. Mechanical, thermal and electrical properties of Al_2O_3 (http://accuratus.com/alumox.html).

3.2.3 Oxidation of PM2000

The oxidation resistance of high-temperature alloys and metallic coatings is dependent on the formation of a protective surface oxide. In an ideal case, the oxide layer should be highly stable, continuous, slow growing, free from cracks or pores, adherent and coherent. α-Al_2O_3 is an oxide which comes close to satisfying these requirements; the slow growth rate is related to its highly stoichiometric structure and its large band gap which makes electronic conduction difficult.

One of the most crucial factors in the oxidation of alumina-formers is the temperature, which must be high enough to promote the formation of α-Al_2O_3 in preference to the less protective transition alumina. Another critical factor is the aluminum content which must be sufficiently high to develop and maintain an alumina layer and prevent subsequent breakaway oxidation.

The addition of chromium to Fe-Al alloys promotes the formation and maintenance of a complete layer of α-Al_2O_3 by acting as a getter and preventing internal oxidation of the aluminum (Wood, 1970).

Iron and chromium are the major impurities present in Al_2O_3 scales formed on PM2000. Primarily their oxides formed during the transient stage and were incorporated into the α-Al_2O_3 scale. Fe segregated to some α-Al_2O_3 grain boundaries, but not Cr. The Al_2O_3 scale became progressively purer with oxidation time. It is possible that the Fe in the Al_2O_3 scale increases the scaling rate and, in particular, enhances lateral growth that causes scale convolution.

The growth of α-Al_2O_3 scales that form on FeCrAl alloys during high temperature oxidation is generally considered to be controlled by oxygen inward diffusion through oxide grain boundaries (Mennicke et al., 1998; Quaddakkers et al., 1991). Aluminum also diffuses out, which can cause growth within the scale (Golightly et al., 1979). The degree of Al outward transport can be significantly reduced by the presence of reactive elements, such as Y, Hf or Zr (Mennicke et al., 1998; Quaddakkers et al., 1991), which segregate to Al_2O_3 grain boundaries (Przybylski et al., 1987). However, the extent of outward growth seems to differ appreciably among several reactive-element doped Fe based alloys.

The EDS results of the average Fe and Cr concentrations in the scale as a function of scale thickness from different transmission electron microscopy (TEM) specimens are summarized in Table 2.

Oxidation condition	**Scale thickness (μm)**	**Oxide grain size (nm)**	**Average [Cr] (at%)**	**Average [Fe] (at%)**	**Lattice parameter (nm)**
1000 °C, 0.5 h	0.39	107±36	4.12±0.95	4.49±1.67	a = 0.495±0.004 c = 1.353 ± 0.008
1000 °C, 1 h	0.9	191±44	0.34±0.28	1.91±0.49	a = 0.475±0.004
1000 °C, 26 h	1.77	186±53	0.24±0.28	0.60±0.32	c = 1.347 ± 0.008
1200 °C, 2 h	2.94	291±46	0.3±0.64	0.70±0.57	a = 0.471±0.004
1200 °C, 120 h	4-5.5	1546±423	0.27±0.20	0.06±0.06	c = 1.323 ± 0.008

Standard parameters for α-Al_2O_3 are: a = 0.4758 nm, c = 1.2991 nm.

Table 2. Fe and Cr concentrations in α-Al_2O_3 scales and effect on lattice parameters (Hou et al., 2004)

Fig. 4 give SEM images and XRD patterns of PM2000 after oxidation at 1023 K for 76 h and 1223 K for 72 h.

EDX analysis (Fig. 5) of the sample oxidized at 1123 K indicates that the film mainly consists in aluminium and oxygen elements and very small amount of the substrate constituting elements are observed. As the film thickness decreases, like sample oxidized at 1023 K, the signature of the substrate increases due to interactions of the electrons with the underlying substrate mater as the film is thinner than that formed at 1123 K.

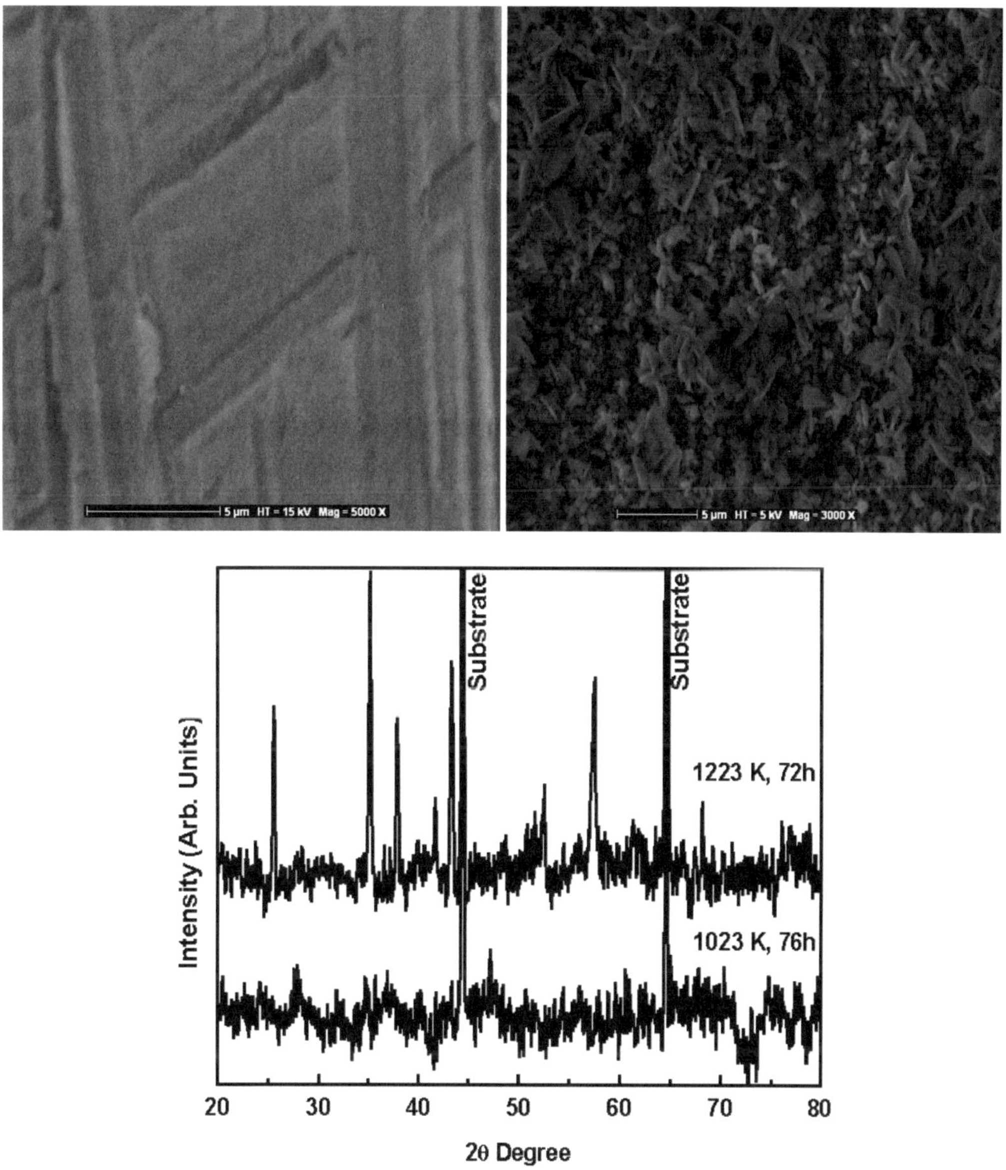

Fig. 4. SEM images of the outer oxidized surface and XRD patterns of α-Al_2O_3 obtained by oxidation of PM2000 at 1023 K for 76 h and 1223 K for 72 h.

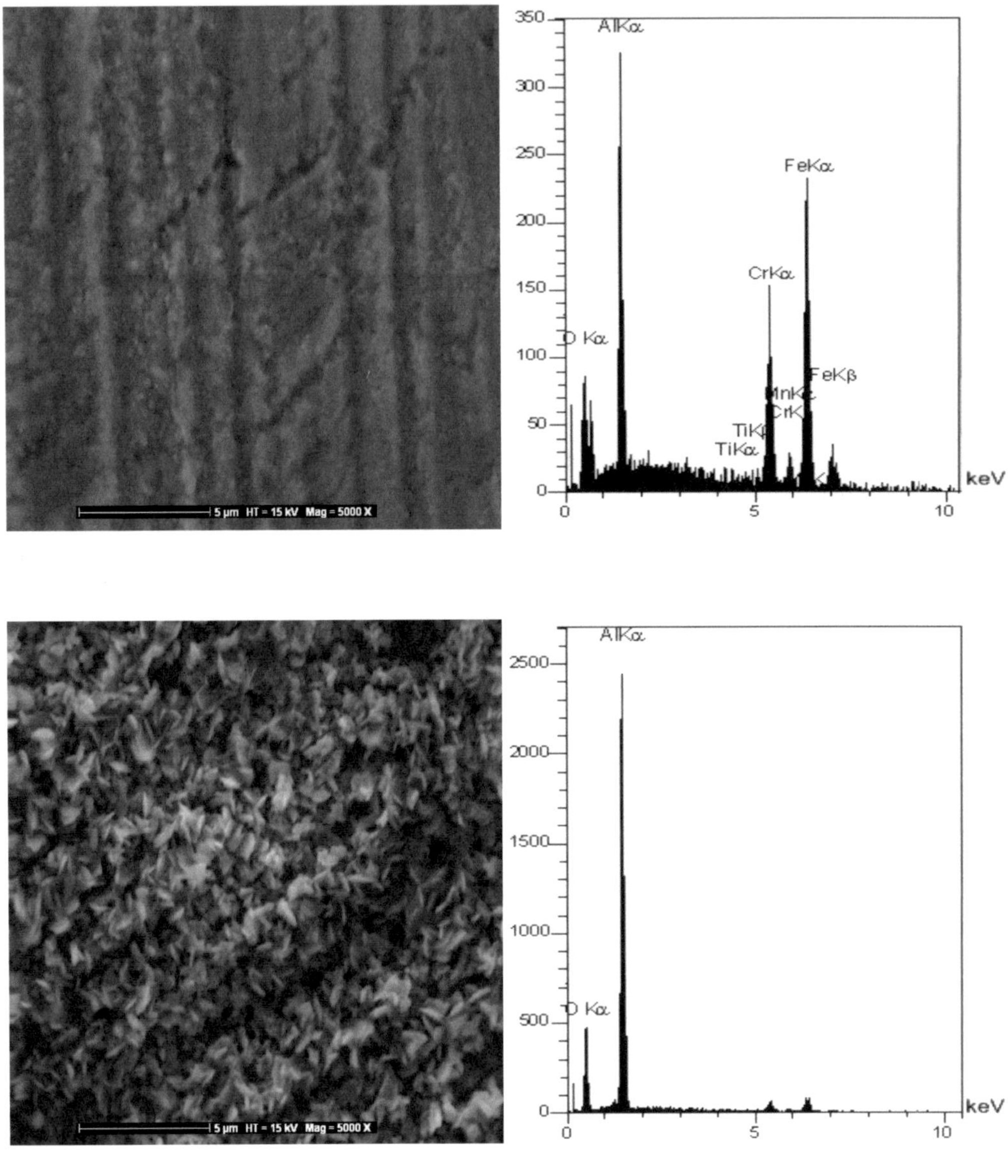

Fig. 5. SEM images of the outer surface and EDX analyses of α-Al_2O_3 films obtained by oxidation of PM2000 in air at 1023 K for 76 h (top) and 1123 K for 95 h (bottom).

3.2.4 Rietveld refinement of the structures

The Rietveld refinement of the structures was performed using the WinPlotr/FullProf suite package (Rodríguez-Carvajal, 1993). The peak shape was described by a pseudo-Voigt function, and the background level was modeled using a polynomial function. The profiles were refined using the space group and structure models of Al_2O_3 ($R\bar{3}C$, JCPDS 46-1212) (in this structure cations occupy the 12c sites and oxygen ions the 18e sites).

The XRD data for the 2θ regions between 20° and 80° was used for the refinement. The observed, Rietveld refined and difference patterns are shown in Figs. 6 and 7.

It was obvious that the agreement between the experimental data and the simulations was excellent since the R_{wp} (weighted residual error) factor was small (≤ 14.1%). The Rietveld results (cell parameters, atom position, reliability factors and crystallite size (*D*)) are given in Table 3 and 4. The size of α-alumina crystallites (*D*=36 nm) obtained using PM2000 alloy was smaller than that (*D*=43 nm) obtained using gibbsite precursor. Furthermore, the α-Al_2O_3 from PM2000 has a greater lattice parameter (*a*=0.4763nm, *c*=1.3047nm) than that of JCPDS file 46-1212 (*a*=0.4758 nm, *c*=1.2991 nm, *c*/*a*=2.730) and a greater *c*/*a* ratio (*c*/*a* =2.739) than that of JCPDS file 46-1212. However, the α-Al_2O_3 from gibbsite has a smaller lattice parameter (*a*=0.4752nm, *c*=1.2980nm) and a similar c/a ratio (*c*/*a* =2.731) than that of JCPDS file 46-1212. Furthermore, the α-Al_2O_3 from gibbsite has a greater calculated density (3.996 g/cm³) than that of α-Al_2O_3 from PM2000 (3.961 g/cm³). Phase transformations are frequently accompanied by microstructural changes. This fact could explain the crystallographic parameters differences between a two α-alumina.

Lattice parameters (nm)	atom	Wyck.	Site	*x*	*y*	*z*	Biso	occupancy
a= 0.47523 *c*= 1.29805	**Al**	12c	3.	0	0	0.3521	0.2200	0.3333
	O	18e	.2	0.3065	0	0.2500	0.2400	0.5000

Table 3. The Rietveld refinement results: α-Al_2O_3 from gibbsite powder, R_p= 16.7%, R_{wp}= 6.93%, Calc. density =3.996 g/cm³, *D*=43 nm.

Lattice parameters (nm)	atom	Wyck.	Site	*x*	*y*	*z*	Biso	occupancy
a= 0.47637 *c*= 1.30472	**Al**	12c	3.	0	0	0.3417	4.7208	0.9602
	O	18e	.2	0.3392	0	0.2500	2.7191	1.1186

Table 4. The Rietveld refinement results: α-Al_2O_3 from PM2000, R_{wp}= 14.1%, Calc. density =3.961 g/cm³, *D*=36nm.

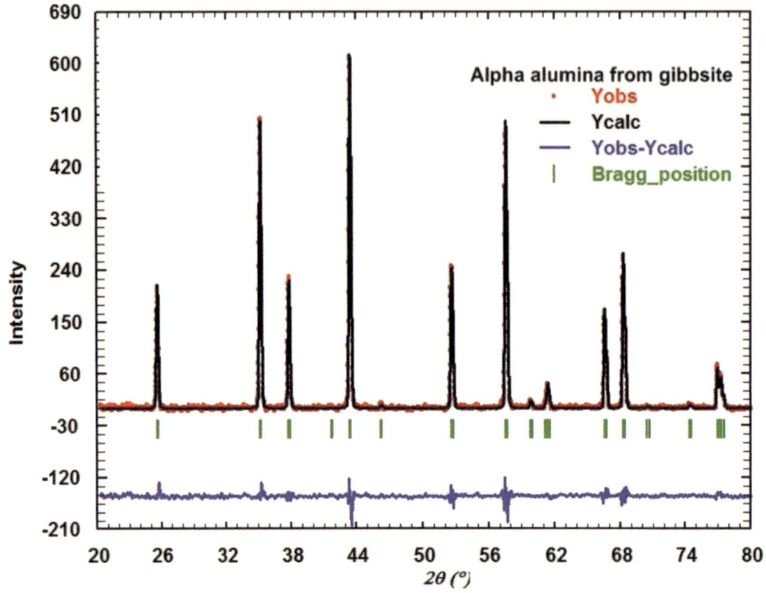

Fig. 6. The observed and calculated diffraction patterns of α-alumina powder prepared by calcination at 1573K for 24h of gibbsite powder. Vertical bars indicate the calculated position of the Bragg peaks. The blue curves in the bottom correspond to the differences between experimental and calculated profiles.

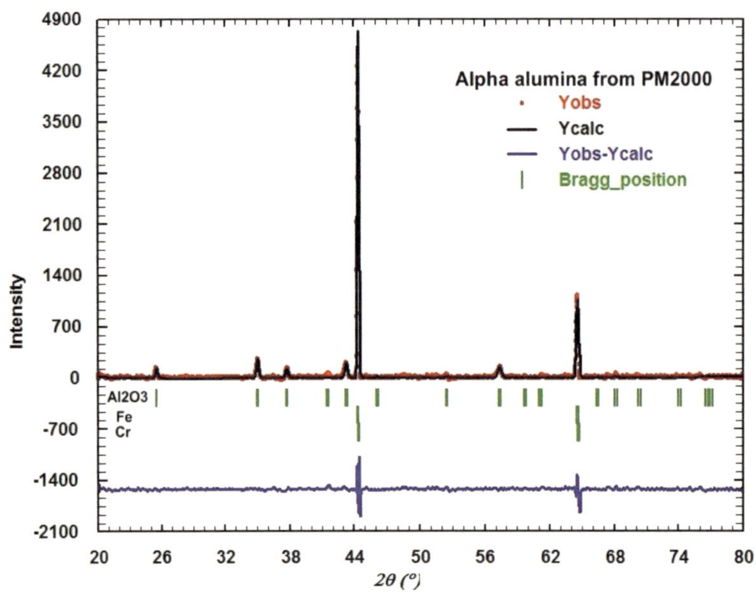

Fig. 7. The observed and calculated diffraction patterns of α-alumina obtained by oxidation of PM2000 at 1223K for 72h.

3.2.5 Fourier transform infrared (FTIR) analysis

Fig. 8 compares the FTIR absorbance spectra of α-Al_2O_3 obtained after calcination of gibbsite and oxidation of PM2000. For the samples calcined at 1573 K, significant spectroscopic bands at ~640, ~594, ~447 cm^{-1} and ~386 cm^{-1} appear which are identified to be the characteristic absorption bands of α-Al_2O_3 (Barker, 1963). This is in good agreement with XRD observations. Common bands exist in all cases, such as the broad OH band centered around 3420 cm^{-1}, and the 1640 cm^{-1} H_2O vibration band (Ma et al., 2008). The very high surface area of these materials results in rapid adsorption of water from the atmosphere because the FTIR samples were kept and grinded in air. Three peaks of very weak intensities at 2850 cm^{-1}, 2920 cm^{-1} and 2960 cm^{-1} are observed which are due to C-H stretching vibrations of alkane groups. The absorption in ~2356 cm^{-1} is due to CO_2 molecular presence in air.

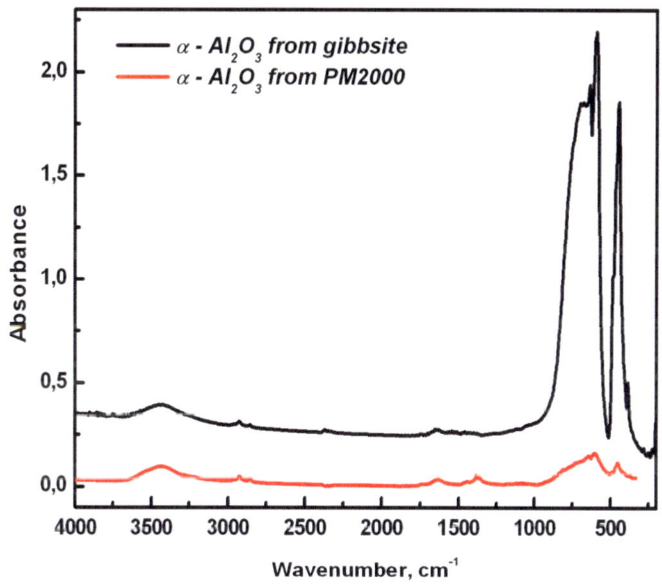

Fig. 8. FTIR absorption spectra of α-alumina powder prepared by calcination of gibbsite (black line) or generated by oxidation of PM2000 (red line).

3.3 Cathodoluminescence

CL is the phenomenon of light emission from specimens as a result of interaction with an electron beam. In insulating crystals, the origin of the luminescence arises from impurity atoms (e.g. transition metals or rare earths) in the crystal lattice. Using an electron microscope to produce the electron beam, the spatial distribution of luminescent sites can be observed with submicron spatial resolution, and correlated with features of the specimen morphology or microstructure.

The mechanisms for CL are similar to those for photoluminescence, but the energy input or excitation source is that of an electron beam rather than a visible or ultraviolet light beam. When an energetic (keV range) electron beam propagates within a semiconductor or insulator, the primary electrons lose energy by the creation of electron-hole pairs. These electron-hole pairs then recombine via radiative and non-radiative processes. Only the radiative recombination process which leads to the creation of a photon is viewed with CL. Radiative recombination may be intrinsic (arising from electronic states of the perfect crystal) or extrinsic (arising from electronic states that are localized at defects or impurities in the crystal). Extrinsic luminescence thus provides information about defects and impurities in the crystal lattice. For conciseness, the defects and impurities that give rise to extrinsic luminescence are often denoted the luminescence centers. Each type of luminescence center in a particular crystal has a characteristic emission spectrum. The spectrum may contain both narrow lines and broad bands, depending on the energy level structure of the luminescence center and the coupling of the center to the host lattice. CL is advantageous compared to photoluminescence. CL can potentially give additional information about local positions in the sample because the electron beam can be focused on several nanometers. In addition, the CL system operates under UHV conditions of less than 10^{-9} Torr. Hence, CL measurements can be performed in a contamination-free environment, which is very effective in detecting weak luminescence. Moreover, the depth dependent emission profiles can be examined in CL by controlling the accelerating voltage.

3.3.1 Oxygen vacancy in Al_2O_3

CL signal is a good signature of the material qualities and is used in this study to characterize the point defects associated to oxygen vacancies in α-Al_2O_3. The CL spectra of α-Al_2O_3 formed from gibbsite and PM2000 are given in figure 9 and 12.

In figure 9, the CL spectra shows that wide band over the interval of (200–600 nm) consists of a series of overlapping bands. The main emission bands located at about 250 nm (4.96 eV), 281 nm (4.41 eV), 325 nm (3.81eV), 373 nm (3.32 eV) and 487 nm (2.54 eV) occur in alpha alumina powder and also in alpha alumina films generated by oxidation of a metallic FeCrAl alloy. We believe that the observed CL peak at 4.96 eV is related to the interband transitions or to defect that is different in origin to the F or F^+ centers in α-alumina. The luminescence band at 4.41 eV is detected only if the excitation density is high and was previously observed in α-alumina by Kortov et al. (Kortov et al., 2008).

In α-Al_2O_3 (corundum structure) each O atom is surrounded by four Al atoms forming two kinds of Al-O bonds of length 0.186 and 0.197 nm. This is why in corundum F-type centers have low C_2 symmetry. Besides, an O vacancy has two nearest neighbor O atoms, forming the basic O triangle with O-O band length of 0.249 nm in perfect corundum. Thus, the F-type centers are surrounded by six nearest atoms which determine mainly their optical properties.

Defects induced in Al_2O_3 may be of various kinds: F centers (oxygen vacancy with two electron), F^+ centers (oxygen vacancy with one electron), F_2 centers (two oxygen vacancies with four trapped electrons), F_2^+ centers (two oxygen vacancies with three electrons) and F_2^{2+} centers (two oxygen vacancies with two electrons) (Ghamdi & Townsend, 1990).

In the case of α-alumina and sapphire, there have been reported a number of F-type centers including the F^+ and F centers (Evans, 1995). As for the luminescence of irradiation defects, it is known that the luminescence of the F^+ center is observed at the UV region around 3.8

eV (325 nm) while that of the F centers is observed at a lower photon energy region around 3.0 eV (410 nm). Thus the presently observed luminescence from α-alumina and sapphire which is centered at 330 and 420 nm can be attributed to the F^+ and F centers. As for the others, it is noted that the luminescence at 250 and 290 nm is also observed in the case of α-alumina. The luminescence intensity at these bands was found to be sensitive to thermal annealing at higher temperatures, and then might be attributed to the effect of some impurities such as OH.

In Ref. (Oster & Weise, 1994), the absorption bands of 220 and 260 nm in pure α-Al_2O_3 crystals have been attributed to F+ absorption bands. Due to the presence of C_2-symmetry in the F^+-center (oxygen vacancy occupied by a single electron) in pure α-Al_2O_3 crystals, the excited state is split into three levels, ^1B, ^2A, and ^2B, according to the theory of La et al. (La et al., 1973), giving three polarized optical absorption bands located at 255, 229 and 200 nm.

The optical properties of these luminescent centres are well known. They possess absorption and emission bands which are produced in the gap as summarised in Fig. 10. F+ centre is characterised by three absorption bands at 6.3, 5.4 and 4.8 eV and emits at 3.8 eV (330 nm). F centre absorbs at 6 eV and emits at 3 eV (415 nm).

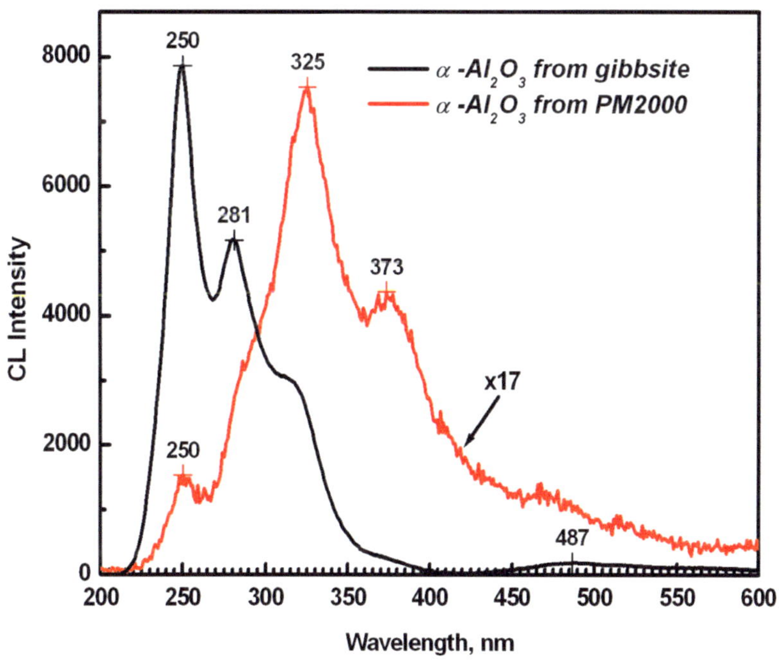

Fig. 9. CL spectra of α-Al_2O_3, formed by calcination of gibbsite powder at 1573 K for 24 h (black line) and by oxidation of PM2000 at 1223K for 72h (red line), obtained in the region of 200 to 600 nm at room temperature.

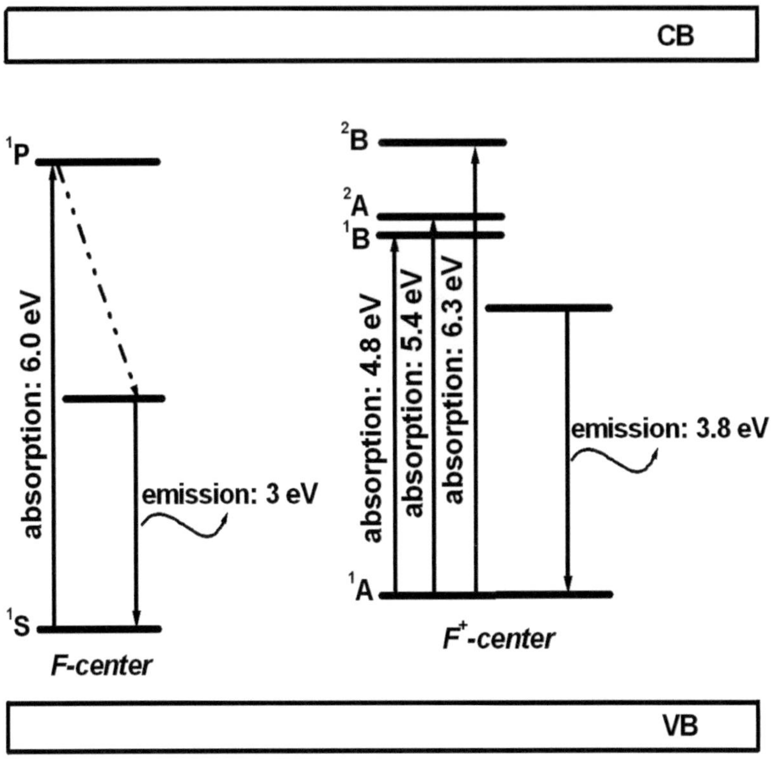

Fig. 10. Schematic energy level diagram for absorption and emission of F- and F^+-centres in α-Al_2O_3 crystal.

In Ref. (Evans, 1994), synthetic sapphire single crystals grown by four different techniques all showed an anisotropic 5.4 eV (230 nm) absorption broad band. There was not any other absorption band presented in his cases. As for the nanometer-sized Al_2O_3 powder apart from normal lattice vacancy-type defects such as the F-, F^+- and F_2-type centers it also has the surface defect because the nanometer powder has a larger specific surface area and there exists a lot of dangling bonds in the surface.

Experimental measurements of absorption and luminescence energies for single-vacancy and for dimer centers in Al_2O_3 are collected in Table 5.

	Absorption	Luminescence
F center	6.0	3.00
F^+ center	6.3, 5.4, 4.8	3.80
F_2	4.1	2.40
F_2^+	3.5	3.26
F_2^{2+}	2.7	2.22

Table 5. Experimental measurements of optical properties for single-vacancy and for dimer centers in Al_2O_3, energies are in eV (Crawford, 1984; Evans et al., 1994).

The band of F-centers is absent in both samples due to the existence of impurity Cr (Aoki, 1996). For instance, the CL intensity at 3.81 eV (F^+-center) of the α-Al_2O_3 formed from gibbsite is approximately 5 times higher than that of the α-Al_2O_3 formed from PM2000 measured under the same excitation conditions. The main specific feature of the CL spectra of the α-Al_2O_3 is the presence of a new emission band with the maximum at 3.32 eV. It is a possible that the new emission band is related to surface F_s-centers concentrating on nanoparticle boundaries (Evans, 1995).

Luminescence spectra obtained for anion-deficient aluminum corundum exposed to different types of excitation and stimulation exhibit luminescence with nanosecond (F^+-centers), microsecond (Ti^{3+} and $Al^+{}_i$) and millisecond (F and Cr^{3+}) decay times (Surdo et al., 2005). Significantly, the aforementioned centers, which actively participate in relaxation processes, have considerably different decay times t and emission band maxima $h\nu$ (Table 6).

Parameter	F^+	F	Al_i^+	Cr^{3+}	Ti^{3+}
$h\nu$ (eV)	3.8	3.0	2.4	1.79	1.75
τ	2 ns	34 ms	56 μs	4 ms	3.5 μs

Table 6. Basic parameters of the emission of most active centers (Surdo et al., 2001; Springis & Valbis, 1984)

The band 487 nm (2.54 eV) can be related to the aggregate F_2-centers produced by double-oxygen vacancies and the centers formed by interstitial aluminum ions. It is known that these centers are responsible for the green luminescence in highly disordered crystals of aluminum oxide (Tale et al., 1996; Springis & Valbis, 1984).

3.3.2 Cr^{3+} in Al_2O_3

A classic example of the isolated luminescent centre is Cr^{3+} in Al_2O_3 (ruby) when the excited electronic energy levels of the host are at much higher energy than those of the dopant ion. The dopant ion colours the colourless host lattice red. If the concentration of the dopant ion is low, the interaction between the dopant ions can be neglected. This is what we consider here as an isolated luminescent centre.

Different impurities in corundum (α-Al_2O_3) produce different color varieties. All colors of corundum are referred to as sapphire, except for the red color, which is known as ruby. Corundum has a trigonal lattice $D^6{}_{3d}$ structure. The crystals have an approximately hexagonal closed packing structure of oxygen and metal atoms. The six oxygen ions are octahedrally coordinated cations; and only two-thirds of the octahedral sites are filled. If corundum has more than 1000 ppm Cr^{3+} ions as impurities, it is referred to as a ruby. Rubies can be used in solid-state lasers (Soukieh et al., 2004), and they fetch high prices in gem markets. Chromium can be substituted for the aluminum in corundum and is present as chromium oxide. The Cr^{3+} ion is slightly larger than Al^{3+}; therefore, it naturally enters easily into the corundum structure. As a result, Cr^{3+} ions form $3d^3$, with only three unpaired electrons in the 3d orbitals. If the Cr^{3+} ion is located in the Al^{3+} site in corundum, it coordinates the six oxygens into a distorted octahedral configuration (Nassau, 1983).

According to the ligand field theory (Figgis et al., 2000), splitting of the $3d^3$ (Cr^{3+}) orbital should result in the spectroscopic terms $^4A_2$ (*A*: no degeneracy) , $^4T_2$, $^4T_1$ (*T*: three fold degeneracy), and ^2E (*E*: two fold degeneracy).

For Cr^{3+} in Al_2O_3 crystal, Cr^{3+} substitutes for some of Al^{3+}, and adopts octahedral ligand coordination. The 3d levels are extremely host sensitive. The strong crystal field in Al_2O_3 leads to the splitting of 3d electron orbits of Cr^{3+} and produces the ground level: $^4A_2$, and the excited states: ^2E, $^4T_2$, and $^4T_1$, etc. the transitions from $^4A_2$ to $^4T_2$, and $^4T_1$ are spin-allowed, so these energy levels act as broad pumping levels. The ^2E is the narrow lowest excited band, acting as emitting level. The unusual magnitude of this crystal field splitting extends the lowest ^2E state 14400 cm^{-1} (694 nm) above the ground state. Thus the ^2E-$^4A_2$ transition of Cr^{3+}: Al_2O_3 crystal lies in visible spectral region. Exciting any of the pumping bands of $^4T_2$, and $^4T_1$ results in fast relaxation to lowest ^2E excited state. At room temperature, the fluorescence emitting from ^2E state appears as a sharp band with a peak at 694 nm corresponding to the transition to the ^2E terminal state. The Cr^{3+} ion has two strong absorption bands in the visible part of the spectrum, which explain the red color, i.e., 2.2 eV light can be absorbed to raise the chromium from the $^4A_2$ ground level to the $^4T_2$ excited level as absorption in the yellow-green, and 3.0 eV light raises it to the $^4T_1$ level as violet absorption. In addition, the absorption decreases to zero in the red region below 2.0 eV. Therefore, rubies have a red color with a slight purple overtone.

Chromium impurity in α-Al_2O_3 lattice is characterised by two bands of absorption (3.1 and 2.2 eV) and one fine emission structure peaked at 1.8 eV (693 nm) as summarised in Fig. 11.

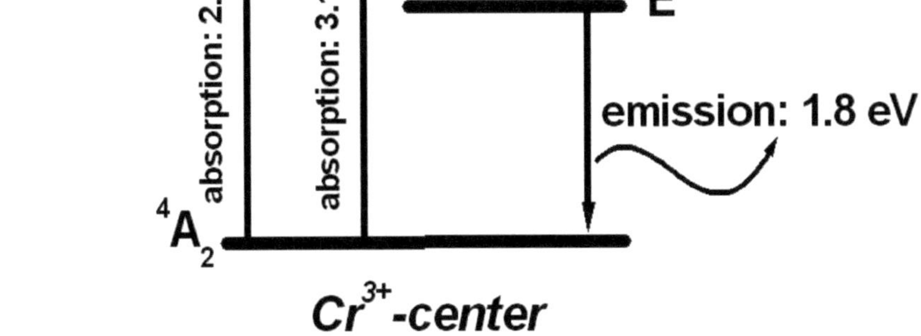

Fig. 11. Schematic energy level diagram for absorption and emission of Cr^{3+}-center in α-Al_2O_3 crystal.

In oxide insulators, a number of transition metal and rare earth impurities act as luminescence centers. The trivalent chromium ion (Cr^{3+}), with electronic configuration $3d^3$, is an efficient luminescence center in many light-metal oxides, including Al_2O_3 and MgO. The trivalent chromium ion enters substitutionally and is surrounded by an octahedron of oxygen ions. In aluminum oxide, the surroundings of the chromium ion are not quite cubic, as the oxygen octahedron is stretched along its trigonal symmetry axis C_3.

Chromium in α-Al_2O_3 lattice gives a luminescence in the visible domain. In CL, the narrow band at 693 nm is attributed to chromium impurity (Ghamnia et al., 2003). In Al_2O_3:Cr^{3+} (ruby) the apparent lifetime of the R-line emission may increase from the intrinsic value of 3.8 ms up to 12 ms (Auzel & Baldacchini, 2007).

The typical CL spectra of α-Al_2O_3, formed by calcination of gibbsite powder at 1573 K for 24 h (black line) and by oxidation of PM2000 at 1223K for 72h (red line), obtained in the region of 600 to 800 nm at room temperature are shown in Fig. 12. The sharp band at 693 nm (1.79 eV), with a radiative lifetime τ_R~4 ms (de Wijn, 2007), as well as features at 706 nm (1.76 eV) and 713 nm (1.74 eV) undoubtedly belongs to Cr^{3+} emission in α-alumina, and the subband at 677 nm (1.83 eV) is attributed to the ^2E–$^4A_2$ transition of Mn^{4+} ions in Al_2O_3 (Jovanic, 1997; Geschwind et al., 1962; Crozier, 1965).

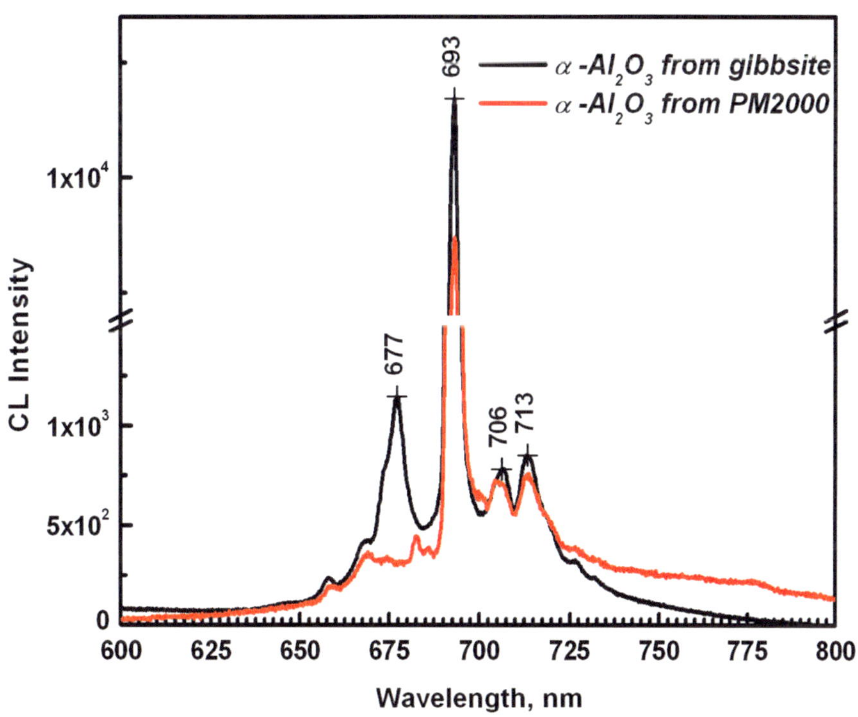

Fig. 12. CL spectra of α-Al_2O_3, formed by calcination of gibbsite powder at 1573 K for 24 h (black line) and by oxidation of PM2000 at 1223K for 72h (red line), obtained in the region of 600 to 800 nm at room temperature.

Mn^{4+} is known to emit doublet lines at 672 and 676 nm in α-Al_2O_3 (Kulinkin et al., 2000).

A similar emission was recently reported in α-Al_2O_3 microcones (Li et al., 2010). Thus, it can be concluded that the incorporation of Mn^{4+} ions in α-Al_2O_3 observed in the experiment is irreversible and occurs during its formation from the Mn^{4+} ions dissolved quite uniformly in the bulk of the low-temperature polymorphic modifications of alumina (gibbsite →χ-→κ-→α-Al_2O_3). The Mn^{4+} impurity emission at 677 nm is absent in α-alumina films obtained by oxidation of a metallic FeCrAl alloy. Taking into account high sensitivity of the method, this indicates very low concentration of such ions.

4. Conclusion

α–Al_2O_3 was prepared either by calcination of gibbsite and also generated by oxidation of a metallic FeCrAl alloy. The Mn^{4+} impurity emission at 1.83 eV is absent in α-alumina thin films obtained by oxidation of a metallic FeCrAl alloy. The band of F-centers is absent in both samples due to the existence of impurity Cr. The difference in oxygen vacancies (F^+-centers) amount between α–Al_2O_3 from gibbsite and from PM2000 was confirmed by CL spectra.

5. References

Aoki, Y.; My, N. T.; Yamamoto, S. & Naramoto, H. (1996). Luminescence of sapphire and ruby induced by He and Ar ion irradiation. *Nuclear Instruments and Methods in Physics Research Section B*, Vol.114, No.3-4, (July 1996), pp. 276-280, ISSN 0168-583X

Auzel, F. & Baldacchini, G. (2007). Photon trapping in ruby and lanthanide-doped materials: Recollections and revival. *Journal of Luminescence*, Vol.125, No.1-2, (July-August 2007), pp. 25-30, ISSN 0022-2313

Barker, Jr, A.S. (1963). Infrared lattice vibrations and dielectric dispersion in corundum. *Physical Review*, Vol.132, No.4, (November 1963), pp. 1474–1481, ISSN 0031-899X

Boumaza, A. & Djelloul, A. (2010). Estimation of the intrinsic stresses in a-alumina in relation with its elaboration mode. *Journal of Solid State Chemistry*, Vol.183, No.5, (May 2010), pp. 1063–1070, ISSN 0022-4596

Boumaza, A.; Djelloul, A. & Guerrab, F. (2010). Specific signatures of α-alumina powders prepared by calcination of boehmite or gibbsite. *Powder Technology*, Vol.201, No.2, (July 2010), pp. 177–180, ISSN 0032-5910

Brewer, J.D.; Jeffries, B.T. & Summers, G.P. (1980). Low-temperature fluorescence in sapphire. *Physical Review B condensed matter and Materials Physics* Vol.22, No.10, (November 1980), pp. 4900–4906, ISSN 1098-0121

Brindley, G.W. & Choe, J.O. (1961). Reaction series gibbsite→chi alumina→kappa alumina→corundum. *American Mineralogist*, Vol.46, No.7-8, (July-August 1961), pp. 771-785, ISSN 0003-004X

Crawford Jr., J.H. (1984). Defects and defect processes in ionic oxides: Where do we stand today?. *Nuclear Instruments and Methods in Physics Research Section B*, Vol.1, No.2-3, (February 1984), pp. 159-165, ISSN 0168-583X

Crozier, M.H. (1962). Optical Zeeman effect in the R_1 and R_2 lines of Mn^{4+} in Al_2O_3. *Physics Letters*, Vol.18, No.3, (September 1965), pp. 219-220, ISSN 0031-9163

de Wijn, H.W. (2007). Phonon physics in ruby studied by optical pumping and luminescence. *Journal of Luminescence*, Vol.125, No.1-2, (July-August 2007), pp. 55-59, ISSN 0022-2313

Evans, B.D.; Pogatshnik, G.J. & Chen, Y. (1994). Optical properties of lattice defects in α-Al_2O_3. *Nuclear Instruments and Methods in Physics Research Section B*, Vol.91, No.1-4, (June 1994), pp. 258-262, ISSN 0168-583X

Evans, B.D. (1994). Ubiquitous blue luminescence from undoped synthetic sapphires. *Journal of Luminescence*. Vol.60-61, (April 1994), pp. 620-626, ISSN 0022-2313

Evans, B.D. (1995). A review of the optical properties of anion lattice vacancies, and electrical conduction in α-Al_2O_3: their relation to radiation-induced electrical degradation. *Journal of Nuclear Materials*, Vol.219, (March 1995), pp. 202-223, ISSN 0022-3115

Figgis, B.N. & Hitchman, M.A. (2000). *Ligand Field Theory and its Applications*, John Wiley & Sons Inc, ISBN 978-0-471-31776-0, New York

Geschwind, S.; Kisliuk, P.; Klein, M.P.; Remeika, J.P. & Wood, D.L. (1962). Sharp-Line Fluorescence, Electron Paramagnetic Resonance, and Thermoluminescence of Mn^{4+} in *a*-Al_2O_3. *Physical Review*, Vol.126, No.5, (June 1962), pp. 1684-1686, ISSN 0031-899X

Ghamdi, A. Al. & Townsend, P.D. (1990). Ion beam excited luminescence of sapphire. *Nuclear Instruments and Methods in Physics Research B*, Vol.46, No.1-4, (February 1990), pp. 133-136, ISSN 0168-583X

Ghamnia, M.; Jardin, C. & Bouslama, M. (2003). Luminescent centres F and F^+ in α-alumina detected by cathodoluminescence technique. *Journal of Electron Spectroscopy and Related Phenomena*, Vol.133, No.1-3, (November 2003), pp. 55-63, ISSN 0368-2048

Golightly, F.A.; Stott, F.H. & Wood, G.C. (1979). The Relationship Between Oxide Grain Morphology and Growth Mechanisms for Fe-Cr-Al and Fe-Cr-Al-Y Alloys. *Journal of The Electrochemical Society*, Vol.126, No.6, (June 1979), pp. 1035-1042, ISSN 0013-4651

Hou, P.Y.; Zhang, X.F. & Cannon, R.M. (2004). Impurity distribution in Al_2O_3 formed on an FeCrAl alloy. *Scripta Materialia*, Vol.50, No.1, (January 2004) pp. 45–49, ISSN 1359-6462

http://accuratus.com/alumox.html/ Aluminum Oxide, Al_2O_3 Material Characteristics

Jheeta, K.S.; Jain, D.C.; Fouran Singh; Ravi Kumar & Garg, K.B. (2006). Photoluminescence and UV–vis studies of pre- and post-irradiated sapphire with 200 MeV Ag^{8+} ions. *Nuclear Instruments and Methods in Physics Research B*, Vol.244, No.1, (March 2006), pp. 187–189, ISSN 0168-583X(02)

Jovanic, B.R. (1997). Shift under pressure of the luminescence transitions of corundum doped with Mn^{4+}. *Journal of Luminescence*, Vol.75, No.2, (September 1997), pp. 171-174, ISSN 0022-2313

Kislov, A.N.; Mazurenko, V.G.; Korzov, K.N. & Kortov, V.S. (2004). Interionic potentials and localized vibrations in Al_2O_3 crystals with vacancies. *Physica B: Condensed Matter*, Vol.352, No.1-4, (October 2004), pp. 172–178, ISSN 0921-4526

Kogure, T. (1999). Dehydration sequence of gibbsite by electron beam irradiation in a TEM. *Journal of the American Ceramic Society*, Vol. 82, No.3, (March 1999), pp. 716-720, ISSN 0002-7820

Kortov, V.S.; Ermakov, A.E.; Zatsepin, A.F. & Nikiforov, S.V. (2008). Luminescence properties of nanostructured alumina ceramic. *Radiation Measurements*, Vol.43, No.2-6, (February-June 2008), pp. 341-344, ISSN 1350-4487

Kortov, V. & Milman, I. (1996). Some New Data on Thermoluminescence Properties of Dosimetric Alpha-Al_20_3 Crystals. *Radiation Protection Dosimetry*, Vol.65, No.1-4, (June 1996), pp. 179-184, ISSN 0144-8420

Kulinkin, A. B.; Feofilov, S. P. & Zakharchenya, R. I. (2000). Luminescence of impurity 3*d* and 4*f* metal ions in different crystalline forms of Al_2O_3. *Physics of the Solid State*, Vol.42, No.5, (May 2000), pp. 857-860, ISSN 1063-7834

La, S.; Bartram, R.H. & Cox, R.T. (1973). The F^+ center in reactor-irradiated aluminum oxide . *Journal of Physics and Chemistry of Solids*, Vol.34, No.6, (June 1973), pp. 1079-1086, ISSN 0022-3697

Li, P.G.; Lei, M. & Tang, W.H. (2010). Raman and photoluminescence properties of α-Al_2O_3 microcones with hierarchical and repetitive superstructure. *Materials Letters*, Vol.64, No.2, (January 2010), pp. 161-163, ISSN 0167-577X

Ma, C.; Chang, Y.; Ye, W.; Shang, W. & Wang, C. (2008). Supercritical preparation of hexagonal γ-alumina nanosheets and its electrocatalytic properties. *Journal of Colloid and Interface Science*, Vol.317, No.1, (January 2008), pp. 148–154, ISSN 0021-9797

MacKenzie, K.J.D.; Temuujin, J. & Okada, K. (1999). Thermal decomposition of mechanically activated gibbsite. *Thermochimica Acta*, Vol.327, No.1-2, (March 1999), pp. 103–108, ISSN 0040-6031

Mennicke, C.; Schumann, E.; Ruhle, M.; Hussey, R.J.; Sproule, G.I. & Graham, M.J. (1998). The Effect of Yttrium on the Growth Process and Microstructure of α-Al_2O_3 on FeCrAl. *Oxidation of metals*, Vol.49, No.5-6, (June 1998), pp. 455-466, ISSN 0030-770X

Michizono, S.; Saito, Y.; Suharyanto; Yamano, Y. & Kobayashi, S. (2007). Surface characteristics and electrical breakdown of alumina materials. *Vacuum*, Vol.81, No.6, (February 2007), pp. 762–765, ISSN 0042-207X

Monteiro, T.; Boemare, C.; Soares, M.J.; Alves, E.; Marques, C.; McHargue, C.; Ononye, L.C. & Allard, L.F. (2002). Luminescence and structural studies of iron implanted α-Al_2O_3. *Nuclear Instruments and Methods in Physics Research Section B: Beam Interactions with Materials and Atoms*, Vol.191, No.1-4, (May 2002), pp. 638-643, ISSN 0168-583(2)

Nassau, K. (1983). *The Physics and Chemistry of Color: The Fifteen Causes of Color,* John Wiley & Sons Inc, ISBN 0471867764, NewYork

Ollivier, B.; Retoux, R.; Lacorre, P.; Massiot, D. & Férey, G. (1997). Crystal Structure of Kappa Alumina an X-Ray powder diffraction, TEM and NMR study. *Journal of Materials Chemistry*, Vol.7, No.6, (1997), pp. 1049- 1056, ISSN 0959-9428

Oster, L.; Weise, D. & Kristiapoller, N. (1994). A study of photostimulated thermoluminescence in C-doped alpha -Al_2O_3 crystals. *Journal of Physics D: Applied Physics*, Vol.27, No.8, (August 1994), pp. 1732-1736, ISSN 0022-3727

Przybylski, K.; Garrett-Reed, A.J.; Pint, B.A.; Katz, E.P. & Yurek, G.J. (1987). Segregation of Y to Grain Boundaries in the Al_2O_3 Scale Formed on an ODS Alloy. *Journal of The Electrochemical Society*, Vol.134, No.12, (December 1987), pp. 3207-3208, ISSN 0013-4651

Quadakkers, W.J.; Elschner, A.; Speier, W. & Nickel, H. (1991). Composition and growth mechanisms of alumina scales on FeCrAl-based alloys determined by SNMS. *Applied Surface Science*, Vol.52, No.4, (December 1991), pp. 271-287, ISSN 0169-4332

Rodríguez-Carvajal, J. (1993). Recent Advances in Magnetic Structure Determination by Neutron Powder Diffraction. *Physica B Condensed Matter*, Vol. 192, No. 1-2, (October 1993), pp. 55–69, ISSN 0921-4526

Saalfeld, H. & Wedde, M. (1974). Refinement of the structure of gibbsite, Al(OH)3. *Zeitschrift für Kristallographie*, Vol.139, No.1-2, (April 1974), pp. 129–135, ISSN 0044-2968

Saalfeld, H.N. (1960). Strukturen des Hydrargillitis und der Zwis- chenstufen biem Entwassern. *Neues Jahrbuch für Mineralogie,* Vol.95, No.7-8, (1960), pp. 1–87, ISSN 0077-7757

Soukieh, M.; Ghani, B.A. & Hammadi, M. (2004). Numerical calculations of intracavity dye Q-switched ruby laser. *Optics and Lasers in Engineering*, Vol.41, No.1, (January 2004), pp. 177-187, ISSN 0143-8166

Springis, M.J. & Valbis, J.A. (1984). Visible luminescence of colour centres in sapphire. *Physica status solidi. B*, Vol.123, No.1, (May 1984), pp. 335 343, ISSN 0370-1972

Stumpf, H.C.; Russel, A.S.; Newsome, J.W. & Tucker, C.M. (1950). Thermal transformations aluminas and alumina hydrates - Reaction with 44% Technical Acid, *Industrial & Engineering Chemistry Research*, Vol.42, No.7, (July 1950), pp. 1398-1403, ISSN 0888-5885

Surdo, A.I.; Kortov, V.S. & Pustovarov, V.A (2001). Luminescence of F and F^+ centers in corundum upon excitation in the interval from 4 to 40 eV. *Radiation Measurements*, Vol.33, No.5, (October 2001), pp. 587-591, ISSN 1350-4487

Surdo, A.I.; Pustovarov, V.A.; Kortov, V.S.; Kishka, A.S. & Zinin, E.I. (2005). Luminescence in anion-defective α-Al_2O_3 crystals over the nano-, micro- and millisecond intervals. *Nuclear Instruments and Methods in Physics Research A*, Vol.543, No.1, (May 2005), pp. 234–238, ISSN 0168-9002

Tale, I.; Piters, T.M.; Barboza-Flores, M.; Perez-Salas, R.; Aceves, R. & Springis, M. (1996). Optical Properties of Complex Anion Vacancy Centres and Photo-Excited Electronic Processes in Anion Defective Alpha-Al_2O_3. *Radiation Protection Dosimetry*, Vol.65, No.1-4, (June 1996), pp. 235-238, ISSN 0144-8420

Whittington, B. & Ilievski, D. (2004). Determination of the gibbsite dehydration reaction pathway at conditions relevant to Bayer refineries. *Chemical Engineering Journal*, Vol.98, No.1-2, (March 2004), pp. 89-97, ISSN 1385-8947

Wood, G. C. (1970). High-temperature oxidation of alloys. *Oxidation of Metals*, Vol.2, No.1, (March 1970), pp. 11-57, ISSN 0030-770X

Yu, Z.Q.; Li, C. & Zhang, N. (2002). Size dependence of the luminescence spectra of nanocrystal alumina. *Journal of Luminescence*. Vol.99, No.1, (August 2002), pp. 29-34, ISSN 0022-2313

9

Cathodoluminescence Properties of SiO_2: Ce^{3+},Tb^{3+}, SiO_2:Ce^{3+}, Pr^{3+} and SiO_2: PbS

Odireleng M. Ntwaeaborwa[1], Gugu H. Mhlongo[1,2], Shreyas S. Pitale[1], Mokhotswa S. Dhlamini[1,3], Robin E. Kroon[1] and Hendrik C. Swart[1]
[1]*Department of Physics, University of the Free State, Bloemfontein*
[2]*National Centre for Nanostructured Materials, CSIR, Pretoria*
[3]*Department of Physics, University of South Africa, Pretoria*
South Africa

1. Introduction

Silicon dioxide (SiO_2), also known as silica, is an oxide of silicon (Si) that is found in nature in two different forms, namely amorphous and crystalline. Traditionally, amorphous SiO_2 is used in many applications such as semiconductor circuits, microelectronics and optical fibers for telecommunication. In modern age research, amorphous glassy SiO_2 has emerged as a potential host lattice for a variety of rare-earth ions to prepare light emitting materials or phosphors that can be used in different types of light emitting devices. SiO_2 based phosphors can also be prepared by encapsulating semiconducting nanocrystals of zinc oxide (ZnO) and lead sulphide (PbS). In addition, recent studies have demonstrated that when semiconducting nanocrystals are incorporated in glassy SiO_2 activated with trivalent rare earth ions (Ce^{3+},Tb^{3+},Eu^{3+},Pr^{3+}) light emission from the rare-earth luminescent centres could be increased considerably as a result of energy transfer from the nanocrystals to the rare-earths, i.e. semiconducting nanocrystals act as sensitizers for radiative relaxation processes in these centres.

Like any other oxide, amorphous SiO_2 is expected to be a good host matrix for rare-earth ions for preparation of phosphors because oxides are more chemically stable than traditional sulphide matrices. In addition, SiO_2 has been reported to have advantages such as high transparency, dopant solubility and ease of production (Nogami et al., 2007; Ntwaeaborwa et al., 2007,2008). In recent studies, the preparation of phosphors by incorporation of semiconducting nanocrystals and rare-earths ions in amorphous SiO_2 has commonly been achieved by using a sol-gel method. This method has been reported to have more advantages over other wet chemistry and conventional glass processing methods because of its potential to produce materials with high purity and homogeneity at low temperatures (Hench & West, 1990; Ding, 1991). It extends the traditional glass melting processes by allowing incorporation of semiconductor nanocrystals and rare-earth activators at low temperatures and predetermined concentrations in such a way that the size and shape of the particles can be controlled (Reisfeld, 2000) during nucleation and growth processes. The sol-gel method was used in this study to prepare respectively red, green and blue cathodoluminescent phosphors by incorporating trivalent rare-earth ions such as prasedomium (Pr^{3+}), terbium (Tb^{3+}) and cerium (Ce^{3+}) in amorphous SiO_2. Orange-red cathodoluminescence was also observed when PbS nanocrystals were incorporated in the sol-gel derived SiO_2. Cathodoluminescence intensity of these phosphors was shown to increase either by co-doping with a different rare-earth ion or encapsulating ZnO nanocrystals. The CL properties and the intensity degradation of these SiO_2 based phosphors were evaluated for their possible application in low voltage cathodoluminescent flat panel displays such as field emission displays (FEDs) and plasma display panels (PDPs).

This chapter is a review of cathodoluminescent properties of phosphor materials prepared by incorporating Ce^{3+}, Tb^{3+}, Pr^{3+} and PbS nanocrystal in amorphous SiO_2 host. The effects of rare-eaths co-doping and encapsulating of ZnO nanocrystals on cathodoluminescent intensity of these materials is also presented. In addition, a review of cathodoluminescence intensity degradation when these phosphors were irradiated with a beam of electron is discussed. It was demonstrated that when these phosphors were irradiated with the beam of electrons for a long period of time they lose their CL intensity and this occured simultaneously with desorption of oxygen (O) from the phosphor surfaces. In the process, an oxygen deficient non-luminescent layer was formed on the surface whose formation could be explained by an electron stimulated surface chemical reaction (ESSCR) model proposed by Holloway and co-workers (Holloway et al., 1996, 2000). The desorption of atomic species was explained by Knotek-Fiebelman electron stimulated desorption (ESD) proposed by Knotek and Feibelman (Knotek and Feibelman, 1978). The objective of the study was to investigate the effects of a prolonged electron beam irradiation, accelerating voltage, and oxygen gas on the CL properties and the chemical stability of these phosphors.

2. Cathodoluminescence intensity degradation

Cathodoluminescence (CL) degradation can be defined as a process by which cathodoluminescent phosphors, used mainly in cathode ray tubes (CRTs) or field emission display (FEDs), lose their brightness (luminescence intensity) as a result of prolonged irradiation by a beam of electrons during normal operation. The CL intensity degradation of CRT/FED phosphors, with special reference to traditional sulphide based phosphors, is a well documented phenomenon. When a beam of electrons is incident on a phosphor, a number of physical processes can occur. These include emission of secondary electrons, Auger electrons and back-scattered electrons. Hundreds of free electrons and free holes are produced along the path of the incident electron (primary excitation electrons). As illustrated in figure 1, the free electrons and holes may couple and produce electron-hole (e-h) pairs that can diffuse through the phosphor and transfer their energy to luminescent centres resulting in emission of visible photons (Stoffers, 1996; Raue, 1989). This process is referred to as radiative recombination. Unwanted processes in which the e-h pairs recombine non-radiatively by transferring their energy to killer centers (incidental impurities and/or inherent lattice defects) are also possible. The e-h pairs can also diffuse to the surface of the phosphor and recombine non-radiatively (Stoffers, 1996). In this case, a non-luminescent oxide layer, which is known to reduce the CL intensity of the CRT/FED phosphors, may form on the surface. For examples, it was demonstrated that when zinc sulfide (ZnS) based phosphors were exposed to a prolonged irradiation by energetic beam of electrons, the ZnS host dissociated into reactive ionic Zn^{2+} and S^{2-} species, which in turn combined with ambient vacuum gases such as O_2 and H_2O to form non-luminescent ZnO or

$ZnSO_4$ layers or H_2S gas (Swart el al., 1998; Itoh et al., 1989) as explained by the ESSCR model (Holloway et al., 1996, 2000). In the case of oxide based systems, the electron beam induced dissociation of atomic species is followed by desorption of oxygen from the surface. For example, Knotek and Fiebelman observed desorption of surface oxygen when SiO_2 or TiO_2 was irradiated with a beam of electrons and they developed a model called Knotek-Fiebelman electron stimulated desorption (ESD) mechanism to explain the desorption process. According to this model, O^+ species produced in the valence band of SiO_2 (TiO_2) desorb from the surface following the creation of an electron-hole pair in the $L_{2,3}$ level of Si, Auger relaxation and emission (Knotek and Feibelman, 1978).

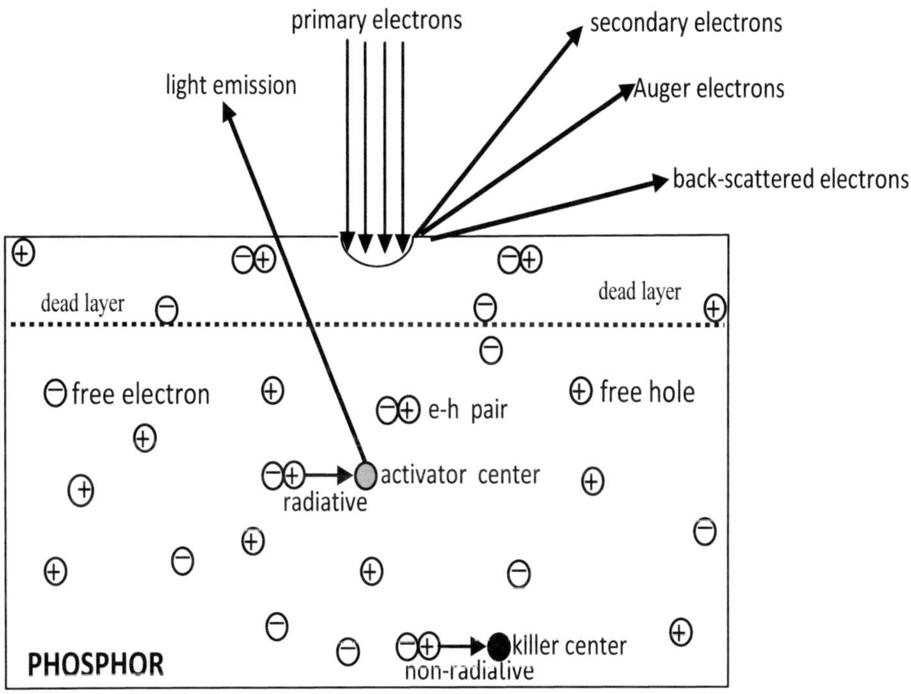

Fig. 1. Cathodoluminescence process in a phosphor grain.

3. Experiments

A sol-gel process was used to prepare rare-earths (Ce^{3+},Tb^{3+},Pr^{3+}) doped and semiconducting nanocrystals (ZnO and PbS) encapsulated SiO_2 based phosphors. The sol-gel is a wet chemical technique commonly used to prepare glassy and ceramic materials at low temperatures. It involves the preparation of a viscous solution (sol) from one or more precursor materials and the conversion of the sol into a gel. There are two major chemical reactions integral to the sol-gel process, namely hydrolysis (reaction of metal alkoxides with water to form oxides) and condensation/polymerization (the formation of a three dimensional network). In the presence of a catalyst, the conversion of the sol to gel can be completed in a short time (within ~1-2 hrs). For more information on the sol-gel process, the reader is referred to the literature cited (Hench & West, 1990; Klein, 1985; Huang et al., 1985).

3.1 Preparation of PbS and ZnO nanocrystals

Lead sulphide (PbS) nanocrystals were prepared by dissolving anhydrous lead acetate ($Pb(CH_3COO)_2$) in boiling ethanol (C_2H_5OH) and the resulting solution was cooled and then combined with the ethanol solution of sodium sulphide (Na_2S) in ice water. The PbS nanocrystals were formed during hydrolysis and condensation of dissolved species according to the following reaction equation:

$$Pb(CH_3COO)_2 + Na_2S \rightarrow PbS + 2CH_3COO^- + 2Na^+$$

Similarly, ZnO nanocrystals were prepared by dissolving anhydrous zinc acetate ($Zn(CH3COO)_2$) in boiling ethanol and the resulting solution was cooled and then combined with the ethanol solution of sodium hydroxide (NaOH) in ice water. The ZnO nanocrystals were formed during hydrolysis and condensation of dissolved species according to the following reaction equation:

$$Zn(CH_3COO)_2 + 2NaOH \rightarrow ZnO + 2CH_3COO^- + 2Na^+ + H_2O(\uparrow)$$

The unwanted CH_3COO^- and Na^+ impurity ions were removed by centrifuging repeatedly in a mixture of ethanol and heptane. The resulting PbS and ZnO precipitates were either dispersed/suspended in ethanol for mixing with SiO_2 or were dried at 90°C in an oven for characterization.

3.2 Preparation of rare-earths, PbS and ZnO nanocrystals incorporated SiO_2

A silica (SiO_2) sol was prepared by hydrolyzing tetraethylorthosilicate ($Si(OC_2H_5)_4$) or TEOS with a solution of water, ethanol and dilute nitric acid (HNO_3). The mixture was stirred vigorously at room temperature. The SiO_2 was formed during the condensation reaction according to the following equation:

$$Si(OC_2H_5)_4 + 2H_2O \rightarrow SiO_2 + 4C_2H_5OH$$

The SiO_2 sol was divided into two parts and one part was mixed with the ethanol suspension of PbS nanocrystals and the other part was mixed with $RE(NO_3) \cdot 6H_2O$ (RE = Ce/Tb/Pr) dissolved in ethanol before adding the ethanol suspension of ZnO nanocrystals and the mixtures were stirred vigorously until thick viscous gels were formed. The gels were dried at room temperature for 3 – 8 days, were ground using a pestle and mortar, and were finally annealed at 600°C in air for 2 hours. Samples that were prepared include SiO_2; $SiO_2:Pr^{3+}/Ce^{3+}/Tb^{3+}$; $SiO_2:Ce^{3+},Tb^{3+}$; $ZnO\text{-}SiO_2:Pr^{3+}$; and SiO_2:PbS powders. The molar concentrations of rare-earth ions and ZnO in SiO_2 were varied while that of the PbS nanocrystals was fixed at 0.34 mol%.

4. Results and discussions

4.1 X-ray diffraction

Figure 2 shows the X-ray diffraction patterns of ZnO and PbS nanocrystals. The patterns are consistent with the hexagonal and cubic phases of ZnO and PbS referenced in JCPDS cards number 36-14551 and 05-0592 respectively. The broadening of the diffraction peaks is attributed to smaller particles. The average crystallite sizes estimated from the broadened

XRD peaks were 5 and 6 nm in diameters for ZnO and PbS respectively. Figure 3 shows XRD patterns of SiO_2 and ZnO-SiO_2:Pr^{3+}, all annealed at 600°C for two hours. The concentrations of ZnO and Pr^{3+} in SiO_2 host were 5 and 1 mol% respectively. All the spectra are characterized by the well known broad diffraction of the amorphous SiO_2 peak at $2\theta = \sim 20 - 25^{o}$. The absence of x-ray diffraction peaks from encapsulated 5 mol% of ZnO nanocrystals and 1 mol% of Pr^{3+} is probably due to their relatively low concentration and/or high scattering background from amorphous SiO_2.

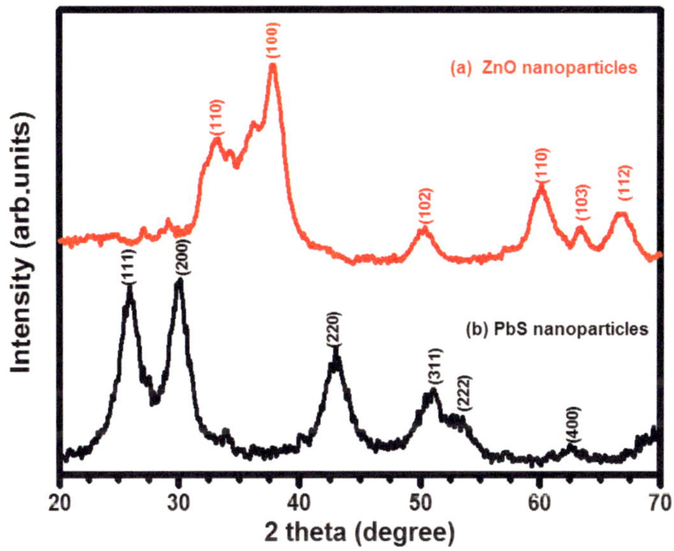

Fig. 2. XRD patterns of (a) ZnO and (b) PbS nanoparticles.

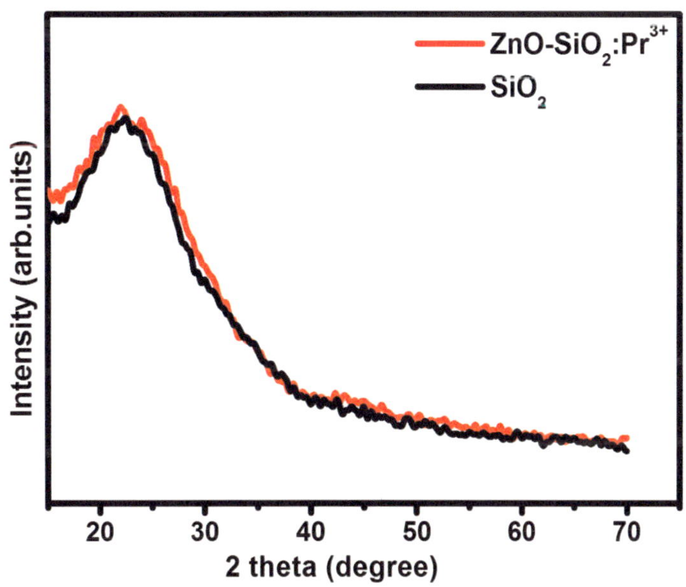

Fig. 3. XRD patterns of SiO_2 and ZnO-SiO_2:Pr^{3+}

4.2 Cathodoluminescence: Properties and intensity degradation

Cathodoluminescence data were recorded using S2000 Ocean Optics CL spectrometer attached to a vacuum chamber of the PHI 549 Auger electron spectrometer (AES) either at base pressure or after backfilling with oxygen gas. Figure 4 presents a simplified schematic diagram of the AES system whose major components are vacuum chamber, housing for the cylindrical mirror analyzer, and electron gun. Attached to the chamber are the fiber optics CL spectrometer and the residual gas analyzer (RGA). The AES and CL spectrometers are connected to two separate computers equipped with programs for recording the Auger and CL data during electron beam irradiation. The Auger and CL data were recorded when the samples were irradiated with a beam of electrons using different accelerating voltages and beam currents.

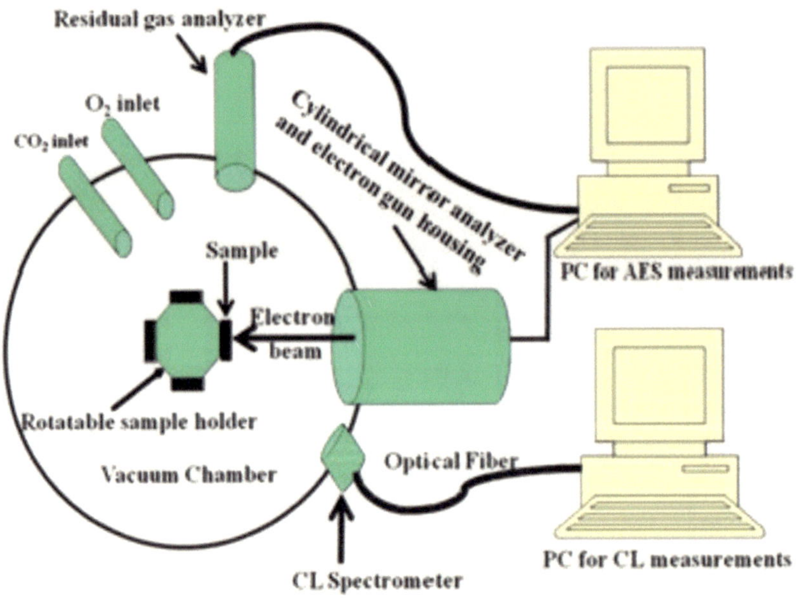

Fig. 4. Schematic diagram of the PHI 549 Auger system.

4.2.1 $SiO_2:Ce^{3+},Tb^{3+}$

$SiO_2:Ce^{3+},Tb^{3+}$ (Ce^{3+} =Tb^{3+}= 0.5 mol%) powders were irradiated with a 54 mA/cm^2 beam of electrons accelerated at 2 kV in a vacuum chamber of the Auger spectrometer maintained at either 1×10^{-7} or 1×10^{-8} Torr O_2 for 10 hours. Figure 5 compares the normalized CL emission spectra of SiO_2 and $SiO_2:Ce^{3+}$ (0.5 mol%). The inset shows the CL spectrum of $SiO_2:Tb^{3+}$ (0.5 mol%). The SiO_2 spectrum is characterized by a broadband blue emission with a maximum at 445 nm and a satellite peak at 490 nm. The visible emission from SiO_2 is attributed to carrier trapping by structural defects (Lin and Baerner., 2000; Han et al., 2002; Gu et al., 1999) or charge transfer between Si and O atoms (García et al., 1995). Upon incorporation of Ce^{3+} and Tb^{3+} the defects emission at 445 nm from SiO_2 was suppressed and blue and green emissions from Ce^{3+} and Tb^{3+} were observed. The blue emission with a maximum at 489 nm can be attributed to 4f→5d transitions of Ce^{3+} while the line emissions with the main emission at ~550 nm in the inset can be attributed to the 5D→ 7F transitions of Tb^{3+}. Figure 6 shows the CL emission spectra of the co-activated SiO_2:Ce^{3+},Tb^{3+} before and after electron irradiation. The insets are the photographs of the irradiated area in the beginning and at the end of irradiation. The emission peaks are associated with radiative transitions of Tb^{3+} ions. The main emission peak was observed at ~550 nm and it is associated with $^5D_4 \rightarrow {}^7F_5$ transitions of the Tb^{3+} ions. The emission intensity of this peak was reduced by ~50% after 10 hours of irradiation as shown by the fading of green luminescence in the inset. The synergies between the CL intensity degradation and the changes in the surface chemistry were determined using the Auger electron and the X-ray photoelectron spectroscopy.

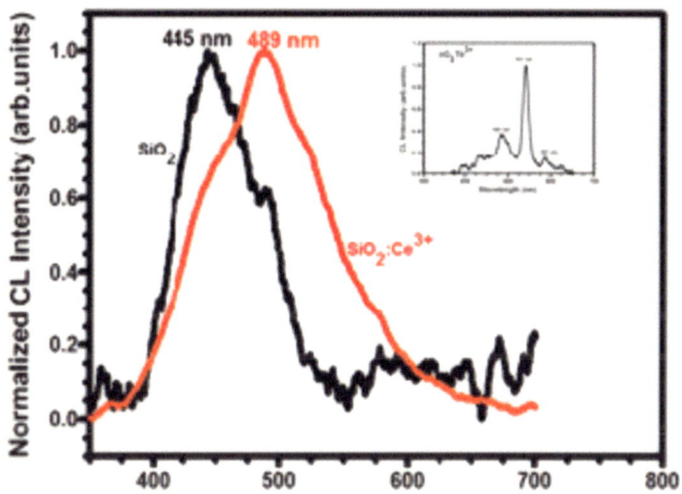

Fig. 5. Normalized CL intensity of SiO_2 and SiO_2:Ce^{3+}. The inset is the CL intensity versus wavelength spectrum of SiO_2:Tb^{3+}.

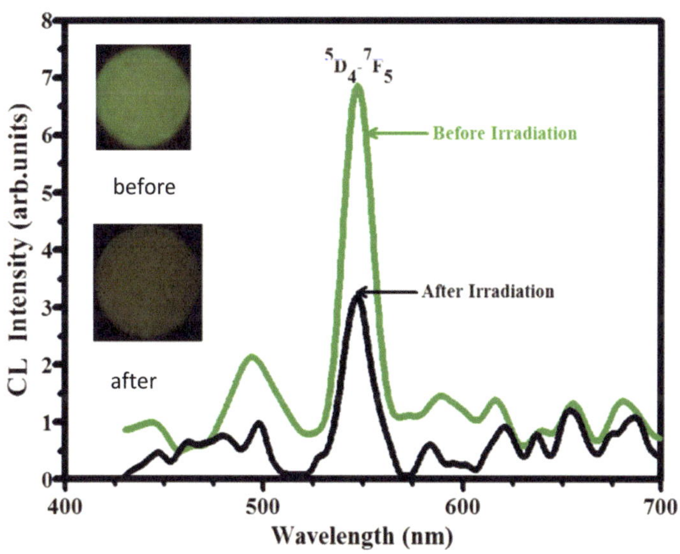

Fig. 6. CL intensity spectra of SiO_2:Ce^{3+},Tb^{3+} before and after electron irradiation.

Figure 7 shows the Auger spectra of the SiO_2 powder before and after electron irradiation. The main features in both spectra are the Si (76.9 eV) and O (505 eV) peaks. Note the decrease in the O Auger peak intensity after 10 hours of continuous irradiation. The Auger peak at 76.9 eV is associated with Si in SiO_2. As a result of the prolonged electron irradiation this peak shifted to 82.7 eV and its intensity was reduced slightly. Thomas (Thomas; 1974) attributed the shift to change in the density of state in the valence band rather than the shift in the binding energies of Si. The XPS data showed that with continuous irradiation an Auger peak associated with elemental Si developed at 98.2 eV and there was also a subsequent change in colour of the chathodoluminescence at the irradiated area.

Figure 8 shows the decrease/degradation of the CL intensity and the Auger-peak-to-peak-heights (APPHs) of oxygen (O), silicon (Si) and adventitious carbon (C) as a function of electron dose for the data recorded when the chamber was maintained at 1×10^{-7} Torr O_2. During electron irradiation, the decrease in the CL intensity was simultaneous with rapid desorption of oxygen from the surface as shown in the figure. Ce^{3+} and Tb^{3+} ions were not detected probably due to their relatively low concentration. While the C peak was almost unchanged, the Si peak was shown to decrease marginally but steadily. Similar trend was observed when the chamber pressure was 1×10^{-8} Torr O_2.

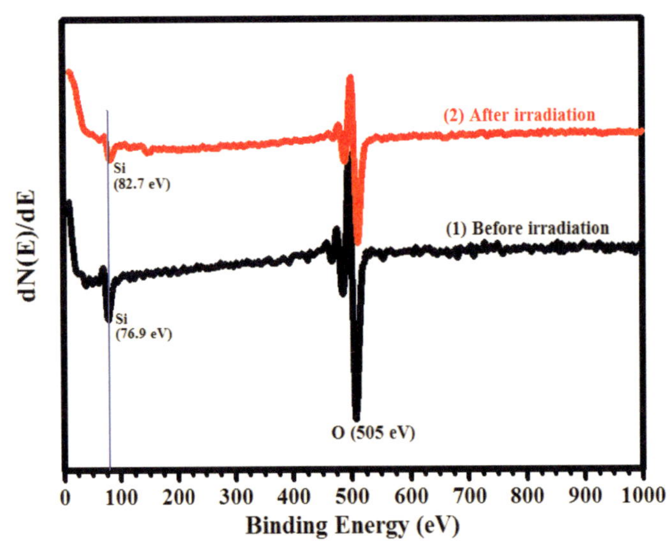

Fig. 7. AES spectra of SiO_2:Ce^{3+},Tb^{3+} before and after electron irradiation.

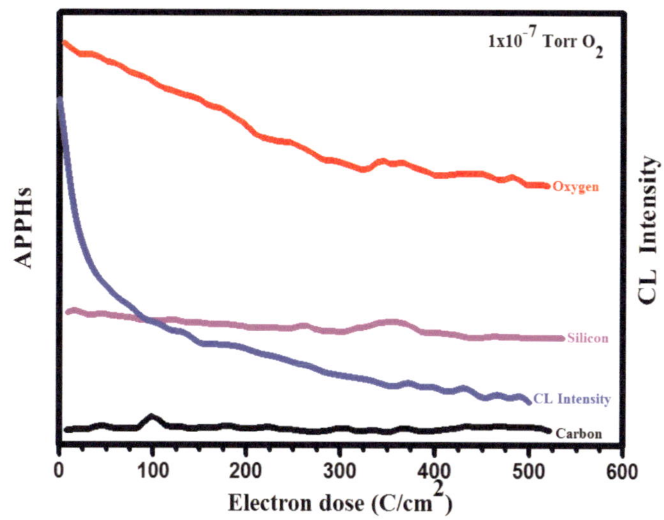

Fig. 8. Auger peak-to-peak-heights of O, Si, C; and the CL intensity as a function of electron dose (Ntwaeaborwa et al., 2006).

Figure 9 and 10 compare respectively the normalized O APPHs and the CL intensity degradation as functions of electron dose at the chamber pressures of 1×10^{-7} and 1×10^{-8} Torr O_2. As shown in figure 9, the rate of O desorption was faster at the low flux (10^{-8} Torr O_2) of O_2 while that of the CL intensity degradation in figure 10 was faster at higher oxygen pressure (10^{-7} Torr O_2). The data in figure 9 and 10 suggest that there is a correlation between O desorption and the CL intensity degradation.

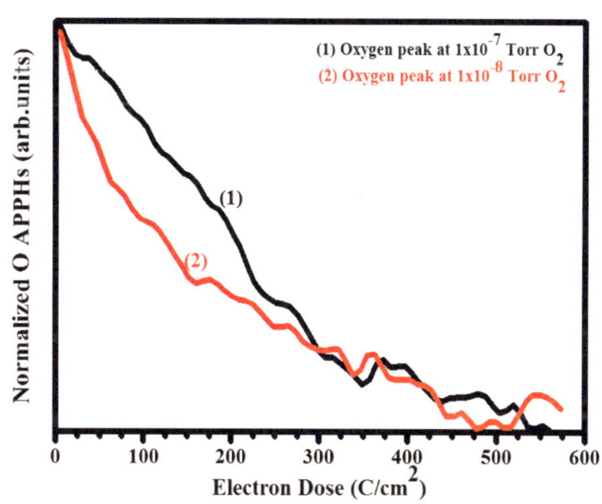

Fig. 9. Normalized O APPHs as a function of electron dose at 1×10^{-7} and 1×10^{-8} Torr O_2 (Ntwaeaborwa et al., 2007).

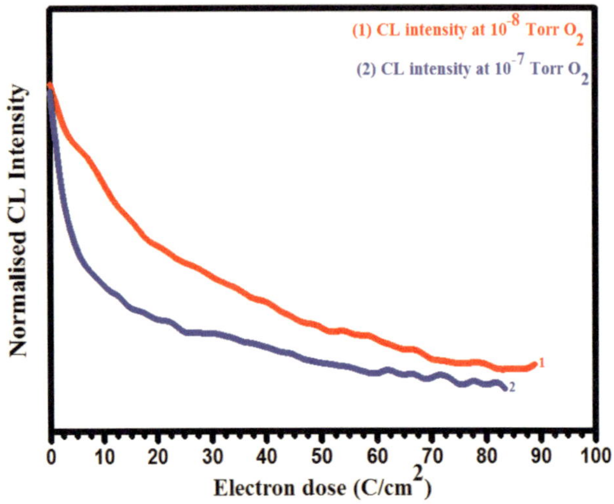

Fig. 10. Normalized CL intensity as a function of electron dose at 1×10^{-7} and 1×10^{-8} Torr O_2 (Ntwaeaborwa et al., 2006).

X-ray photoelectron spectroscopy (XPS) was used to investigate the correlation between the CL intensity degradation and the changes on the surface chemistry. The XPS data were recorded from the $SiO_2:Tb^{3+},Ce^{3+}$ powders before and after electron irradiated. Figure 11 compares the high resolution Si 2p XPS peaks before and after irradiation. Note the shift of 0.2 eV to the right in the peak position, development of the new peak at 98.2 eV and the narrowing of the spectrum after electron irradiation. While the peak shift and narrowing can be attributed to surface charging, among other things, the development of the new peak can be attributed to the chemical changes that occurred on the surface during irradiation.

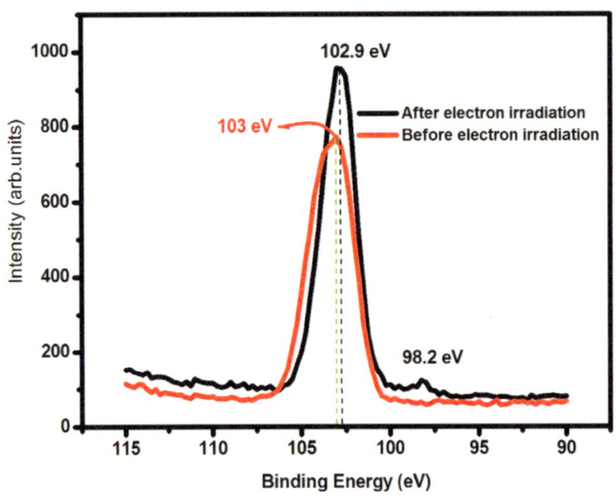

Fig. 11. High resolution Si XPS peaks before and after electron irradiation.

Figures 12 and 13 show respectively the fitted high resolution Si 2p XPS peaks before and after irradiation. The peaks at 102.5, 103.7 and 104.8 eV in figure 12 can be assigned to SiO_x

(x < 2), SiO_2 and chemisorbed species (from atmospheric oxygen/moisture) respectively (Nagpure et al.; 2011). Note the absence of the peak from chemisorbed species and the appearance of two additional peaks at 100.6 eV and 98.2 eV in figure 12 that can be assigned to SiC (Nagpure et al.; 2011) and elemental Si (Moulder et al.; 1992) respectively. The well known effect of the prolonged electron beam irradiation of SiO_2 is the breaking (dissociation) of Si-O bond and the subsequent desorption of oxygen. As shown in the XPS spectrum after electron beam irradiation in figure 11, the depletion of oxygen was accompanied by a growth of free (elemental) silicon (Si) at 98.2 eV. As a result of oxygen desorption the surface was rich in Si and the remaining structure was probably an oxygen deficient SiO_x ($0<x<2$). The mechanism of desorption of oxygen from SiO_2 is beyond the scope of this chapter and the readers are referred to the literature cited for further reading (Carriere and Lang, 1977; Knotek and Feibelman, 1978; Thomas, 1974; Fiori and Devine, 1984). It is most likely that the oxygen deficient SiO_x was non-luminescent and it therefore contributed to the CL intensity degradation (Ntwaeaborwa et al., 2006, 2007).

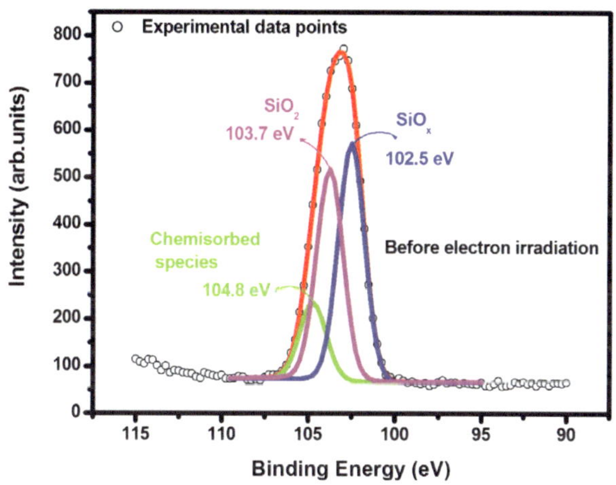

Fig. 12. Fitted high resolution Si 2p XPS peak before electron irradiation.

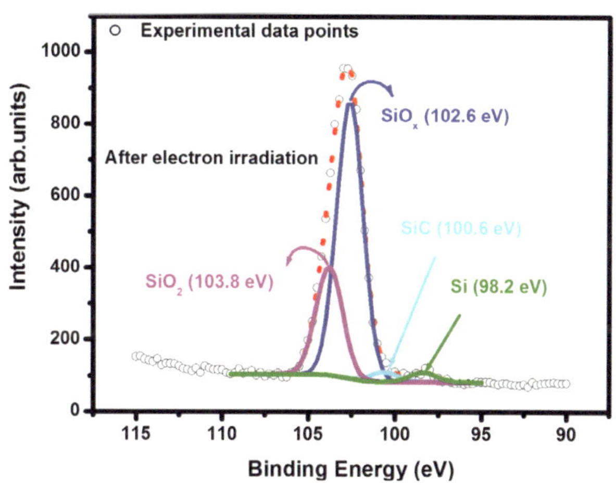

Fig. 13. Fitted high resolution Si 2p XPS peak after electron irradiation.

4.2.2 ZnO-SiO_2:Pr^{3+} and SiO_2:Ce^{3+},Pr^{3+}

In this section the effects of ZnO nanoparticle, Ce^{3+} and Pr^{3+} concentrations on cathodoluminescence intensities of ZnO-SiO_2:Pr^{3+} and SiO_2:Ce^{3+},Pr^{3+} are discussed. Figure 14 shows the CL spectra of SiO_2:Pr^{3+}, with different concentrations (0.05 - 0.25 mol%) of Pr^{3+}, measured in the AES vacuum chamber at a base pressure of 10^{-8} Torr. Note that relatively low concentrations of Pr^{3+} were used to avoid quenching at higher concentrations. The CL intensity increased with concentration from 0.05 to 0.2 mol% and it decreased when the concentration was increased to 0.25 mol% due to concentration quenching effect. It is well known that at higher concentrations of light emitting ions, luminescence can be quenched as a result of clustering of or cross relaxation between the ions (Blasse and Grabmaier, 1994; Solé et al., 2005). The CL emission spectra are characterized by multiple line emissions in the visible region of the electromagnetic spectrum. The main emission associated with $^3P_0 \rightarrow ^3H_6$ transitions of Pr^{3+} was observed at 616 nm and minor emissions due to transitions from $^3P_0$ and $^1D_2$ to $^3H_{(J=6,5,4)}$ and $^3F_{(J=2,3,4)}$ were also observed. These transitions were not measured but were assigned according to the literature cited (Mhlongo et al., 2011a, Sokólska et al., 2000). The inset of figure 13 shows the maximum CL intensity of the 616 nm peak as a function of Pr^{3+} concentration illustrating the increase in intensity with concentration from 0.05 to 0.2 mol% and a sudden decrease when the concentration was increased to 0.25 mol%.

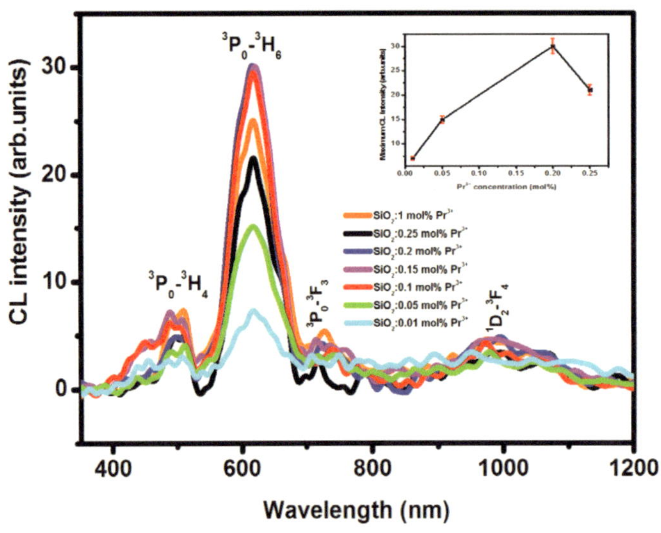

Fig. 14. CL spectra of SiO_2:Pr^{3+} with different concentrations of Pr^{3+}.

Figure 15 compares the CL emission spectra of SiO_2, SiO_2:1mol%Pr^{3+} and ZnO-SiO_2:1mol%Pr^{3+} (ZnO = 5mol%) powders recorded when the powders were irradiated with a beam of 2 keV electrons and a beam current of 20 μA in the chamber kept at a base pressure of 1.6×10^{-8} Torr. With incorporation of Pr^{3+}, the blue emission of SiO_2 at 445 nm was suppressed and all the observed emissions were due to transitions in Pr^{3+}. The main emission at 616 nm was enhanced considerably by the incorporation of ZnO nanopareticles. This emission is 2× more intense than the emission from SiO_2:Pr^{3+} suggesting that ZnO contributed to this enhancement by transferring excitation energy to Pr^{3+} ions. Note that because of severe charging upon irradiation with electrons, cathodoluminescence of ZnO nanoparticles could not be measured. However it is a well known phenomenon that ZnO nanoparticles display dual emission in the UV and visible region irrespective of the type of excitation (i.e. UV photons or electrons) (Mhlongo et al., 2011b). Since both the UV and defects emissions were not detected when ZnO nanoparticles were incorporated in SiO_2:Pr^{3+} and there was a subsequent increase in the red emission from Pr^{3+}, it is reasonable to attribute the increase to energy transfer from ZnO to Pr^{3+}. The reader is referred to the literature cited to read about the mechanism of energy transfer from ZnO to rare-earth ions in glassy SiO_2 host (Bang et al., 2006; Ntwaeaborwa and Holloway, 2005; Mhlongo et al., 2010)

Figure 16 shows the CL emission spectra of Ce^{3+}-Pr^{3+} co-activated SiO_2 phosphors. The concentration of Pr^{3+} was fixed at 0.2 mol% while that of Ce^{3+} was varied from 0.2 - 2 mol%. The spectra were recorded when the powders were irradiated with 2 keV electrons and a beam current of 8.5 μA in the AES vacuum chamber maintained at a base pressure of 1.2 $\times 10^{-8}$ Torr. The CL spectrum of Pr^{3+} singly doped SiO_2 consists of a major emission peak at 616 nm and minor peaks at 510 and 489. The CL spectrum of Ce^{3+} consists of broad emission peak with a maximum at 489 nm. Notice the suppression of the Pr^{3+} emission in co-activated SiO_2:Ce^{3+},Pr^{3+} and the change in PL intensity of the 489 nm peak with Ce^{3+} concentration. The maximum intensity was observed when 1 mol% of Ce^{3+} was co-doped with 0.2 mol% of Pr^{3+} as shown in the inset of figure 16. The intensity was quenched, probably as a result of concentration quenching effects, when the Ce^{3+} was increased to 1.5 and 0.2 mol%.

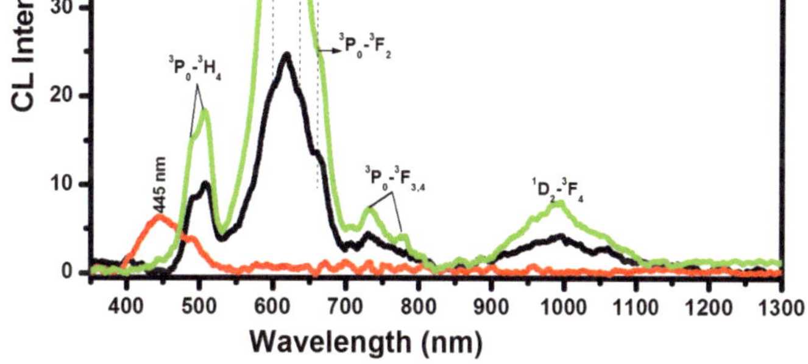

Fig. 15. CL spectra of SiO_2, SiO_2:Pr^{3+} and ZnO-SiO_2:Pr^{3+} (Mhlongo et al., 2010)

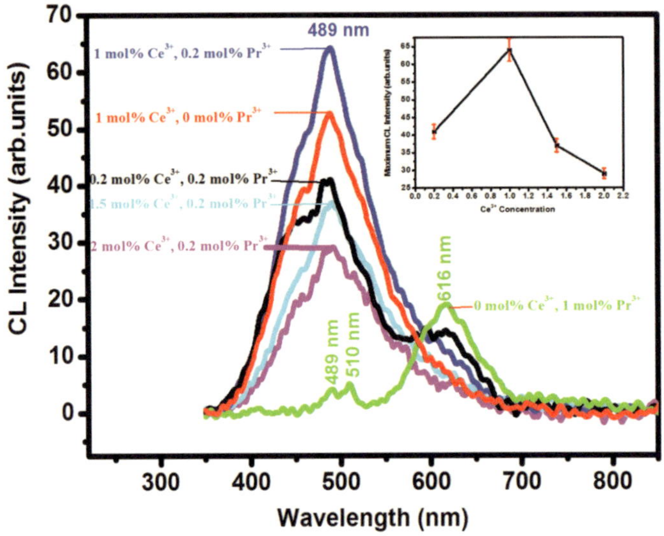

Fig. 16. CL spectra of SiO_2:Ce^{3+}, SiO_2:Ce^{3+},Pr^{3+} and SiO_2:Pr^{3+}. The inset is the maximum CL intensity of the 489 peak as a function of Ce^{3+} concentration.

The fact that the increase in the blue emission at 489 nm was simultaneous with the quenching of Pr^{3+} emission from SiO_2:Ce^{3+},Pr^{3+} (Ce^{3+} = 1 mol% and Pr^{3+} = 0.2 mol%) suggests that energy was transferred from Pr^{3+} to Ce^{3+}. Note that the energy associated with the $^1P_0 \rightarrow {}^3H_6$ transition of Pr^{3+} is 2.0 eV (616 nm), which is too low to transfer to Ce^{3+} at 2.5 eV (489 nm). On the other hand, the energy associated with the $^3P_0 \rightarrow {}^3H_4$ transition of Pr^{3+} occurs at the same energy position of 2.0 eV (489 nm) as the blue emission from Ce^{3+}. It is therefore reasonable to speculate that energy was transferred resonantly or by phonon mediated processes from the $^3P_0$ state of Pr^{3+} to $5d^1$ states of Ce^{3+}. Before discussing the proposed mechanism of energy transfer, it is important to discuss briefly the transitions responsible for blue emission in Ce^{3+}. Shown in figure 17 is the deconvoluted CL emission spectrum of SiO_2:Ce^{3+},Pr^{3+} (Ce^{3+} = 1 mol%) recorded during irradiation with 2 keV beam of electrons and a beam current of 8.5 μA at a base pressure of 1.2×10^{-8} Torr. The 489 nm peak is deconvoluted into two Gaussian peaks, one at 480 nm and the other one at 525 nm. The 4f ground state of the Ce^{3+} is split into two components ($^2F_{5/2}$ and $^2F_{7/2}$) due to spin-orbit interaction while the excited $5d^1$ state can be split into 2-5 components by the crystal field (Ntwaeaborwa et al., 2008; Mhlongo et al., 2011a). Here we consider a case where the $5d^1$ is split into two components, namely the upper $^2D_{5/2}$ and the lower $^2D_{3/2}$. We speculate that the absorbed excitation energy causes transition from the ground state to the upper $^2D_{5/2}$ state followed by radiationless transition to the lower $^2D_{3/2}$ and a subsequent radiative transition to the $^2F_{7/2}$ (480 nm) and $^2F_{5/2}$ (525 nm) energy states.

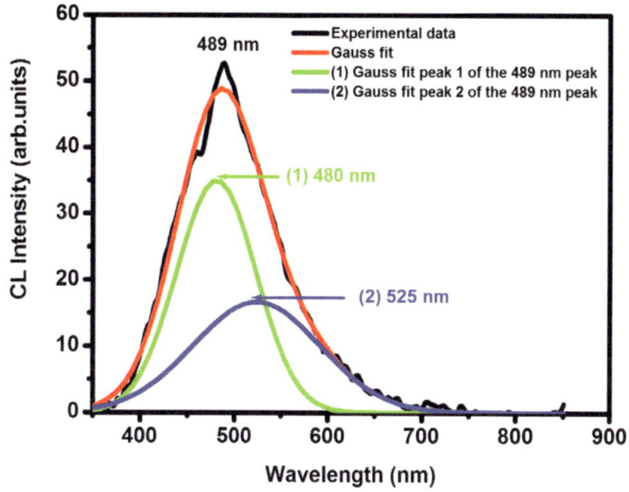

Fig. 17. Deconvoluted CL emission spectrum of SiO_2:Ce^{3+}.

The proposed mechanism of energy transfer from Pr^{3+} to Ce^{3+} is presented in figure 18. According to this mechanism, the absorbed excitation energy causes transition from the ground state ($^3H_4$) to higher energy states (4f5d) of Pr^{3+}. This is followed by a radiationless transition to the $^3P_{2,1,0}$ states of Pr^{3+}. In a Pr^{3+} singly doped SiO_2 the $^3P_{1,0}$ states will be followed by radiative transitions to the $^3H_{6,5,4}$ states. However, in the Ce^{3+}-Pr^{3+} co-activated SiO_2 the $^3P_0$ state of Pr^{3+} will transfer energy resonantly or by phonon-assisted processes to the $5d^1$ states of Ce^{3+} resulting in an enhanced blue emission during radiative transition to the ground state. This is only possible if the transfer rate is faster than the radiative transition to the $^3H_{6,5,4}$ states. Similar mechanism was reported by Ntwaeaborwa et al. (Ntwaeaborwa et al., 2008) for energy transfer between Eu^{3+} and Ce^{3+}.

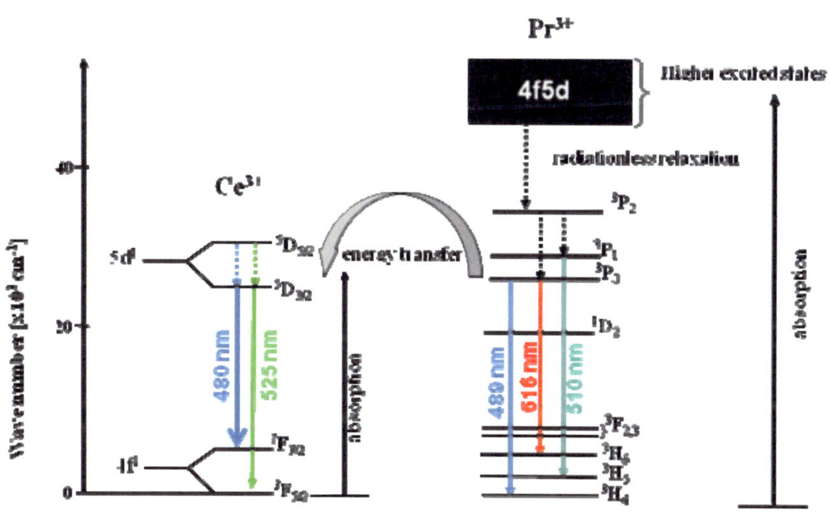

Fig. 18. Simplified energy level diagrams of Ce^{3+} and Pr^{3+} illustrating energy transfer from Pr^{3+} to Ce^{3+}.

The effect of accelerating voltage on the brightness of $SiO_2:Ce^{3+},Pr^{3+}$ (Ce^{3+} = 1 mol% and Pr^{3+}=0.2 mol%) was evaluated by varying the voltage from 1 - 5 kV when the beam current was fixed at 8.5 μA. Figure 19 shows the CL intensity as a function of accelerating voltage. As shown in the inset, the CL intensity of the 489 nm peak is increasing linearly with voltage from 1 to 5 kV. It is well known that the depth of penetration by incident electrons is proportional to the accelerating voltage (Kumar et al., 2010), i.e. high the accelerating voltage deeper will be the penetration. The proportional increase in the CL intensity (brightness) with accelerating voltage suggests that the rate of generation of free electron and holes was high at greater penetration depths and a subsequent increase of radiative recombination. A lack of luminescence saturation at voltages up to 5 kV is a sign of good prospect for this phosphor to be used in low voltage (1 - 10kV) field emission display technology.

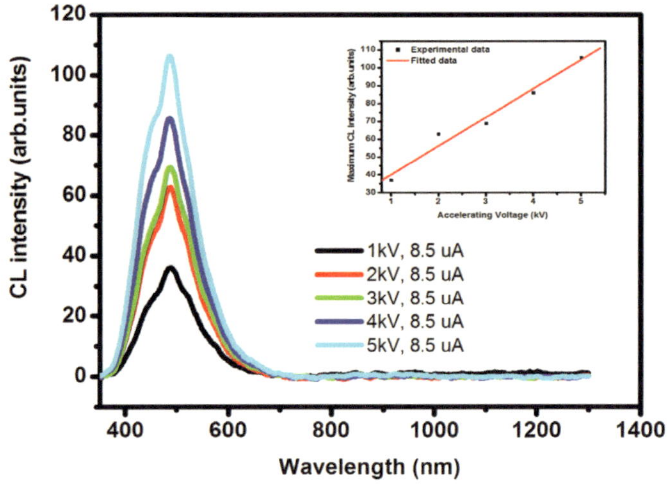

Fig. 19. The effect of accelerating voltage on the CL intensity of $SiO_2:Ce^{3+},Pr^{3+}$.

4.2.3 SiO_2:PbS

In this section, cathodoluminescent properties and intensity degradation of SiO_2:PbS phosphor are discussed. The preparation of the samples was discussed in section 1.2. The concentration of PbS nanoparticles in SiO_2 was 0.34 mol%. The CL data were recorded when the SiO_2:PbS powders were irradiated with a 2 keV beam of electrons in the Auger spectrometer chamber at different oxygen pressures. The Auger-peak-to-peak-heights (APPHS) of O, Si and C and CL intensity as a function of electron dose resembled that of $SiO_2:Ce^{3+},Tb^{3+}$ in figure 8 suggesting that the decrease in the CL intensity was simultaneous with desorption of O. Figure 20 shows broadband CL emission spectra, before and after electron irradiation, with a maximum at ~700 nm. While the CL emission from pure SiO_2 was observed at 445 nm (figure 5), the CL emission from pure PbS nanoparticles could not be recorded due to excessive surface charging. It is however well known that light emission from PbS nanoparticles associated with excitonic recombination is usually in the near infrared region at 1100 - 1700 nm (Ntwaeaborwa et al., 2009a, which is blue-shifted from PbS bulk emission at ~3000 nm due to quantum confinement effects. Therefore the orange-red emission in figure 20 is neither coming from SiO_2 nor PbS nanoparticles. This emission is similar to emission associated with transitions in Pb^{2+} ions in ZnS:Pb^{2+} (Bol and Meijerink, 2001) and SiO_2:Pb^{2+} (Ntwaeaborwa et al., 2009b). Although the exact mechamism of this emission is not known yet, it is reasonable to attribute the emission to transisions in Pb^{2+}. This can only be possible if Pb^{2+} ions are selectively excited. Note that blue, green and red emissions from Pb^{2+} have been reported and the red emission, similar to the one in figure 20, was attributed to $^3P_{0,1}\rightarrow{}^1S_0$ transition of Pb^{2+} (Bol and Meijerink, 2001). As shown in the inset of figure 20, the red-orange cathodoluminescence degraded by more than 50% following the prolonged irradiation by 2 keV electrons. Similar mechanism that involves desorption of O and a subsequent formation of less luminescent SiO_x ($x < 2$) discussed in section 2.2.1 can be used to explain the CL intensity degradation of the SiO_2:PbS phosphor.

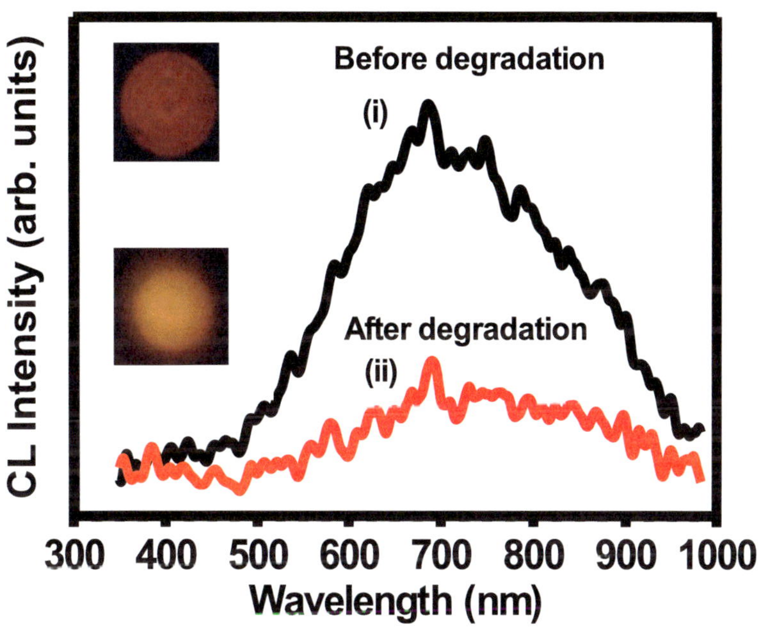

Fig. 20. CL emission spectra of SiO_2:PbS before and after electron irradiation.

5. Conclusion

Green, blue and orange-red cathodoluminescence was observed, respectively, from SiO_2:Ce^{3+},Tb^{3+}; SiO_2:Ce^{3+},Pr^{3+} and SiO_2:PbS phosphors prepared by the sol-gel method. Green emission from Tb^{3+} was enhanced by Ce^{3+} co-doping. The CL intensity of the green emission degraded by 50% when irradiated with a beam of electrons for 10 hours and the CL intensity degradation was simultaneous with desorption of O from the surface. The degradation was attributed to the formation of oxygen deficient SiO_2 ($x<2$) layer. Enhanced blue emission from Ce^{3+} was observed from Ce^{3+}-Pr^{3+} co-doping while the red emission of Pr^{3+} suppressed. Possible mechanism of energy transfer from Pr^{3+} to Ce^{3+} was discussed. Orange-red cathodoluminescence from SiO_2:PbS was attributed to $^3P_{0,1}\rightarrow{}^1S_0$ transition of Pb^{2+}. SiO_2 is a potential host matrix for rare-earths and semiconducting nanocrystals for preparation of phosphors that can be used in light emitting devices including cathode ray tubes and field emission displays.

6. Acknowledgment

The authors would like to acknowledge the financial support from the South African National Research Foundation and the Nanoscience cluster fund of the University of the Free State.

7. References

[1] Nogami, M., Enomoto, T., Hayakawa, T. (2002). Enhanced fluorescence of Eu^{2+}induced by energy transfer from nanosized SnO_2 crystals in glass. *J. Lumin.* Vol. 97, No. 3-4, pp. (147-152)

[2] Ntwaeaborwa, O.M., Swart, H.C., Kroon, R.E., Botha, J.R., Holloway, P.H. (2007). Cathodoluminescence degradation of SiO_2:Ce,Tb powder phosphors prepared by a sol-gel process. *J. Vac. Sci. Technol. A.* Vol. 25, No. 4, pp. (1152-1155).

[3] Ntwaeaborwa, O.M., Swart, H.C., Kroon, R.E., Botha, J.R., Ngaruiya, J.M., Holloway, P.H. (2008). Enhanced photoluminescence of rare-earth activators in sol-gel derived SiO_2 by energy transfer from ZnO nanoparticles and co-activators, In: *Photoluminescence Research Progress*, Wright H.K. and Edwards, G.V., pp. (287-306), Nova publishers, 978-1-60456-538-6, New York.

[4] Hench, L.L., Wesk, J.K. (1990). The Sol-Gel Proces, *Chem. Rev.*, Vol. 90, No. 1, pp. (33-72).

[5] Ding, J.Y., Day, D.E. (1991). Preparation of silica glass microsphere by sol-gel, *J. Mat. Res*, Vol. 6, No. 1, pp. (168-174).

[6] Ntwaeaborwa, O.M.; Swart H.C.; Kroon, R.E.; Holloway P.H.; Botha, J.R. (2006), Enhanced luminescence and degradation of SiO_2:Ce^{3+},Tb^{3+}powder phosphor prepared by a sol-gel process, *J. Phys. Chem. Sol.*,Vol. 67, pp (1749-1753)

[7] Reisfeld, R., Gaft M., Saridarov, T., Panczer, G., Zelner, M. (2000). Nanoparticles of cadmium sulfidewith europium and terbium in zirconia films having intensified luminescence, *Mat. Lett*, Vol. 45, No. 3, pp. (154-156).

[8] Holloway, P.H., Sebastian, J., Trottier, T., Jones, S., Swart, H., Petersen, R.O. Degradation mechanisms and vacuum requirements for FED phosphors. (1996). *Mat. Res. Soc. Symp. Proc*, Vol. 424, pp. (425-431).

[9] Holloway, P.H., Trottier, T.A., Sebastian, S., Jones, S., Zhang, X.-M., Bang, J.-S., Abrams, B., Thomes, W.J., Kim, T.J. (2000). Degradation of field emission display phosphors, *J. Appl. Phys.*, 88(1), pp. (483-488)

[10] Knotek, M.L. and Feibelman, P.J. (1978). Ion desorption by core-hole Auger decay, *Phys. Rev. Lett.*, Vol. 40, No. 14, pp. (964-967)

[11] Stoffers, C.; Yang, S.; Jacobsen, S. M. and Summers, C.J. (1996). Saturation of phosphor under low voltage excitation, *J. Soc. Inf. Display.* Vol. 4, No. 4, pp. (337-341)

[12] Raue, R.; Vink, A.T. and Welker T. (1989). Phosphors screens in cathode ray tubes for projection television *Phillips Tech. Rev*, Vol. 44, No. (11-12), pp. (335 – 347).

[13] Swart, H.C.; Oosthuizen, L.; Holloway, P.H. & Berning G.L.P. (1998). Degradation behavior of ZnS phosphors under different experiment conditions, *Surf. Interface Anal.*, Vol. 26, No. 5, pp (337-342)

[14] Itoh, S.; Kimizuka, T. and Tonegwa T. (1989). Degradation mechanism for low voltage cathodoluminescence sulphide phosphors, *J. Electrochem. Soc.*, Vol. 136, No. 1819-1823

[15] Klein, L.C. (1985). Sol-gel processing of silicates, *Ann. Rev. Mater. Sci.*, 15,pp. (227-248)

[16] Huang, H-H.; Orler, B. and Wilkes, G.L. (1985). Ceramers: Hybrid materials incorporating polymeric/oligomeric species with inorganic glasses by a sol-gel process – 2. Effect of acid content on the final product, *Polymer Bulletin.,* Vol. 14, No. , pp. (557-564)

[17] Nagpure, I.M.; Pitale, S.S.; Tshabalala, K.G.; Kumar, V.; Ntwaeaborwa, O.M., Terblans, J.J. and Swart, H.C. (2011). Luminescence response and CL degradation of combustion synthesized spherical SiO_2:Ce nanophosphor, *Mater. Res. Bull.,* doi:10.1016/j.materresbulls.2011.08.051

[18] Thomas, S. (1973). Electron-irradiation effect in the Auger analysis of SiO_2, *J. Appl. Phys.,* Vol. 45, No. 1, pp (161-166).

[19] Lin, J. and Baerner K. (2000). Tunable photoluminescence in sol-gel derived silica xerogels, *Mat. Lett.,* Vol. 46, No. (2-3), pp (86-92)

[20] Han, Y.; Lin J. and Zhang H. (2002). Photoluminescence of organic-inorganic hybrid SiO_2 xerogels, *Mat. Lett.,* Vol. 54, No. (5-6), pp (389-396)

[21] Gu G.; Ong P.P. and Chu C. (1999). Thermal stability of mesoporous silica molecular sieve, J. *Phys. Chem. Sol.,* Vol. 60, No. 7, pp (943-947)

[22] García M.; Mondragón M.A.; Téllez S.; Campero A. and Castano V.M. (1995). Blue emission and tetraethoxysilane and silica gels, *Mater. Chem. Phys.,* Vol. 41, No. 1, pp (15-17).

[23] Moulder, J.F.; Stickle, W.F.; Sobol, P.E. and Bomben, K. (1992). *Handbook of X-ray photoelectron spectroscopy*, Perkin Elmer Corp., Minnesota (USA).

[24] Carriere, B. and Lang, B. (1977). A study of the charging effect and dissociation of SiO_2 surfaces by AES, *Surf. Sci.,* Vol. 64, pp (209-223)

[25] Fiori, C. and Devine, R.A.B. (1984). Photon-induced oxygen loss in thin SiO_2 films, *Phys. Rev. Lett.,* Vol. 52, No. 23, pp (2081-2083)

[26] Blasse, G.; Grabmaier, B.C. (1994). *Luminescent Materials,* Springer Verlag, ISBN 3-540-58019-0 New York

[27] Solé, J.G.; Bausá, L.E.; Jaque D. (200). *An Introduction to the Optical Spectroscopy of Inorganic Solids,* John Wiley and Sons, Ltd, ISBN 0-470-86886-4,Chichester.

[28] Mhlongo, G.H.; Ntwaeaborwa, O.M., Dhlamini, M.S.; Swart, H.C; Hillie, K.T. (2011a). Effects of Ce^{3+} concentration, beam voltage and current on the cathodoluminescence of intensity of SiO_2:Ce^{3+}-Pr^{3+} nanophosphor, J. Alloy. Compd., Vol. 509, pp (2986-2992)

[29] Sokólska, I.; Golab, S.; Baluka, J.; Ryba-Romanowski, W. (2000), Quenching of Pr^{3+}in single crystals of $K_5Pr_xLa_{1-x}Li_2F_{10}$., *J. Lumin.,* Vol. 91, pp (79-86).

[30] Mhlongo, G.H.; Ntwaeaborwa, O.M., Swart, H.C.; Kroon, R.E.; Solarz, P.Ryba-Romanowski, W., Hillie, K.T. (2011b). Luminescence dependence of Pr^{3+} activated SiO_2 nanophosphor on Pr^{3+}concentration, temperature and ZnO incorporation, *J. Phys. Chem. C.,* Vol. 115, pp (17625 - 17632).

[31] Mhlongo, G.H.; Ntwaeaborwa, O.M.; Dhlamini, M.S.; Swart, H.C.; Hillie, K.T. (2010) Cathodoluminescence properties of SiO_2:Pr^{3+} and ZnO SiO_2:Pr^{3+}, *J. Mater. Sci.,* Vol. 45, pp (5228 - 5236).

[32] Kumar, V.; Mishra, V.; Pitale, S.S.; Nagpure, I.M.; Coetsee, E.; Ntwaeaborwa, O.M.; Terblans, J.J.; Swart, H.C. (2010). Surface chemical reactions during electron beam irradiation of nanocrystalline CaS:Ce^{3+} phosphor, *J. Appl. Phys.,* Vol. 107, pp (123533-1 - 123533-6)

[33] Bol, A.A.; Meijerink, A. (2001). Luminescence of nanocrystalline ZnS:Pb^{2+}, *Phys. Chem. Chem. Phys.*, Vol. 3 (11), pp (830-832).

[34] Ntwaeaborwa, O.M.; Kroon, R.E.; Kumar, V.; Dubroca T.; Ahn, J.-P.; Park, J.-K.; Swart, H.C. (2009a), Ex-situ synthesis and optical properties of ZnO-PbS nanocomposites, *J. Phys. Chem. Sol.*, Vol. 70, pp (1438 – 1442)

[35] Ntwaeaborwa O.M.; Swart H.C.; Kroon, R.E.; Terblans, J.J.; Holloway, P.H. (2009b). Synthesis, characterization and luminescent properties of ZnO-SiO_2:PbS, *J. Vac. Sci. Technol. A.*, Vol. 27(4), pp (767 – 769)

10

Cathodo- and Photo- Luminescence of Silicon Rich Oxide Films Obtained by LPCVD

Rosa López-Estopier[1, 2], Mariano Aceves-Mijares[3] and Ciro Falcony[4]
[1]*Department of Electronics, ITSPR, Poza Rica, Veracruz*
[2]*Department of Applied Physics, ICMUV, University of Valencia, Burjassot, Valencia*
[3]*Department of Electronics, INAOE, Tonantzintla, Puebla*
[4]*Department of Physics, CINVESTAV-IPN, Distrito Federal*
[1,3,4]*México*
[2]*Spain*

1. Introduction

Silicon technology dominates the electronics industry today, so it is highly desirable the development of silicon-based components compatible with silicon technology, allowing integration of electrical and optical components on a single chip. One promising approach to the development of a silicon based light emitter is Silicon Rich Oxide (SRO), also called off-stoichiometric silicon oxide. The interest on the optical properties of this material has grown since it was demonstrated that SRO films subjected to high-temperature annealing exhibit efficient photoluminescence (PL) (Iacona et al., 2000; Shimizu-Iwayama et al., 1996).

At present, different techniques have been employed to produce SRO films, these include plasma enhanced chemical vapor deposition (PECVD) (Pai et al., 1986), low pressure chemical vapor deposition (LPCVD) (Dong et al., 1978), silicon implantation into thermal oxide (SITO) (Pavesi et al., 2000), reactive sputtering (Hanaizumi et al., 2003) and others. An indicator of the Si content in this material is the parameter R_0, which is the ratio of the partial pressure of the precursor gases (in this case, N_2O/SiH_4) when it is prepared by gas phase deposition methods like CVD (Chemical Vapor Deposition). Silicon excess ranging from 17% to 0% can be obtained varying the R_0 from 3 to 100. Among CVD methods, the low pressure chemical vapor deposition (LPCVD) is a very convenient approach for the deposition of SRO films since it allows an exact variation of the Si content and produces higher PL than other methods.

Different applications of SRO have been proposed, such as: visible light emission devices (Shimizu-Iwayama et al., 1996), non-volatile or electrically alterable memory devices (Calleja et al., 1998), surge suppressors (Aceves et al., 1999), microsensors (Aceves et al., 2001), and single electron devices (Yu et al., 2003).

SRO can be considered as a multi-phase material composed of a mixture of stoichiometric silicon oxide (SiO_2), off-stoichiometric oxide (SiO_x, $x<2$) and elemental silicon. After thermal treatment at temperatures above 1000°C, off-stoichiometric oxide and silicon excess are separated into Si nanoClusters (crystalline or amorphous depending on their size), defects (oxidation states) and SiO_2 (Yu et al., 2006b). Transmission electron microscopy (TEM) studies on the PECVD-SRO films with Si excess of 4-9% have confirmed the existence of Si nanocrystals with sizes between 1-2 nm (Iacona et al., 2000). The observation of Si nanocrystals in LPCVD-SRO films with Si excess of ~13% has also been reported (Irene et al., 1980). Although no Si nanocrystals were observed in LPCVD-SRO with less Si excess (<6%) (DiMaria et al., 1984), however, it was still proposed that amorphous Si nanodots exist in the SiO_2 matrix, which could not be distinguished by TEM (Nesbit, 1985). It was found that the growth of the Si nanodots is a diffusion-controlled process with a diffusion coefficient as larger as 1×10^{-16} cm^2/s at 1100°C (Nesbit, 1985); it is estimated that a diffusion length of 10-20 nm can be obtained during the thermal annealing of 3 hours. This diffusion length is long enough for the excess Si to precipitate into Si nanodots during the annealing process.

Depending of the silicon excess, SRO shows various properties, among others, variable conductivity (Aceves et al., 1999), charge-trapping effect (Aceves, et al., 1996) and photoluminescence (PL) (Shimizu-Iwayama et al., 1996). These properties depend on the concentration and the exact nature of the excess Si. For example, the trapped charge density and photoluminescence intensity of SRO increase with decreasing its Si excess. The light emission is one promising property of SRO that makes it a good candidate for future's light-emitting devices. It has been observed that strong visible light can be emitted from SRO films deposited by LPCVD, its emission wavelength and intensity is tuned by the Si excess and thermal annealing parameters (temperature and time). It has also been found that SRO can trap both negative and positive charges, and the trapped charge density depends on the same parameters.

Although this material has been extensively studied there is a considerable uncertainty about the nature of the luminescence mechanisms. A clear understanding of the luminescence mechanisms in SRO is the key for the production of silicon-based light-emitting devices. Some authors attribute the luminescence to quantum confinement effects (QCE) in Si-nCs (Si nanocrystals) (Chen et al., 2006; Iacona et al., 2000), to the emission of several types of defects in the matrix or in the interface of SiO_2/Si-nCs (Kenyon et al.1996; Lockwood et al., 1996), including the decay of donor acceptor pairs, DAD (López-Estopier et al., 2011; Yu et al., 2006a).

Even though SRO has been strongly study by PL, there are only a few works that study the cathodoluminescence (CL) of SRO. CL expands the possibilities to understand the emission in SRO. For example point defects in silicon oxide and silicon could be observed trough CL, allowing studying the distribution of such defects on the surface and inside the structure. An advantage of this method is the possibility to observe optical transitions of electrons from high energy levels (or band to band), which need to be exited with high energies, this is very important for wide-gap materials.

CL and PL are similar techniques with some possible differences associated with the excitation of the electron-hole pairs due to photons or more energetic electrons (Yacobi & Holt, 1990). PL emission depends strongly on the excitation energy, and not all luminescence mechanism could be excited. Cathodoluminescence, in general, leads to emission by all the luminescence mechanism present in the solid due to the high energy excitation.

In this chapter, different techniques, such as Fourier transform infrared spectroscopy (FTIR), X-ray photoelectron spectroscopy (XPS), photoluminescence (PL) and cathodoluminescence

(CL) were used to study SRO films obtained by LPCVD. In section 2, the experimental details of the fabrication of SRO films and their characterization techniques used are described. Section 3 gives the results of the structural and optical measurements. Also, in this section, the discussion and analysis of the results are included. Finally, in Section 4, conclusions are drawn. The structural and optical properties of SRO-LPCVD films were related to better understand the process of light emission in these materials.

2. Experimental

SRO samples were deposited by Low Pressure Chemical Vapor Deposition (LPCVD) at 720 °C on n-type Si substrates. The silicon excess in the SRO films was varied by adjusting the ratio of partial pressure of the reactive gases, Silane (SiH_4) and Nitrous Oxide (N_2O).

$$R_0 = \frac{P_{[N_2O]}}{P_{[SiH_4]}} = 10, 20 \text{ and } 30 \tag{1}$$

With this R_0 variation a silicon excess of 11-5.5% is approximately obtained. The SRO layers thickness is about 300 nm. After the deposition, one part of samples was keep as deposited and other part was annealed at 1100 °C in N_2 atmosphere during 180 minutes.

The PL emission spectra were obtained with a Fluoromax-3 Spectrofluorometer; all the films were excited with 250 nm. Cathodoluminescence (CL) measurements were performed using a Luminoscope equipment model ELM2-144, 0.3 mA current and energies of 2.5, 5, 10 and 15 keV were used. The luminescence spectra (PL and CL) were measured at the room temperature. The IR spectra were measured with a Brucker FTIR spectrometer model V22, which works in a range of 4000-350 cm^{-1} with 5 cm^{-1} resolution. The silicon excess in SRO films was measured with a PHI ESCA-5500 X-ray photoelectron spectrometer (XPS) using a monochromatic Al radiation source with energy of 1486 eV.

3. Results and discussion

3.1 Compositional and structural properties

The IR absorbance spectra of SRO samples are depicted in Fig. 1. Their characteristic absorption bands are enumerated and identified in the Table 1. SRO films show the absorption peaks associated with rocking at ~458 cm^{-1}, bending at ~812 cm^{-1} and stretching at ~1080 cm^{-1} vibration modes of the Si-O-Si bonds in SiO_2 (Ay & Aydinli, 2004; Pai et al., 1986). In addition, the shoulder at ~1150 cm^{-1} corresponds to the out-of-phase Si-O stretching vibration (Pai et al., 1986). The insets of Fig. 1 show Si-N stretching vibration at ~886-990 cm^{-1} (Ay & Aydinli, 2004; Daldosso et al., 2007); in addition Si-H bending and stretching at ~886 and 2258 cm^{-1} respectively (Iacona et al., 2000).

The as deposited samples exhibit a characteristic IR absorption at a lower frequency than that of a thermal SiO_2 film (1080 cm^{-1}), and the frequency of stretching vibration peak decreases as R_0 decreases due to Si atoms replacing O atoms (Pai et al., 1986); this result is related to off-stoichiometric of SRO, frequency of the Si-O stretching decreases with x of SiO_x (silicon excess increases). For all samples, the stretching frequency increases after annealing for all silicon excess, this result suggest phase separation during the thermal annealing.

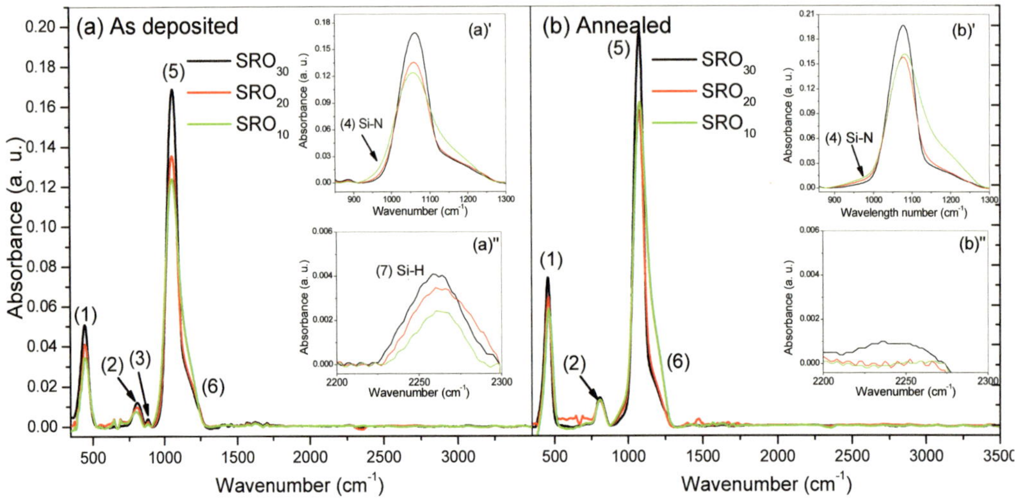

Fig. 1. FTIR absorption spectra of SRO films with $R_0 = 10$ (a) before and (b) after thermal annealing. The inset shows Si-N and Si-H bonding in the 850 to 1300 cm^{-1} and 2200 to 2300 cm^{-1} range.

Vibration mode	Wavenumber [cm^{-1}]					
	As deposited			Annealed		
	SRO_{30}	SRO_{20}	SRO_{10}	SRO_{30}	SRO_{20}	SRO_{10}
(1) Si-O rocking	447	445	455	457	459	460
(2) Si-O bending	810	808	802	808	808	810
(3) Si-H bending	883	883	883	-	-	-
(4) Si-N stretching	980	984	940	984	991	954
(5) Si-O symmetric stretching	1060	1058	1055	1078	1078	1082
(6) Si-O asymmetric stretching	1176	1170	1159	1192	1190	1190
(7) Si-H stretching	2258	2258	2258	-	-	-

Table 1. Infrared vibrations modes observed in the as deposited samples.

It has been reported that SRO films deposited by the mixture of N_2O and SiH_4 display absorption bands associated with Si-H (660, 880, 2250 cm^{-1}), Si-N (870-1030 cm^{-1}), Si-OH (940, 3660 cm^{-1}) and N-H (1150, 1180, 3380 cm^{-1}) vibrations in addition to the three characteristic bands related to the Si-O-Si bonding arrangement (Fazio et al., 2005; Pai et al., 1986). The presence of some nitrogen and hydrogen characteristic peaks was also observed in the IR spectra of the films. Si-H bending and stretching were observed only in as deposited samples; these peaks disappear after annealing. Si-N stretching was also found in as deposited and annealing samples.

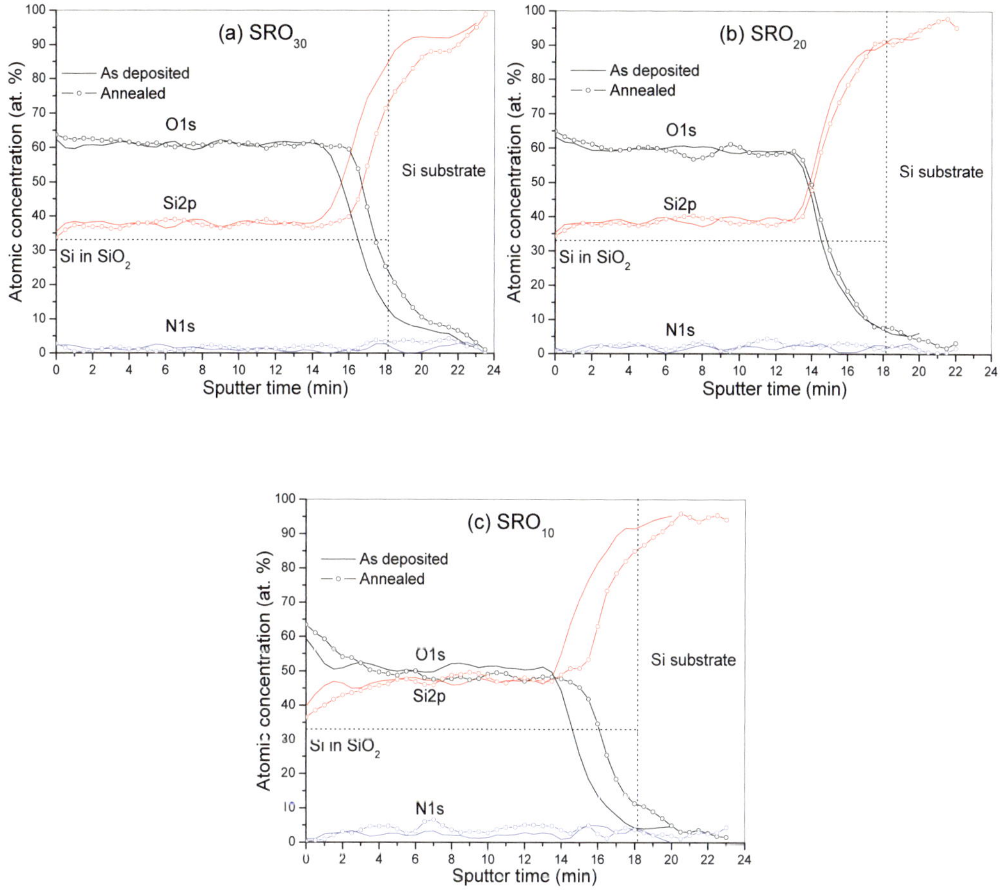

Fig. 2. Silicon, oxygen and nitrogen concentration in SRO films with different flow ratio (R_0) values as measured by XPS.

Fig. 2 indicates the composition measured by XPS of as deposited and annealed SRO films with different flow ratio, R_0 = 30, 20 and 10. As it was expected, the silicon excess varies depending on the flow ratio (R_0). For SRO films with R_0 = 30, the silicon profile stays about 37.5 at.% and it increases until 47 at.% as R_0 value reduces to 10. A thin silicon dioxide layer is formed at the SRO surface; which can be explained by the silicon oxidation at the end of the deposition process. A nitrogen profile was also observed, however is negligible, specially compared with that reported for SRO-PECVD films where the N incorporation is about 10 at.% (Morales et al., 2007; Ribeiro et al., 2003). Table 2 summarizes the silicon excess in the SRO films; silicon (Si), oxygen (O) and nitrogen (N) concentration in SRO is also shown.

R_0	As deposited					Annealed				
	Si excess (at. %)	Concentration (%)			SiOx	Si excess (at. %)	Concentration (%)			SiOx
		O	Si	N	x=O/Si		O	Si	N	x =O/Si
30	4.54	60.90	37.87	1.23	1.61	4.21	61.15	37.54	1.31	1.63
20	5.38	59.67	38.71	1.62	1.54	5.09	59.29	38.42	2.29	1.54
10	13.50	50.89	46.83	2.28	1.09	14.50	48.09	47.83	4.07	1.01

Table 2. Silicon excess and atomic concentration of SRO films obtained by XPS.

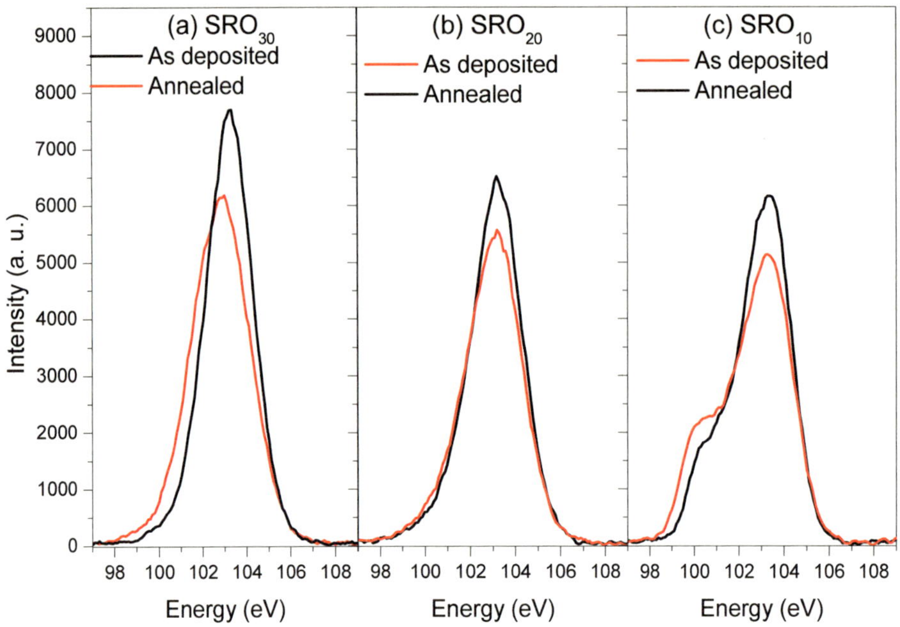

Fig. 3. Si 2p XPS peaks for as deposited and annealed samples with different flow ratio (R_0 30, 20 and 10).

The microstructure of SRO films was studied based on the analysis of the Si 2p spectra, shown in Fig. 3. As can be seen, the effect of silicon excess and thermal treatment is evident. After annealing and as silicon increases the shoulder in low energies becomes apparent, indicating the contribution of different oxidation states according to the random bonding model (RBM) (Chen et al., 2004, 2005).

The Si 2p XPS peaks were deconvoluted considering the five possible oxidation states for the silicon: Si, Si_2O, SiO, Si_2O_3, and SiO_2, noted as Si^{+0}, Si^{+1}, Si^{+2}, Si^{+3} and Si^{+4} respectively, shown in Fig. 4. Each oxidation state has been fitted by using peaks constituted by Gaussians. The energy positions of the different peaks of the Gaussians were centered with those previously reported in the literature at ~99.8, 100.5, 101.5, 102.5 y 103.5 eV (Alfonsetti et al., 1993; Brüesch et al., 1993; Philipp, 1972; Wang et al., 2003; Yang et al., 2005). The full widths at half maximum (FWHM) have been allowed to vary within a small range of values while increasing in the order $Si^{0+}<Si^{1+}<Si^{2+}<Si^{3+}<Si^{4+}$ (Dehan et al., 1995; Liu et al., 2003). Si^{0+} compounds were found in all samples in both before and after annealing due to silicon excess.

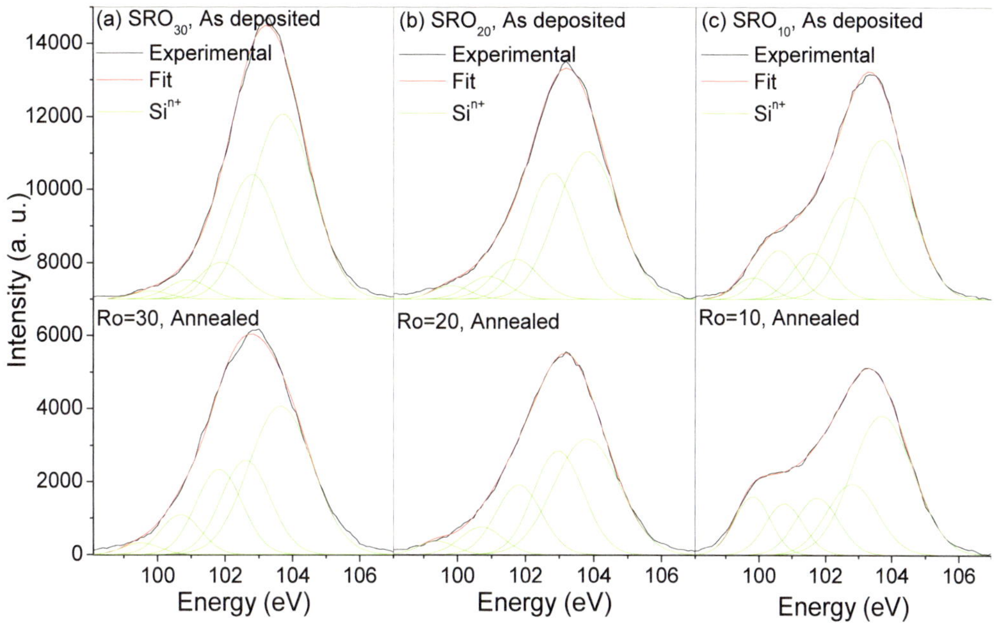

Fig. 4. Illustration of Si 2p peak deconvolution for the SRO films with R_0: (a) 30, (b) 20 and (c) 10.

It would be useful to examine quantitatively the changes of the concentrations of the five oxidation states with annealing. In this regard, as a first-order approximation, the relative concentration (in percentage) of each oxidation state is obtained by calculating the ratio I_{Si}^{n+}/I_{Total} (n = 0, 1, 2, 3, and 4), where I_{Si}^{n+} is the peak area of the oxidation state Si^{n+} and I_{total} is the total area of the Si 2p peaks ($I_{total} = \sum_{i=0}^{4} I_{Si^{i+}}$) (Khriachtchev et al., 2002; Liu et al., 2003).

Fig. 5 shows the effect of annealing in the concentrations of the five oxidation states for R_0 30, 20 and 10. After annealing concentration of Si^{0+}, Si^{1+}, Si^{2+} increases while Si^{3+} and Si^{4+} compounds decrease in all silicon excess indicating the phase separation; this separation is more evident in R_0 10. Then, when SRO films are thermally annealed and the silicon excess is high, the silicon atoms diffuse around of a nucleation site and well defined silicon nanocrystals are observed. However, for lower excess silicon films, the particles are far enough and do not produced agglomeration, therefore the thermal diffusion causes the silicon to be redistributed and form compounds with oxygen. These two mechanisms are not mutually exclusive and both exist simultaneously, but depending on the silicon excess will dominate a mechanism or the other. Because of this effect, concentration of Si^{0+} is much higher in SRO_{10} than SRO_{20} and SRO_{30}; in addition the concentration of Si^{1+} y Si^{2+} is larger for films with low silicon excess (SRO_{20} and SRO_{30}).

On the other hand, it is well known that in SRO with a large excess silicon (higher than 10%), Si-nCs of ~ 9 nm have been found by TEM (Transmission Electron Microscopy), as the silicon excess decreases, the density and size of the nanocrystals decreases until there are no Si-nCs (~ 5.5% silicon excess) (Yu et al., 2006b). Since the presence of Si^{0+} compounds in

$SRO_{20,30}$, agglomeration of amorphous compounds is more likely than elemental silicon agglomeration (nCs), and its possible size is smaller than 2 nm, then they cannot be observed by TEM.

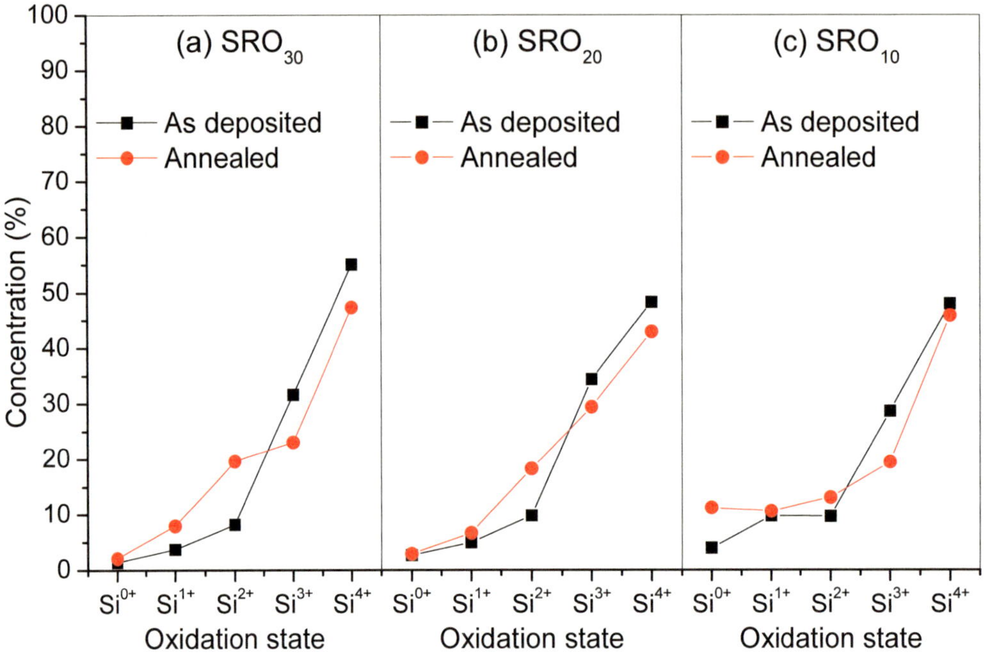

Fig. 5. Changes in the concentration of the five Si oxidation states Si^{n+} (n = 0, 1, 2, 3, and 4) for different R_0 values.

3.2 Photoluminescence

PL spectra of as deposited and annealed films are depicted in Fig. 6. The as deposited samples present a weak emission only in SRO_{30} from ~460 to 650 nm. After annealing all SRO films (R_0= 10, 20 and 30) present a main emission from ~650 to 850 nm. This emission range is characteristic of SRO annealed films deposited by LPCVD (López-Estopier et al., 2011; Yu et al., 2006a).The emission intensity increases when the silicon excess decrease.

The multi-Gaussian deconvolution of PL spectra was performed only for annealed samples, and the set of band positions have been determined. As can be seen in Fig. 7, each spectrum can be well fitted to a superposition of three Gaussian distributions: a main band and two shoulders. Fit distributions are centered at 715, 780, 825 nm with FWHM of 101, 80, 47 nm respectively. Distribution position and FWHM are the same for all R_0 values while intensity varies according to the silicon excess. As silicon excess decreases the intensity of main distribution increase; however, distributions centered at 780 and 825 nm are more intense to SRO_{20} than SRO_{30}; because of this PL emission of SRO_{20} shows a shoulder in this region. Distribution centered at 715 nm is the highest emission (R_0 = 20 and 30), hence this distribution is the main contribution of the PL emission. Since there are different distributions that change with silicon excess can be assumed that PL emission is related with at least three different types of emission centers (or emission mechanism).

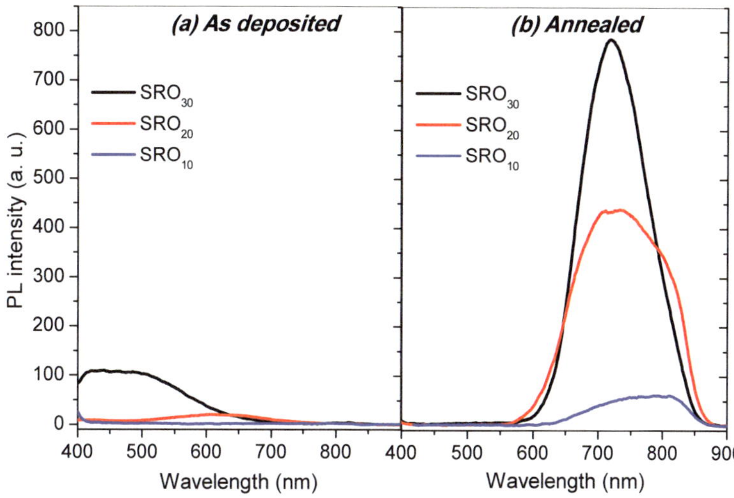

Fig. 6. Photoluminescence of (a) as deposited and (b) annealed SRO films with different ratio flows.

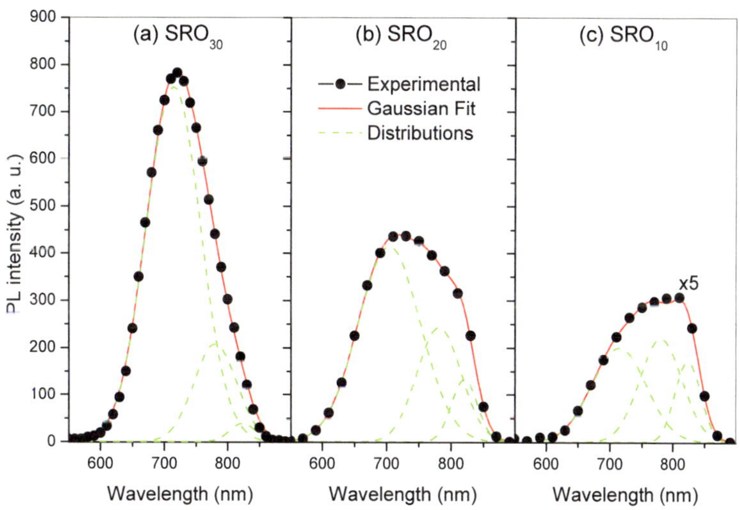

Fig. 7. PL spectra and fits from SRO films with different silicon excess and annealed at 1100° C for 180 min. Symbols are experimental data, lines are the Gaussian fits and dash lines are distributions.

3.3 Cathodoluminescence

CL spectra from SRO films (as deposited and annealed) with different silicon excess are depicted in Fig. 8. In as deposited samples there is CL emission only in SRO_{30}. The CL spectra of SRO with thermal treatment consist of a broad emission in the visible and near infrared region (NIR) from ~400 to 850 nm. After annealing, intensity of the blue band at ~460 nm increases with increasing the R_0. On the other hand, the intensity of the red-NIR CL band seems to have a maximum for R_0=20.

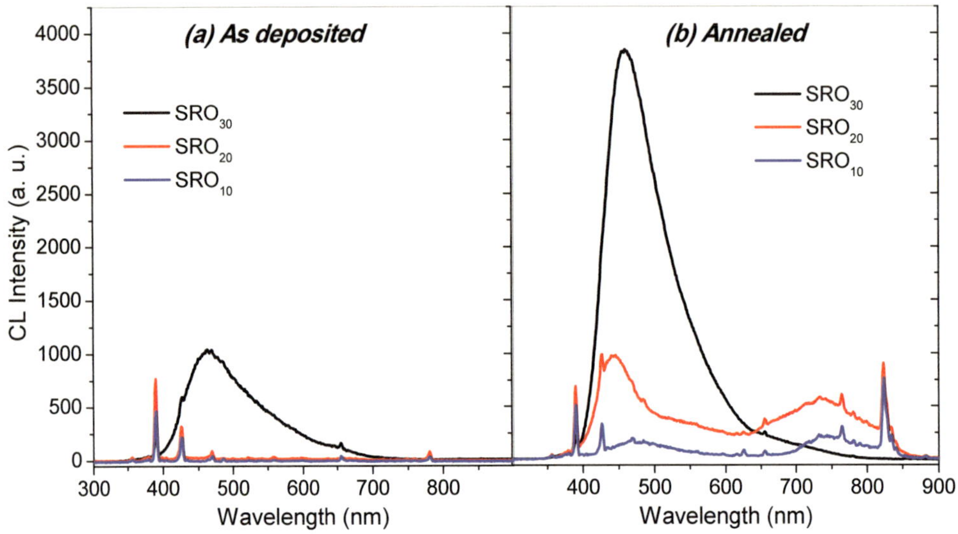

Fig. 8. Cathodoluminescence spectra from SRO films with different silicon excess: (a) as deposited and (b) annealed. Energy excitation of 5 keV and 0.3 mA current was used.

Since CL emission has asymmetrical shape for all SRO samples, it can be assumed that CL emission is due to different causes. Hence, multi-Gaussian deconvolution of CL spectra was also obtained, shown in Fig. 9. The best fit of CL spectra requires 4 and 6 distributions for SRO_{30} and SRO_{20} respectively. Distributions were obtained at about 460, 522, 643, and 714 nm for SRO_{30} and 447, 541, 645, 714, 780 and 823 nm for SRO_{20}. Distributions obtained at 714, 780 and 823 nm in CL are centered in the same position than distributions obtained from PL spectrum in SRO_{20}. Furthermore, distribution centered at 714 nm was obtained for Gaussian fit, in PL and CL in SRO_{30}. Then, the red emission of the CL emission can be ascribed to the same PL emissive centers. PL distributions in higher wavelength are not observed in CL due to either destruction of the emissive centers or emissive center with lower energies do not efficiently emit (Trukhin et al., 1999). The latter one could rise because cathode excited electrons acquire so high energy that they arrive to the higher emissive center where they emit in the blue region (higher energy), however almost none of the excited electrons reach lower energy centers; then, the red emission is not likely to occur. Therefore, there could be several different kinds of emission centers located at different energy levels in SRO.

Depending on the emission wavelength, multiple luminescence centers have been reported to act as radiative recombination centers in SiO_2 films. Luminescent emission at 460 nm (2.7 eV), 520 nm (2.4 eV) and 650 nm (1.9 eV) are mainly related to defects such as Oxygen deficiency-related centers (ODC) or oxygen vacancies (Cervera et al., 2006; Fitting, 2009; Gritsenko et al., 1999), E'δ defect or peroxide radical (Goldberg et al., 1997) and non-bridging oxygen hole centers (NBOHC) (Fitting, 2009; Fitting et al., 2001; Gritsenko et al., 1999), respectively. Since CL measurements have shown luminescent peaks (or distributions) close to those wavelengths, such defects could be inside the SRO films.

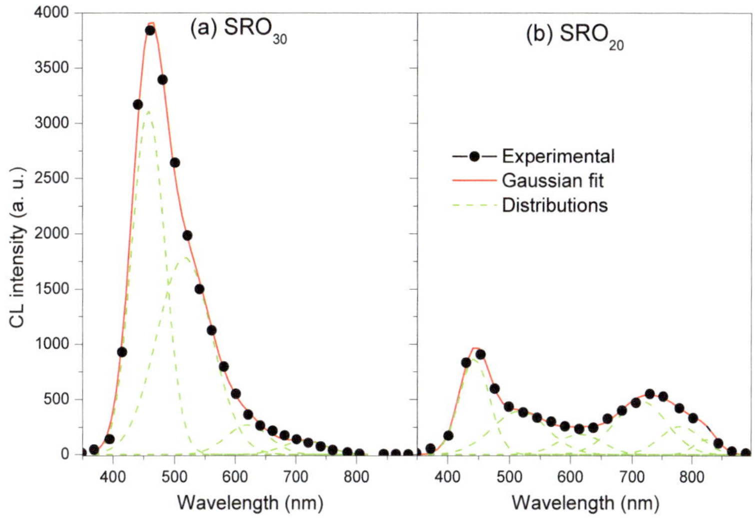

Fig. 9. Gaussian fit of CL experimental spectrum, the best fit requires 4 distributions for (a) SRO_{30} and 6 distributions for (b) SRO_{20}. Symbols are experimental data, lines are the Gaussian fits and dash lines are distributions.

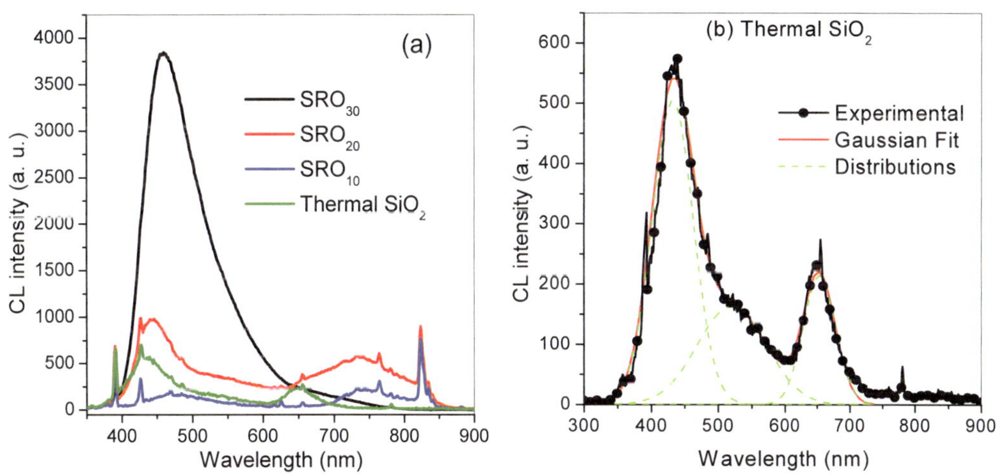

Fig. 10. (a) CL spectra of thermal SiO_2 and SRO annealed samples and (b) Gaussian fit of thermal SiO_2 sample.

In order to compare CL spectra between SiO_2 and SRO, CL spectrum from thermal SiO_2 was obtained, as shown in Fig. 10 (a). CL of SiO_2 results are helpful because a large fraction of the SRO films consists of silicon oxide. As can be seen, spectrum from thermal SiO_2 is similar to SRO_{20} but it does not include the red to infrared emission. To compare emission bands, Gaussian deconvolution was also obtained from thermal oxide; the best fit was centered at 432, 520 and 653 nm, as shown in Fig. 10 (b). Clearly only the 650 nm peak in both SRO_{20} and SiO_2 are comparable. The other emissions are enhanced in both Ro 20 and 30. That could mean that emission in SRO with low silicon excess and without silicon nCs are mainly due to silicon oxidation states.

In summary, luminescence peaks at 460, 520, and 620 nm are mainly due to defects such as Oxygen deficiency-related centers (ODC), E'δ defect, and non-bridging oxygen hole centers (NBOHC), respectively (Cervera et al., 2006; Fitting et al., 2001; Inokuma et al., 1998). Electroluminescence (EL) studies on SRO films with the same silicon excess as in this work exhibit an emission band similar to our CL results with peaks at 450, 500, 550, and 640 nm (Morales-Sánchez et al., 2010).Then the mentioned defects should exist into the SRO films, which are only excited with electron of high energy generated by CL or by EL.

The emission in red and NIR region could be associated to Si-clusters of less of 2 nm and defects interaction. In (Morales-Sánchez et al., 2008), the authors proposed that the band gap is large when the size of the Si cluster is small, and then the energy difference between the defects (localized state) and the Si clusters is big enough to produce the emission. This result is similar to the decay of Acceptor-Donator pair in a crystalline semiconductor. Therefore, we can assume that the absorption and emission processes in SRO films are connected with electrons decay between donor acceptor pairs (localized states) inside the silicon oxide band gap.

3.4 Depth profiling by different electron beam energy

It is well known that depth analysis can be obtained varying the electron energy (excitation range of CL) (Goldberg et al., 1998; Wittry & Kyser, 1967). In this case the CL intensity versus energy would remain constant when assuming a homogeneous depth distribution of luminescence centers. Deviation of such a behavior is easy to detect and should be interpreted.

The depth-resolved CL spectra for SRO annealed samples measured with electron beam energies of 2.5, 5, 10 and 15 keV was obtained, shown in Fig. 11. As can be seen, there is no band shift or any new emission bands by varying the beam energy, indicating specific luminescence correlated with a homogeneous depth impurity distribution in the samples.

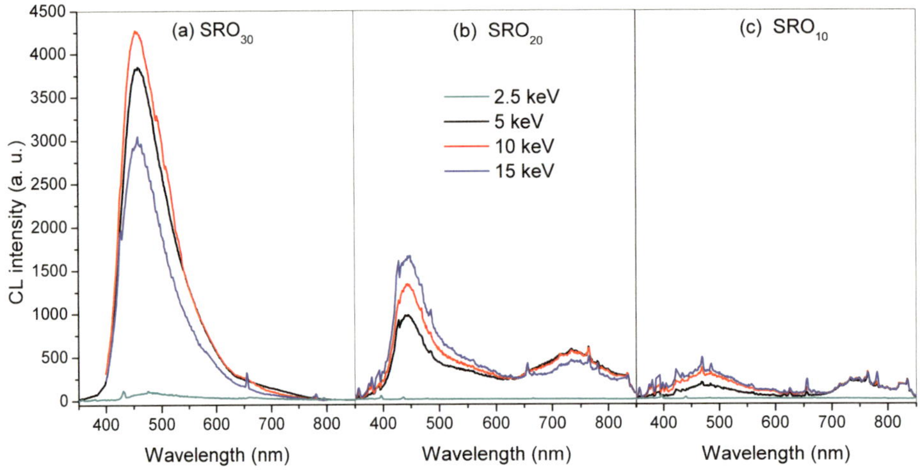

Fig. 11. Cathodoluminescence spectra of SRO annealed films at various electron energies. The spectra were measured at the room temperature.

3.5 Effect of the electron excitation on the emissive centers

In order to study the effect of the cathode excitation on the emissive centers, PL spectra were obtained before and after CL measurements with different energy electron doses; additionally, the effect of time excitation on CL spectra was obtained.

Different SRO_{20} and SRO_{30} annealed samples at 1100°C in N_2 by 180 minutes were used. One sample was not cathode-excitated, and it was used as reference. CL measurements were performed applying energy of 2.5, 5, 10 and 15 keV in each sample, respectively. In all cases the current used in the experiment was fixed at 0.3 mA.

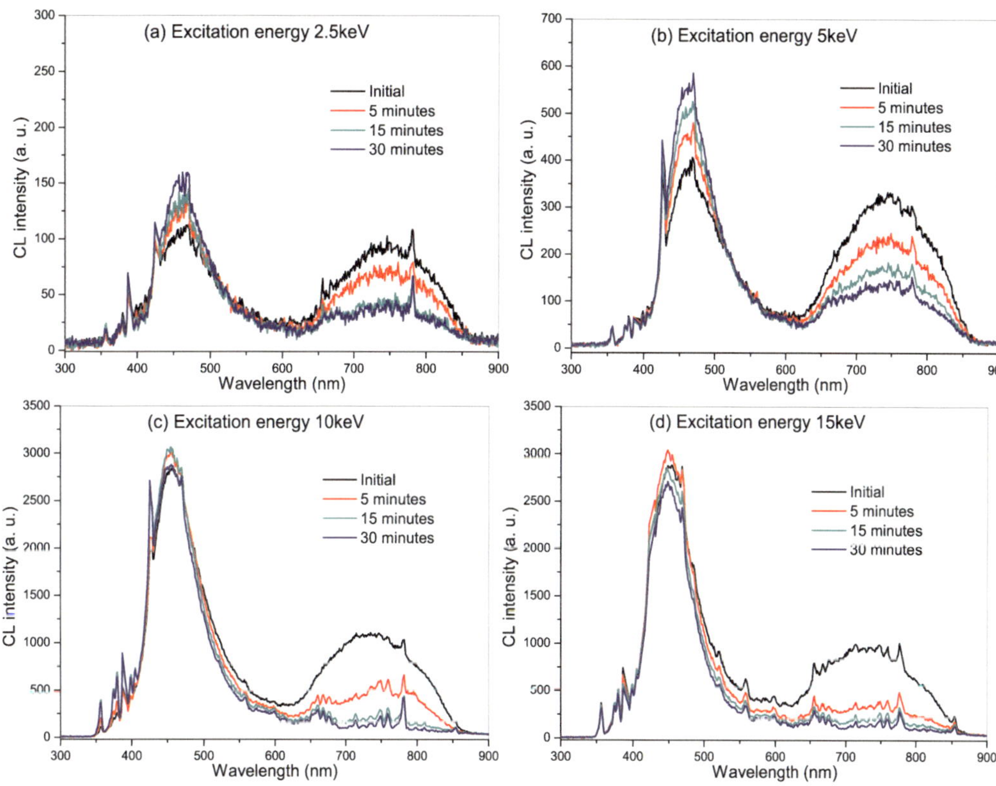

Fig. 12. CL spectra of SRO_{20} after different times of electron bombardment. Different samples were used: (a) sample excited with 2.5 keV, (b) sample excited with 5 keV, (c) sample excited with 10 keV, (d) sample excited with 15 keV.

CL spectra of SRO_{20} samples after different times of electron bombardment and different energy are shown in Fig. 12. CL spectrum consists of a broad emission in the visible and near infrared region (NIR) from ~400 to 850 nm. The different excitation energies produce both characteristic bands of SRO_{20} (red and blue). It can be observe in all the samples and independent of excitation energy that the blue emission does not have a considerable change while red band rapidly decrease or quenched with cathodo-excitation time. Therefore the excitation with high energy electrons destroys or modifies the red emissive states, inclusive with low energy (2.5 keV). Using electron energies greater than or equal to 10 keV, red band practically disappears after 5 minutes of bombardment.

PL spectra obtained after constant electron beam excitation during 30 minutes continually are shown in Fig. 13. After apply excitation energy of 2.5, 5 and 10 keV, the PL emission decreases considerately (50%), and with 15 keV PL emission is reduced 80%. This indicates that electron beam excitation modifies the structure of SRO due to the high energies used, and emission mechanism or emission centers were destroyed. Also it can be observed that after high energy excitation there is no new emission band.

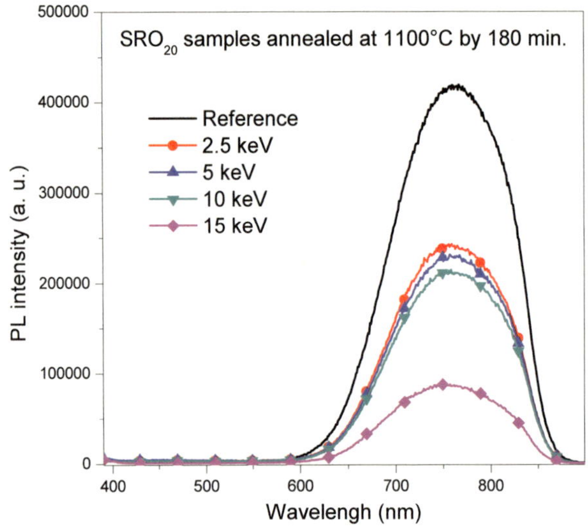

Fig. 13. PL spectra of SRO_{20} samples after continuous cathode-excitation during 30 minutes. Solid line is the reference sample emission (without electron beam irradiation). The symbols represent the PL emission after cathodo excitation.

The evolution of CL spectra for SRO_{30} annealed samples after 10minutes of constant electron bombardment is shown in Fig. 14 (a), (b) and (c) with 5, 10 and 15 keV respectively. There is no CL emission when these samples were excited with 2.5 keV. Samples excited with higher energies (more than 5 keV) exhibit strong blue emission with a shoulder in higher wavelength, characteristic emission of SRO_{30}. Blue emission centered at ~460 nm increases after 10 minutes of constant cathode excitation. The PL spectra of these samples were obtained after constant electron beam excitation during 10 minutes, and shown in Fig. 15. Similar to SRO_{20}, after cathode-excitation the PL emission decreases considerately (~50%), and as energy electron increase the PL emission reduces.

As mentioned before, blue emission centered at ~460 nm has been assigned to oxygen deficiency. Also, it has been proposed that high energy excitation and long exposition time are responsible of this band in SiO_2 (Chen et al., 1999; Goldberg et al., 1997; Inokuma et al., 1998). Since blue emission increases slightly after constant electron excitation in SRO, it is possible that oxygen deficiencies are created by excitation. However, we can assume that the emission in the blue region is due to the non-stoichiometry of the SRO because blue emission is not observed in PL neither before nor after of cathode excitation. Besides, blue emission has also been found in thermoluminescence and electroluminescence in SRO deposited by LPCVD (Morales-Sánchez et al., 2010; Piters et al., 2010). Therefore, the emission is due to band to band recombination through localized states in the SRO.

Degradation of the intensity in red region could be due to the passivation of luminescent centers. It is possible that the electron beam introduces changes around the emission center that affect the emission rate of emission of radiative and non radiative transitions.

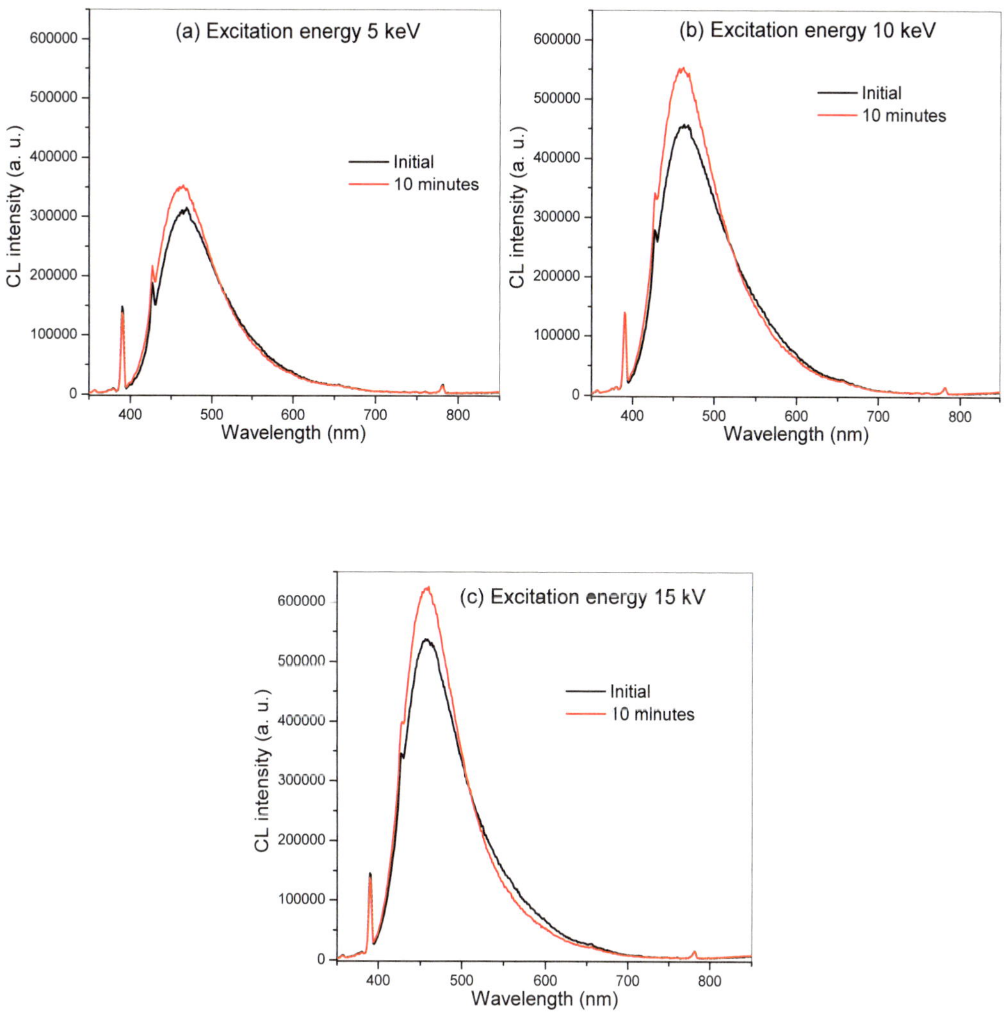

Fig. 14. CL spectra of SRO_{30} annealed samples after different times of electron bombardment. Different samples were used: (a) sample excited with 5 keV, (b) sample excited with 10 keV, (c) sample excited with 15 keV.

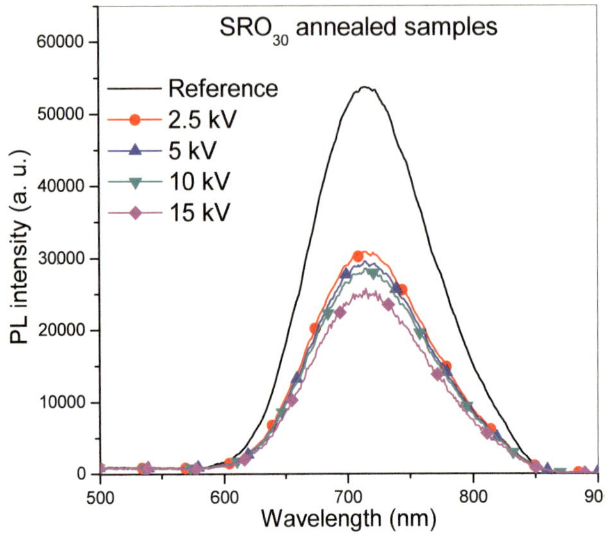

Fig. 15. PL spectra of SRO_{30} after cathode excitation at different beam energies. A sample without electron beam irradiation was used as reference (solid line).

4. Conclusion

SRO films with different silicon excess and with and without thermal treatment were studied using CL and PL. In order to understand their emission FTIR and XPS were used to study the structural composition and the effect of thermal treatment. FTIR and XPS results show the formation of si nanoaglomerates in all samples. Analysis of the XPS Si 2p peaks shows the existence of some chemical structures corresponding to the Si oxidation states in the SRO films; these oxidation states depend on the silicon excess. The concentration of each oxidation state was estimated. According to variation of silicon compounds found in analysis of the Si 2p spectra, we can assume that as silicon excess decrease Si-nCs density decreases and defects (or silicon oxygen compounds) tend to predominate over Si nanoclusters or nanocrystals. In fact, when the Si nanoclusters are very small, the interface will play important role due to the large stress in the interface of the nanoclusters and the localized states will be form.

Photo and Cathodoluminescence (PL and CL) properties in Silicon Rich Oxide (SRO) films with different silicon excess were studied. SRO deposited by LPCVD has shown high visible PL and CL emission at room temperature mainly after thermal treatment at high temperatures. Samples with lower silicon excess show maximum luminescence while samples with higher silicon excess show low luminescence. Strong emission from 650 to 800 nm was found in PL while CL spectra show luminescence emission from 400 to 850 nm in SRO annealed samples. CL spectra are wider than the PL one, because high energy excitation leads the emission of all the emissive centers present in the SRO. Also was found that red and NIR region emission is affected by electron beam.

Luminescent emission in blue-green region is mainly due to defect created by silicon excess, such as oxygen deficiency-related centers (ODC), E'δ defect, and non-bridging oxygen hole centers (NBOHC). Emission in red and near infrared region is associated to some defects that are acting as localized states and also to Si nanoaglomerates and defects interaction.

5. Acknowledgment

The authors express their thanks to the microelectronic laboratory technicians of INAOE for their support and to the CONACYT for providing financial support for this work.

6. References

Aceves, M., Falcony, C., Reynoso-Hernandez, A., C. W., & Torres, A. (1996). The conduction properties of the silicon/off-stoichiometry-SiO_2 diode. *Solid-State Electronics,* Vol. 39 No. 5, (May 1996), pp. 637-644, ISSN 00381101.

Aceves, M., Malik, A., & Murphy, R. (2001). The FTO/SRO/SI structure as a radiation sensor, In: *Sensors & Chemometrics,* M. T. Ramírez-Silva, M. A. Romero Romo, & M. E. Palomar Pardavé (Ed.), 1-25, Research Signpost, ISBN 81-7736-067-1, India.

Aceves, M., Pedraza, J., Reynoso-Hernandez, J. A., Falcony, C., & Calleja, W. (1999). Study on the Al/silicon rich oxide/Si structure as a surge suppressor, DC, frequency response and modeling. *Microelectronics Journal,*Vol. 30, No. 9, (September 1999), pp. 855-862, ISSN 00262692.

Alfonsetti, R., Lozzi, L., Passacantando, M., Picozzi, P., & Santucci, S. (1993). XPS studies on SiOx thin films. *Applied Surface Science,* Vol. 70-71, No. part 1,(June 1993), pp. 222-225, ISSN 01694332.

Ay, F., & Aydinli, A. (2004). Comparative investigation of hydrogen bonding in silicon based PECVD grown dielectrics for optical waveguides. *Optical Materials,* Vol. 26 No. 1, (June 2004), pp. 33-46, ISSN 09253467.

Brüesch, P., Stockmeier, T., Stucki, F., & Buffat, P. A. (1993). Physical properties of semi-insulating polycrystalline silicon. I. structure, electronic properties, and electrical conductivity. *Journal of Applied Physics,* Vol. 73, No. 11, (n. d.), pp. 7677-7689, ISSN 00218979.

Calleja, W., Aceves, M., & Falcony, C. (1998). EEPROM transistor fabricated with stacked SiOx LPCVD films. *Electronics Letters,* Vol. 34, No. 13, (June 1998), pp. 1294-1296, ISSN 00135194.

Cervera, M., Hernández, M. J., Rodríguez, P., Piqueras, J., Avella, M., González, M. A., & Jimenez, J. (2006). Blue-cathodoluminescent layers synthesis by high-dose N^+, C^+ and B^+ SiO_2 implantation. *Journal of Luminescence,* Vol. 117, No.1, (March 2006), pp. 95-100, ISSN 00222313.

Chen, C. M., Pan, H. C., Zhu, D. Z., Hu, J., & Li, M. Q. (1999). Cathodoluminescence and ion beam analysis of ion-implanted combinatorial materials libraries on thermally grown SiO_2. *Nuclear Instruments and Methods in Physics Research.* Vol. 159 No.1-2, pp. 81-88, ISSN 0168583X.

Chen, X. Y., Lu, Y. F., Tang, L. J., Wu, Y. H., Cho, B. J., Xu, X. J., Dong, J. R & Song, W. D. (2005). Annealing and oxidation of silicon oxide films prepared by plasma-enhanced chemical vapor deposition. *Journal of Applied Physics,* Vol. 97 No. 1, (January 2005), pp. 014913-1-014913-1, ISSN 00218979.

Chen, X. Y., Lu, Y. F., Wu, Y. H., Cho, B. J., Song, W. D., & Dai, D. Y. (2004). Optical properties of SiOx nanostructured films by pulsed-laser deposition at different substrate temperatures. *Journal of Applied Physics,* Vol. 96, No. 6, (September 2004), pp. 3180-3186, ISSN 00218979.

Chen, X. Y., Lu, Y. F., Wu, Y. H., Cho, B. J., Tang, L. J., Lu, D., & Dong, J. R. (2006). Correlation between optical properties and si nanocrystal formation of si-rich si oxide films prepared by plasma-enhanced chemical vapor deposition. *Applied Surface Science,* Vol. 253, No. 5, (December 2006), pp. 2718-2726, ISSN 01694332.

Daldosso, N., Das, G., Larcheri, S., Mariotto, G., Dalba, G., Pavesi, L., Irrera, A., Priolo, F., Iacona, F. & Rocca, F. (2007). Silicon nanocrystal formation in annealed silicon-rich silicon oxide films prepared by plasma enhanced chemical vapor deposition. *Journal of Applied Physics,* Vol. 101, No. 11, (n. d.), ISSN 00218979.

Dehan, E., Temple-Boyer, P., Henda, R., Pedroviejo, J. J., & Scheid, E. (1995). Optical and structural properties of SiOx and SiNx materials. *Thin Solid Films,* Vol. 266, No. 1, (September 1995), pp. 14-19, ISSN 00406090.

DiMaria, D. J., Kirtley, J. R., Pakulis, E. J., Dong, D. W., Kuan, T. S., Pesavento, F. L., Theis, T., & Cutro, J. A. (1984). Electroluminescence studies in silicon dioxide films containing tiny silicon islands. *Journal of Applied Physics,* Vol. 56, No. 2, pp. 401-416, ISSN 00218979.

Dong, D., Irene, E. A., & Young, D. R. (1978). Preparation and some properties of chemically vapor-deposited Si-rich SiO_2 and Si_3N_4 films. *Journal of the Electrochemical Society,* Vol. 125, No. 5, (May 1978), pp. 819-823, ISSN 00134651.

Fazio, E., Barletta, E., Barreca, F., Neri, F., & Trusso, S. (2005). Investigation of a nanocrystalline silicon phase embedded in SiOx thin films grown by pulsed laser deposition. *Journal of Vacuum Science and Technology B,* Vol. 23, No. 2, (n. d.), pp. 519-524, ISSN 10711023.

Fitting, H. J. (2009). Can we make silica luminescent?. *Optical Materials,* Vol. 31, No. 12, (October 2009), pp. 1891-1893, ISSN 09253467.

Fitting, H. J., Barfels, T., & Trukhin, A. N. (2001). Cathodoluminescence of crystalline and amorphous SiO_2 and GeO_2. *Journal of Non-Crystalline Solids,* Vol. 279, No. 1, (January 2001), pp. 51-59, ISSN 00223093.

Goldberg, M., Barfels, T., & Fitting, H. J. (1998). Cathodoluminescence depth analysis in SiO_2-Si-systems. *Fresenius' Journal of Analytical Chemistry,* Vol. 361 No. 6-7, (n. d.), pp. 560-561, ISSN 09370633.

Goldberg, M., Fitting, H. J., & Trukhin, A. (1997). Cathodoluminescence and cathodoelectroluminescence of amorphous SiO_2 films. *Journal of Non-Crystalline Solids,* Vol. 220, No. 1, (October 1997), pp. 69-77, ISSN 00223093.

Gritsenko, V. A., Shavalgin, Y. G., Pundur, P. A., Wong, H., & Lau, W. M. (1999). Cathodoluminescence and photoluminescence of amorphous silicon oxynitride. *Microelectronics Reliability,* Vol. 39, No. 5, (May 1999), pp. 715-718, ISSN 00262714.

Hanaizumi, O., Ono, K., & Ogawa, Y. (2003). Blue-light emission from sputtered Si:SiO_2 films without annealing. *Applied Physics Letter,* Vol. 82, No. 4, (January 2003), pp. 538-540, ISSN 00036951.

Iacona, F., Franzo, G., & Spinella, C. (2000). Correlation between luminescence and structural properties of Si nanocrystals. *Journal of Applied Physics,* Vol. 87, No. 3, (February 2000), pp. 1295–1303, ISSN 00218979.

Inokuma, T., Kurata, Y., & Hasegawa, S. (1998). Cathodoluminescence properties of silicon nanocrystallites embedded in silicon oxide thin films. *Journal of Luminescence,* Vol. 80, No. 1-4, (December 1998), pp. 247-251, ISSN 00222313.

Irene, E. A., Chou, N. J., Dong, D. W., & Tierney, E. (1980). On the Nature of CVD Si-Rich SiO_2 and Si_3N_4 Films. *J. Electrochem. Soc.*, Vol. 127, No. 11, pp. 2518-2521.

Kenyon, A. J., Trwoga, P. F., Pitt, C. W., & Rehm, G. (1996). The origin of photoluminescence from thin films of silicon-rich silica. *Journal of Applied Physics*, Vol. 79, No. 12, (June 1996), pp. 9291-9300, ISSN 00218979.

Khriachtchev, L., Novikov, S., & Lahtinen, J. (2002). Thermal annealing of Si/SiO_2 materials: Modification of structural and photoluminescence emission properties. *Journal of Applied Physics*, Vol. 92, No. 10, (November 2002), pp. 5856-5862, ISSN 00218979.

Liu, Y., Chen, T. P., Fu, Y. Q., Tse, M. S., Hsieh, J. H., Ho, P. F., & Liu Y. C. (2003). A study on si nanocrystal formation in Si-implanted SiO_2 films by x-ray photoelectron spectroscopy. *Journal of Physics D*, Vol. 36, No. 19, (October 2003), pp. L97-L100, ISSN 00223727.

Lockwood, D. J., Lu, Z. H., & Baribeau, J. (1996). Quantum confined luminescence in Si/SiO_2 superlattices. *Physical Review Letters*, Vol. 76, No. 3, (n. d.), pp. 539-541, ISSN 00319007.

López-Estopier, R., Aceves-Mijares, M., Yu, Z., & Falcony, C. (2011). Determination of the energy states of the donor acceptor decay emission in silicon rich oxide prepared by low-pressure chemical vapor deposition. *Journal of Vacuum Science and Technology B: Microelectronics and Nanometer Structures*, Vol. 29, No. 2, (March 2011), ISSN 10711023.

Morales, A., Barreto, J., Domínguez, C., Riera, M., Aceves, M., & Carrillo, J. (2007). Comparative study between silicon-rich oxide films obtained by LPCVD and PECVD. *Physica E: Low-Dimensional Systems and Nanostructures*, Vol. 38, No. 1-2, (April 2007), pp. 54-58, ISSN 13869477.

Morales-Sánchez, A., Aceves-Mijares, M., González-Fernández, A. A., Monfil-Leyva, K., Juvert, J., & Domínguez-Horna, C. (2010). Blue and red electroluminescence of silicon-rich oxide light emitting capacitors. *Proceedings of SPIE - the International Society for Optical Engineering*, Vol. 7719, ISBN 978-081948192-4, Brussels, Belgium, April 12, 2010.

Morales-Sánchez, A., Barreto, J., Domínguez-Horna, C., Aceves-Mijares, M., & Luna-López, J. A. (2008). Optical characterization of silicon rich oxide films. *Sensors and Actuators, A: Physical*, Vol. 142, No. 1, (March 2008), pp. 12-18, ISSN 09244247.

Nesbit, L. A. (1985). Annealing characteristics of si-rich SiO_2 films. *Applied Physics Letters*, Vol. 46, No. 1, (n. d.), pp. 38-40, ISSN 00036951.

Pai, P. G., Chao, S. S., Takagi, Y., & Lucovsky, G. (1986). Infrared spectroscopic study of SiOx films produced by plasma enhanced chemical vapor deposition. *Journal of Vacuum Science & Technology A: Vacuum, Surfaces, and Films*, Vol. 4, No. 3, pp. 689-694.

Pavesi, L., Dal Negro, L., Mazzoleni, C., & Franzo, G. (2000). Optical gain in silicon nanocrystals. *Nature*, Vol. 408, No. 6811, (November 2000), pp. 440-444, ISSN 00280836.

Philipp, H. R. (1972). Optical and bonding model for non-crystalline SiOx and SiOxNy materials. *Journal of Non-Crystalline Solids*, Vol. 8-10, No.C, (June 1972), pp. 627-632, ISSN 00223093.

Piters, T., Aceves-Mijares, M., D., B. M., Berriel-Valdos, L., & Luna-López, J. A. (2010). Dose dependent shift of the TL glow peak in a silicon rich oxide (SRO) film. *Revista mexicana de fisica*, Vol. S 57, No. 2, pp. 26-29.

Ribeiro, M., Pereyra, I., & Alayo, M. (2003). Silicon rich silicon oxynitride films for photoluminescence applications. *Thin Solid Films*, Vol. 426, No. 1-2, (February 2003), pp. 200-204, ISSN 00406090.

Shimizu-Iwayama, T., Yoichi, T., Atsushi, K., Motonori, K., Setsuo, N., & Kasuo, N. (1996). Visible photoluminiscence from nanocrystal formed in silicon dioxide by ion implantation and thermal processing. *Thin Solid Films*, Vol. 276, No. 1-2, (April 1996), pp. 104-107, ISSN 00406090.

Trukhin, A. N., Fitting, H. J., Barfels, T., & Czarnowski, V. (1999). Cathodoluminescence and IR absorption of oxygen deficient silica - influence of hydrogen treatment. *Journal of Non-Crystalline Solids*, Vol. 260, No. 1-2, (December 1999), pp. 132-140, ISSN 00223093.

Wang, Y. Q., Chen, W. D., Liao, X. B., & Cao, Z. X. (2003). Amorphous silicon nanoparticles in compound films grown on cold substrates for high-efficiency photoluminescence. *Nanotechnology*, Vol. 14, No. 11, (November 2003), pp. 1235-1238, ISSN 09574484.

Wittry, D. B., & Kyser, D. F. (1967). Measurement of diffusion lengths in direct-gap semiconductors by electron-beam excitation. *Journal of Applied Physics*, Vol. 38, No. 1, (n. d.), pp. 375-382, ISSN 00218979.

Yacobi, B. G., & Holt, D. B. (1990). *Cathodoluminescence Microscopy of inorganic solids*, Springer, ISBN 0-306-43314-1, New York:, USA.

Yang, D. Q., Gillet, J. N., Meunier, M., & Sacher, E. (2005). Room temperature oxidation kinetics of si nanoparticles in air, determined by x-ray photoelectron spectroscopy. *Journal of Applied Physics*, Vol. 97, No. 2, (January 2005), pp. 024303-1-024303-6, ISSN 00218979.

Yu, Z., Aceves, M., Carrillo, J., & Flores, F. (2003). Single electron charging in si nanocrystals embedded in silicon-rich oxide. *Nanotechnology*, Vol. 14, No. 9, (September 2003), pp. 959-964, ISSN 09574484.

Yu, Z., Aceves-Mijares, M., Luna-López, A., Du, J., & Bian, D. (2006b). Formation of silicon nanoislands on crystalline silicon substrates by thermal annealing of silicon rich oxide deposited by low pressure chemical vapour deposition. *Nanotechnology*, Vol. 17, No. 1, (October 2006), pp. 4962-4965, ISSN 09574484.

Yu, Z., Mariano, A., Luna-López, A., Quiroga, E., & López-Estopier, R. (2006a). Photoluminescence and single electron effect of nanosized silicon materials, In *Focus on Nanomaterials Research*, B. M. Caruta (Ed.), pp. 233-273, Nova Science Publishers, ISBN 1-59454-897-8, New York, USA.

11

Textural Characterization of Sedimentary Zircon and Its Implication

Yixiong Qian[1], Wen Su[2], Zhong Li[2] and Shoutao Peng[1]
[1]*Research Institute of Exploration & Production, Sinopec, Beijing*
[2]*Institute of Geology and Geophysics, Chinese Academy of Sciences, Beijing, China*

1. Introduction

Zircon with highly refractory, as accessory, detrital minerals occurs in virtually all sediments and sedimentary rocks, has played a prominent and complex role in interpreting the history of sediments and source history of a deposit, paleogeography, and tectonic reconstructions (Qian Y.X.,et al.,2007a). The main use of CL imaging of Detrital zircon has been as an adjunct to U-Pb dating of zircons, which allows identification of different types of zircon domains ,then may be dated in situ within a spatial resolution of about 15~30μm. By using a SHRIMP(Sensitive High-Resolution Ion Microprobe) , this data analysis would be a powerful tool in understanding origin of the zircon , that is , the various geological processes(magmatism, deposition , metamorphism, hydrothermal alteration, metasomatic leaching of Th, U, and Pb) , their surrounding source terrane of known age may be traced and a geological evolution history of sedimentary basins may be established. In fact, the analyzed sample wouldn't completely represent geological history by including evidence of all the possible provenances and their relationships to each other due to natural complexity of sample. It was mentioned that the analysis techniques involve the sampling protocol and interpretation of data and then show the application of detrital zircon studies to: (1) analysis of origin characteristics such as U-Pb isotopic analysis and composition, (2) determine on the using age of stratigraphic successions and to help recognize time gaps in the geologic records, (3) test regional paleogeographic reconstructions by origin analysis, and (4) reveal geological history relative with the mineral chemistry of detrital zircon, and reveal complexity in order to gain insight into natural processes.

2. Analysis of origin characteristics such as U-Pb isotopic data and composition

The distribution of heavy minerals in sediments is affected by the following factors: provenance, uplift and erosion, paleo-topographic feature, palaeo-climate and palaeo-environment. The analysis of heavy minerals can be applied in explanation of the sedimentary response to tectonic cycles based on the provenance and sedimentary environment.

The heavy minerals assemblage can be act as indicators of provenance. The assemblage of zircon, tourmalie, apatite and little biotite with the well-development crystal shapes is generally believed to come from granitic rocks; the assemblage of a great quantity of Garnet, zircon, epidote, chlorite is derived from metamorphic rocks; while the assemblage of magnetite, ilmentite, anatase, augite and hornblende is commonly be readily traced to balt-igneous rocks. In addition, the characteristic of heavy minerals are not only indictor of the composite of provenance, but also of physical-selection, mechanical abrasion of granular and chemical dissolution of sedimentary rocks during the process of sedimentary and transportation. Based on the primary analysis of sandstones with regard to sedimentary environment and provenance, five heavy minerals assemblages can be reasonable classified as follows: ①stable minerals assemblage: includes the most of Ti oxide , zircon and Tourmalie; ②the relative stable minerals assemblage: predominantly composes of garnet and apatite; ③un-stable minerals assemblage: be dominated by epidote、augite and Hornblende; ④the assemblage of the indication of initial depositional environment :be associated with hematite, pyrite, glauconite, barite, and carbonate minerals; ⑤ore minerals present in hydrothermal mineralization. In addition, the index of ZTR, which shows the percent of zircon, toumalie, rutile, and oblique carbonate minerals in heavy minerals assemblage, the high value ZTR, the great maturity of heavy mineral.

3. Reveal geological history relative with the mineral chemistry of detrital zircon, and reveal complexity in order to gain insight into natural processes

The mineral zircon is extremely variable both in terms of external morphology and internal textures. These features reflect the geologic history of the mineral, especially the relevant episode(s) of magmatic or metamorphic crystallization (and recrystallization), strain imposed both by external forces and by internal volume expansion caused by metamictization, and chemical alteration. One of the major advantages of zircon is its ability to survive magmatic, metamorphic and erosional processes that destroy most other common minerals. Zircon-forming events tend to be preserved as distinct structural entities on a pre-existing zircon grain. Because of this ability, quite commonly zircons consist of distinct segments, each preserving a particular period of zircon-formation(or consumption).

In common rocks, zircon ranges in size from about 20 to 200μm .The elongation (length-to-width) ratios is ranging from 1 to 5. which is commonly believed to reflect crystallization velocity. Indeed, needle-shaped acicular zircon crystals are common in rapidly crystallized, porphyritic, sub-volcanic intrusions, high-level granites, and gabbros. In addition, newly-grown zircon crystals can themselves exhibit evidence for multiple stages of growth and corrosion.

Sedimentary rocks may also contain a significant fraction of zircon. Although authigenic zircon has been reported, sedimentary zircon is predominantly derived from weathered igneous and metamorphic rocks. Detrital zircon in sedimentary rocks and sediments is highly durable and records age information of crustal units that contributed to the sediment load.

The External morphologies of zircons in heavy minerals are included the followings: colors, opaque, lustre, roundness, coarse, dissolution, abrasion and elongation, based on the its roundness of grains, which is to unravel the transportation distance of sediments, which has been attributed to its provinces, two subgroups are classified as:

1. Zircons with a high degree roundness: which are mostly consist of dark pink-rose with few of yellow-pink colors ,sub-opaque , coarse glass luster, and sub-rounded and rounded grains with obvious relics of abrasion , with elongation (length-to-width) ratios ranging from 1 to 2, and preferential growth of grains 0.05mm~0.2mm , in response to transportation of sediments
2. Zircons with a medium degree roundness: : which are dominantly composed by sub-opaque or opaque, weak-diamond to coarse glass luster, pink-shallow pink colors and sub-rounded and rounded grains, euhedral with obvious relics of abrasion of mostly grains with few smooth surface of grains , with elongation (length-to-width) ratios ranging from 1.5 to 3.5 and with a size from 0.05mm~0.35mm in diameter of grains , the grains with a rather high round would be taken a great parts among all of grains with a well selection, indicating the obvious transportation of sediments 。

In simple way, three different types of detrital zircons would be termed :

1. the igneous zircon with six- prismatic concentric zone (Fig.1);
2. Metamorphic zircon with a internal structure of core-mantle-rimmed; either the core-mantle-rimmed or the core- rimmed commonly appears to developed in metamorphic zircon with a different component of core, mantle and rim(Fig.1). The origin and inherit zircons may be observed in the core part of grain of zircon, while the metamorphic zircons grows in the mantle or rimmed(Hermann et al. , 2001) ; equant and weak zoning zircons are not likely to occur in granulite rocks; it is not uncommon to find zircons with a fan-shape and oscillatory zoning in the process of serpentinization of basic and ultra basic rocks , the zircons can retain a plane zoning in response to the metamorphic re-crystallization either in re-melting migmatite or in a neocryst of ovoid shape(Fig.1);
3. re-cycle sedimentary zircons with a feature of cathodoluminescence (CL) or no growth zoning and metamict zones(Fig.1).

The zircons in the Lower Silurian sandstones, well Shun1 also shows the difference among the samples, for example, 15 percent, the most lowest content of six- prismatic concentric zone, 71 percent, the highest content of non-zoning and 33 percent, the rather high cathodoluminescence (CL) are presented in Shun1-21 ; indicating the stable re-cycle sedimentary environment; while 20 percent of six-prismatic concentric zone , 33 percent of structure of core-mantle-rimmed ,48 percent of non-zoning, and 25 percent of cathodoluminescence (CL) in Shun1-22, illustrated that igneous and metamorphic zircon has been majorly attributed to the detrial zircons(Fig.2).

It would be reasonable predicted that unaltered igneous zircon generally contains the highest contents of U 、Th and Pb, which is a comparable to that of Metamorphic zircon and Re-cycle sedimentary zircons, and the re-cycle sedimentary zircons would be various in contents between the igneous zircon and Metamorphic. U as a un-compatible element, would be accumulated in anatectic melting crystallization and partial differentitation; while the contents for U and Th decreased in the metamorphic process due to being dispelled from the crystal lattices during the re- crystallization in solid. In addition, the contents for Pb is supposed to relate the contents of U and Th and age of rocks, the higher of contents of U and Th and the older of age of rocks; the greater of the accumulated content of radiogenic lead in zircons.

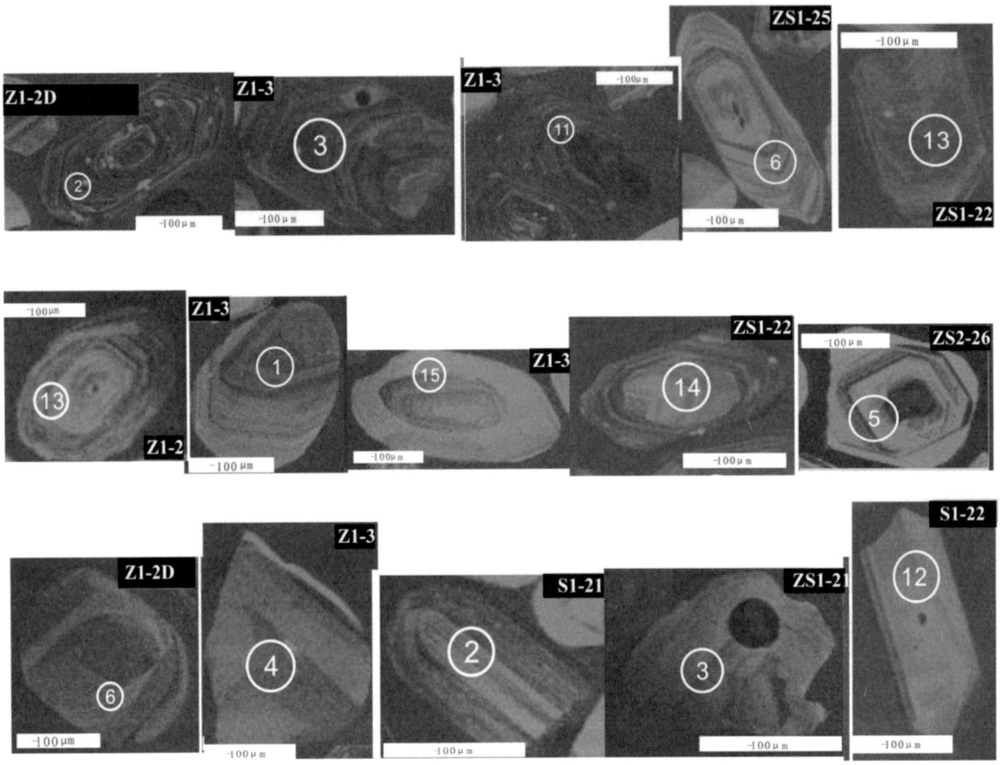

1. Alteration developed along concentric fractures parallel to the boundary between the low-U interior part of the zircon and the high-U outer shell; an outermost low-U rim has radial cracks, which have allowed the access of the fluids(Z1-2(D); Partially preserved growth zoned zircon penetrated by trangressive zones of recrystallization and with local development of recrystallization or convolute zoning (ZS2-25);
2. Recrystallization and new growth of zircon in high-grade metamorphic rocks, Bands or other large segment of homogeneously textured zircon (ZS2-26); strong variations in the relative development of zoned domains, large uniform zone external and much finer oscillatory-zoned bands in internal structure (Z1-3)
3. Late to post-magmatic recrystallization of zircon, Variable appearance of xenocrystic cores in magmatic (Z1-2D,ZS-2)and high-grade meta-morphic rocks(Z1-3,ZS2-26),Z1-2(D),Z1-3); Variations in growth zoning in magmatic zirconAppearance and texture of zircon in meteorite impact structures(ZS1-21);

Fig. 1. Typical Zircon CL structures for different origin and SHRIMP U-Pb Concordia plots with spot locations, from Donghe sandstones of DongheTang formation, the Upper Devonian in western Tazhong area , or the Lower Silurian sandstones ,Tarim basin: igneous zircons(upper),Metamorphic zircon(Middle),Re-cycle sedimentary zircons(down),different scales in zircon megacryst,ranging from 0.5 mm down to a few 50um.

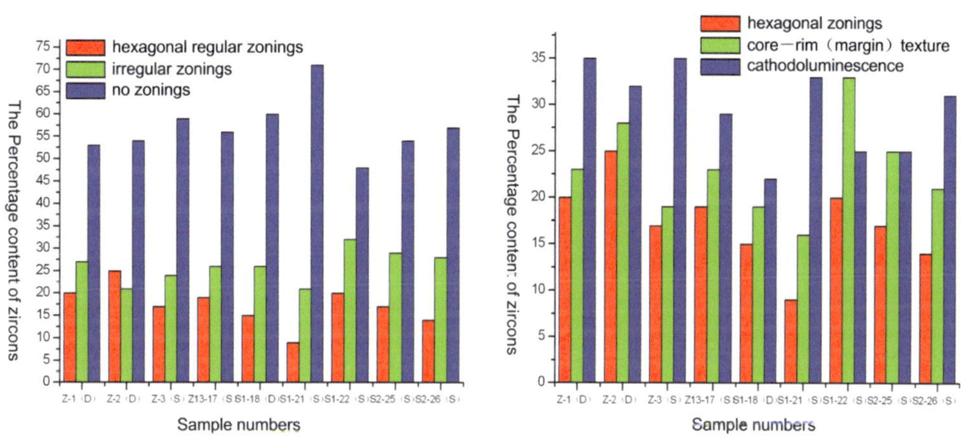

Fig. 2. Diagram showing the statistics of typical Zircon grains based on their sources from Donghe(lift)and the Lower Silurian sandstones(right) in wells of Zhong1,11,12,13 and Shun1,2. Tarim basin (Qian Y.X.,et al.,2007b)

4. Determine maximum age of stratigraphic successions and to help recognize time gaps in the geologic records

The most reliable means of directly determining depositional sedimentary ages is through the dating of interstratified volcanic rocks, or dating time-of-deposition authigenic xenotime overgrowths on detrital zircon grains (Mc Naughton *et al.* 1999). Under certain circumstances, the age of the youngest detrital zircon in a population can approach the age of deposition (Nelson, 2001). Disconformities reveal information about coeval tectonics/uplift and sea-level fluctuations. In the absence of fossils to identify gaps in the geologic record.

The results are typically the U-pb isotopic data of zircon from sensitive high resolution ion microprobe(SHRIMP)analyses can provide the accurate age constrains for record of events associated with the major geodynamic evolution of stage from Proterozoic to the Early Paleozoic in Tarim basin and its surrounding orogenic belts.

The U-pb isotopic data of zircons can be divided into three groups , i.e. 1.8Ga,2.2Ga and 2.6Ga from Meso-Proterozoic to Neo- Proterozoic; 0.84Ga in Cambrian; and 477~439.8Ma and 431.6~421.1Ma from Ordovician to The Lower Silurian(Fig.3), which have a concordance with the records of rocks derived from that of wells or outcrops of Tarim basin and its surrounding orogenic belts.

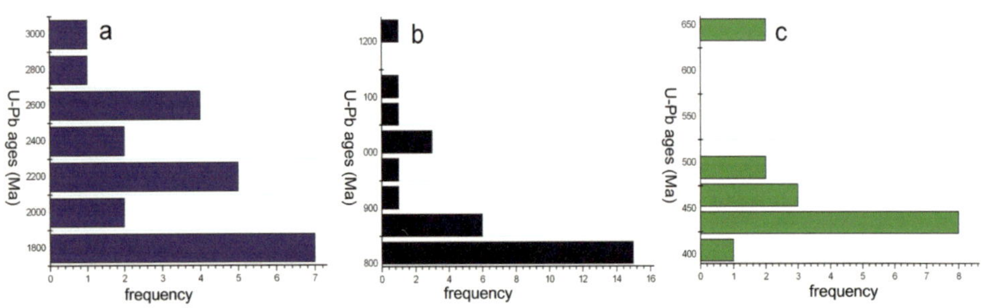

Fig. 3. Frequency histograms for Distribution of $^{206}Pb/^{238}U$ ages of Zircon from Proterozoic(a) to Neo-Meso Proterozoic(b) and Paleozoic(c) in western Tazhong area ,Tarim basin

5. Test regional paleogeographic reconstructions by origin analysis

Previously published research from the Kuqa Sub-basin along northern margin of the Tarim Basin shows five tectonic-depositional phases from Triassic to Neogene time(Li *et al.*,2010). In order to reveal more detailed information on the nature of provenance terrains and tectonic attributes since late Mesozoic time, five typical sandstone samples from Jurassic Neogene strata were collected for U-Pb dating of detrital zircons.

Geochronological constitution of detrital zircons of the Middle Jurassic sample is essentially unimodal and indicates major contributions from the SouthTian Shan, wherein most 370~450Ma zircons probably resulted from tectonic accretion events between the Central Tian Shan block and SouthTian Shan Ocean during Silurian and Devonian time, with sandstone provenance tectonic attributes of passive continental margin. The Lower Cretaceous sample shows a complicated provenance detritalzircon signature, with new peak ages of 290~330Ma as well as 370(or 350)~450Ma showing evidence of arc orogenic provenance tectonic attribute, probably refecting a new provenance supply that resulted from denudation process within the SouthTian Shan and SouthTian Shan suture. There are no obvious changes within age probability spectra of detrital zircons between the Cretaceous and early Paleogene samples, which suggests that similar provenance types and basin-range framework continued from Cretaceous to Early Paleogene time. However, unlike the Cretaceou sand early Paleogene samples, an age spectra of the Miocene sample is relatively unimodal and similar to that of the Pliocene sample, with peak ages ranging between 1392 and 1458Ma older than the comparable provenance ages (peak ages about 370~450Ma) of the Middle Jurassic and Lower Cretaceous samples. Therefore, we can conclude that the SouthTian Shan was rapidly exhumated and the southern South Tian Shan had become the main source of clastics for the Kuqa Sub-basin since the Miocene epoch. and the corresponding age-probability plots in Fig.4.

We contrast the detrital zircon age spectra from the Kuqa Subbasin with those of potential provenance areas, discussing the implications for prove nance and paleogeographical changes. Detrital zircons of the Middle Jurassic sample mostly range from 370 to 450Ma, with a small number of Proterozoic-Archean ages.This geochronological constitution is comparatively unimodal and indicates that the dominant source of detritus was likely from the northern South Tian Shan and southern Central Tian Shan (Fig.5), probably resulting from tectonic accretion events between the Central Tian Shan Block and the South Tian Shan Ocean that occurred during Silurian and early Devonian time, with dominant sandstone provenance tectonic attributes of passive continental margin discriminated by major element composition of whole-rock samples.The Lower Cretaceous sample shows a provenance with complicated detrital zircon age spectra, with new peak ages of 290-330Ma as well as 370 (or 350)-450Ma. Refecting a new provenance supply produced by denudation of the South Tian Shan and South Tian Shan suture. These grains likely refect Carboniferous-Permian volcanism and the Silurian-Devonian tectonic events between the South Tian Shan-Tarim and Central Tian Shan blocks, with dominant sandstone provenance tectonic attributes of active continental margin or island arc (arc orogenic belt). In addition, several clusters of Proterozoic-Archean ages from this sample probably re£ect that some provenance regions may have been deeply exhumated.The lack of obvious changes in detrital zircon age probability spectra between the Cretaceous and early Paleogene samples suggests that similar provenance types and basin-range framework likely continued from Cretaceous to early Paleogene time(Fig. 5).

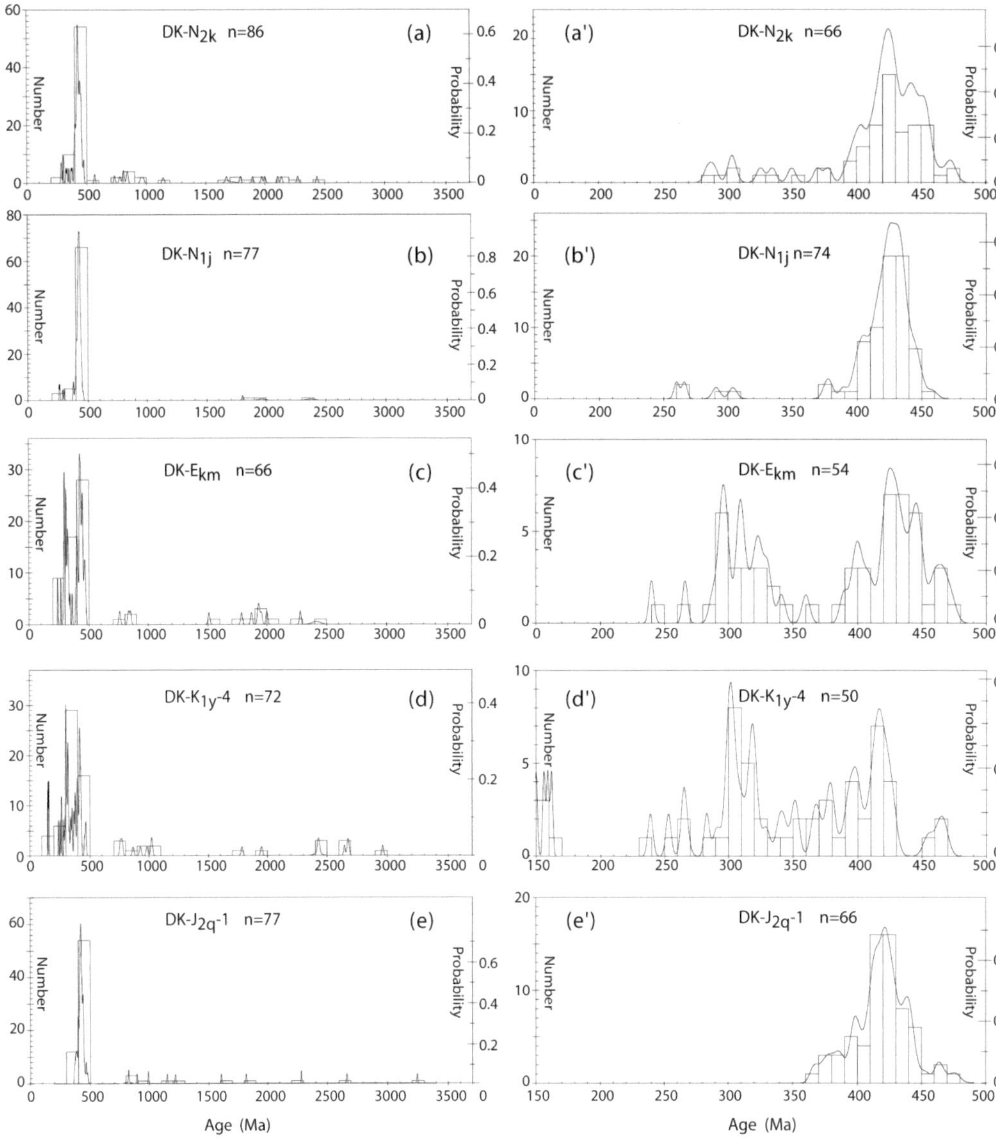

Fig. 4. Probability plots and number histograms of U-Pb ages of detrital zircons from Middle Jurassic-Neogene sandstone samples in Kuqa Subbasin. (a)-(e) 0-3700Ma grains; (a0)-(e0) 150-500Ma grains(after Li *et al.*,2010).

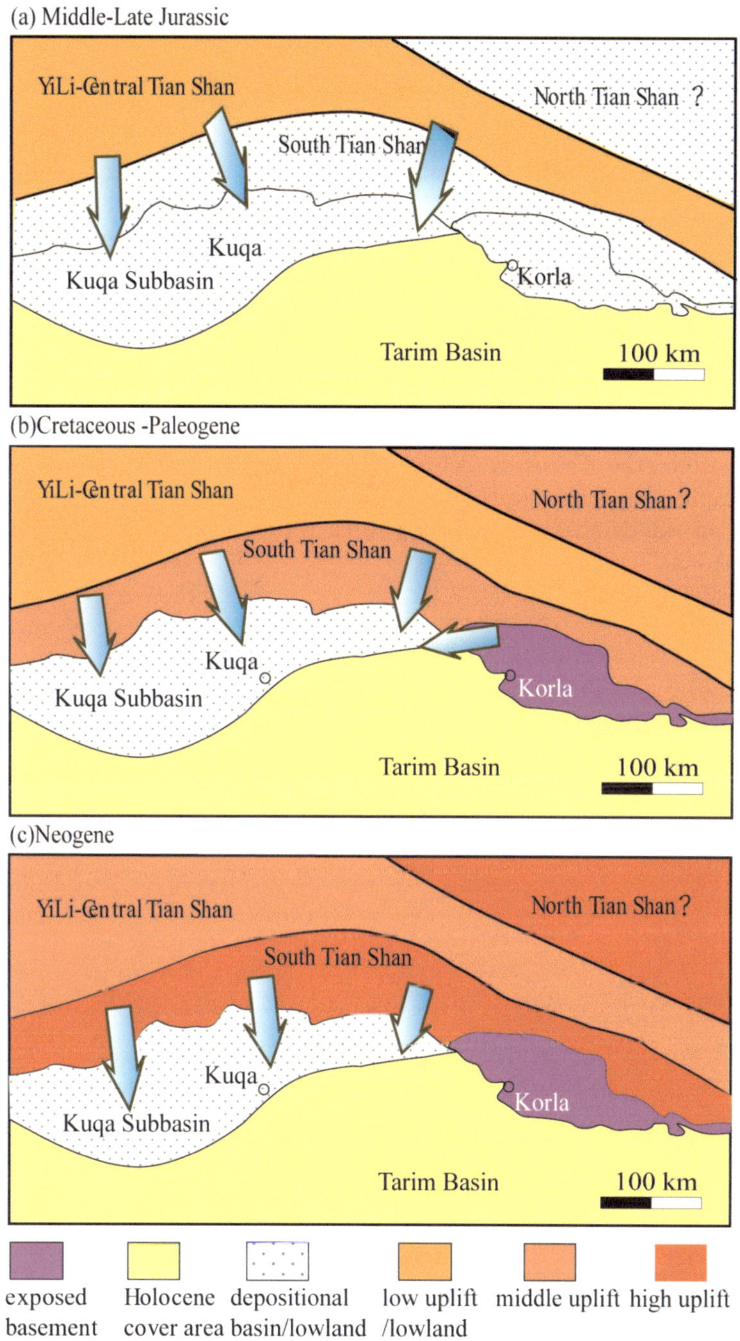

Fig. 5. Schematic map showing major provenance area changes from middle Jurassic (a), through Cretaceous-Paleogene (b), to Neogene (c). Arrowheads, with width, indicate directions, power and ranges of major inferred provenance supply. Structural deformation and intra-continental shortening that occurred in the study area from Jurassic to Neogene is ignored on the above maps(after Li *et al.*,2010)

6. References

Li, Z., Peng, S.T. (2010) Detrital zircon geochronology and its provenance implications: responses to Jurassic through Neogene basin-rangeinteractions alongnorthernmargin of the Tarim Basin,Northwest China. Basin Research,22 : 126~138

Mc Naughton NJ, Rasmussen B, Fletcher IR (1999) SHRIMP uranium-lead dating of diagenetic xenotime in siliciclastic sedimentary rocks. Science, 285:78-80

Nelson DR (2001) An assessment of the determination of depositional ages for Precambrian clastic sedimentary rocks by U-Pb dating of detrital zircon. Sed geol, 141-142

Qian Y.X., He Zh.L., Cai X.Y., *et al.* (2007a) Characteristics of heavy minerals from Upper Devonian Donghe sandstone and Silurian sandstone in western Tazhong area , Tarim Basin , and their geological implications. Acta Petrologica et Mineralogica, 26(2) : 147~154 (in Chinese)

Qian Y.X., He Zh.L., Cai X.Y., *et al.* (2007b) Characteristics of and U-Pb ages of zircons and its implication from zircon of Donghe sandstones of The upper Devonian and Silurian sandstones in Tazhong area,Tarim basin. Acta Petrologica Sinica, 23(11) : 3003~3014 (in Chinese)

Part 4

Related Luminescence

12

Novel Problems in the Solid State Cathodoluminescence of Organic Materials

Zheng Xu[1] and Suling Zhao[2]
[1]*Institute of Optoelectronic Technology, Beijing Jiaotong University, Beijing,*
[2]*Key Laboratory of Luminescence and Optical Information, Ministry of Education, Beijing,*
P.R. China

1. Introduction

Luminescence is the unique efficient supplementary technology to make up a constant deficit of solar light. Among all kinds of luminescence the electric field induced luminescence is especially attractive due to its feasibility of manipulation and direct transformation of electrical energy into light. For this trend cathodoluminescence is used for display already approximately 100 years. But the trend of contemporary display is to make flat panel display. For this reason many efforts were made to use cathodolumunescence, e.g., VFD and FED although the latter achieves still no practical success. At the same time EL finds applications in a certain field. In order to obtain blue color inorganic EL we proposed layered optimization scheme, and very soon we convinced the energy of hot electrons from the acceleration layer is high enough to excite luminescence directly. We realized this idea by using insulator phosphors. But the probability is very low because the environment of luminescent center is not regular near surface. In the layered optimization scheme we used the secondary property of electrons (hot electrons) of inorganic materials in order to raise the electron energy before they are accelerated in EL layer. In this acceleration layer, both the number of electrons and the energy of electrons are increased. The correctness and reasonableness of this idea was fully realized [1]. We tried to use organic materials to substitute the inorganic materials in the layered optimization scheme [2-3], thus discovered the solid state cathodoluminescence of organic luminescent materials. We call it as solid state cathodoluminescence (SSCL) because the hot electrons are accelerated in solids instead of in vacuum. SSCL is an independent branch of electric field induced luminescence. Soon we found also the possibility of integrating SSCL with OEL[4], i.e., using a single electrical source and a single luminescent material we can get two kinds of excitations for SSCL and OEL, which is the unique integrated excitation among all kinds of luminescence.

2. Classification of electric field induced luminescence according to mode of excitation

SSCL is a new kind of electric field induced luminescence. Up to now, electric field induced luminescence play an import role in the field of flat panel display, for example field emission display (FED), vacuum Fluorescent Display (VFD), electroluminescence (EL), light-emitting diode(LED) and organic electroluminescence(OEL), even though some technologies such as FED are still not practical at present. Their excitation mechanism is basically classified as two types, impact excitation and carrier injection.

2.1 Impact excitation

In the impact excitation mechanism, the luminescence center is excited by hot electrons collision which provides an efficient means of producing electronic transition to excited states of the luminescence centers. Based on impact excitation, some technologies of field emission display, vacuum fluorescence display, solid state cathodoluminscence, inorganic electroluminescence are produced.

2.1.1 Field emission display (FED)

A field emission display (FED) is a low power, flat cathodoluminescence display that uses a matrix-addressed cold cathode. Field emission display is a high-voltage display with a triode structure consisting of anode, cathode, and gate electrodes to achieve high illumination by applying a high voltage and a low current. In the field emission display, a strong electric field is formed between a field emitter and gate electrodes disposed on a cathode at a constant interval, so that electrons are emitted from the field emitter so as to impact on phosphors of an anode, thereby emitting light. Field emission display technology makes possible the thin panel as today's liquid crystal displays (LCD), offers a wider field-of-view, provides the high image quality of today's cathode ray tube (CRT) displays, and requires less power than today's CRT displays. However, FED display requires a high level vacuum which is difficult maintain.

2.1.2 Vacuum fluorescent display (VFD)

The VFD is composed of three basic electrodes: the cathode (filaments), anodes (phosphor) and grids under a high vacuum condition in a glass envelope. Electrons emitted from the cathode are accelerated with positive potential applied to both grid and anode, which upon collision with the anode excites the phosphor to emit light. The desired radiative patterns can be achieved by controlling the positive or negative potentials on each grid and anode. This voltage can be as low as 10V DC. The principle of operation is identical to that of a vacuum tube triode. Electrons can only reach (and "illuminate") a given plate element if both the grid and the plate are at a positive potential with respect to the cathode. This allows the displays to be organized as multiplexed displays where the multiple grids and plates form a matrix, minimizing the number of signal pins required. Compared to LCDs, they have relatively high power consumption. Other problems include some segments of a VFD display gradually becoming brighter or dimmer than others (caused by the phosphors glowing less brightly as they get older) and flickering.

2.1.3 Solid state cathodoluminescence (SSCL)

Organic materials for example, Alq_3 [5-6] , MEH-PPV [2], $Ir(ppy)_3$[7], etc, were used to substitute the inorganic materials in the layered optimization scheme and discovered two peaks of luminescence under AC bias. One is the typical organic electroluminescence peak with longer wavelength. Another one is a new luminescence peak with short wavelength. Both peaks constitutes the solid state cathodoluminescence (SSCL). The excitation mechanism of two emission peaks is impact excitation by hot electrons. The emission is not due to the typical recombination electroluminescence of injected carriers because no holes are injected to the organic materials in this kind of device. After primary electrons tunneled into the insulator layer, they are accelerated in the insulator layer and then have the high energy. These hot electrons collide organic materials to excite electrons from HOMO to LUMO like the excitation by photons. These excited electrons relax fast to form excitons with holes in HOMO. Then formed excitons recombine and emit light. Under high electric field, parts of formed excitons should be dissociated and then the light should be emitted directly recombined from electrons in LUMO and holes in HOMO. The excitation mechanism of SSCL is same to that of typical vacuum cathodoluminscence except that the hot electrons are accelerated not in vacuum but in solids.

2.1.4 Electroluminescence (EL)

Electroluminescence is the result of radiative recombination of electrons and holes in a material, usually a semiconductor. The excited electrons release their energy as photons - light. Electroluminescent devices are fabricated using either organic (OEL) or inorganic electroluminescent (IEL) materials. It is well known that the sandwich type thin-film IEL. The phosphor in these devices is not a powder but a thin continuous film prepared by sputtering or vacuum evaporation. The luminescence activators are manganese or rare-earth ions, atomic species doped in ZnS or other host materials with internal electronic transitions that lead to characteristic luminescence. A thin film luminescence (TFEL) device generally has a double insulator structure of electrode/insulator layer/active layer/insulator layer/electrode. For excitation process of luminescence centers in TFEL, the hot electron has been known to be the dominant carrier. Under AC bias, the accumulated electrons in the interface region of the electrode and the insulator layer are accelerated to become hot electrons. These hot electrons impact luminescence centers and EL results. A TFEL device acts like a pure capacitor at low applied voltage; no light is emitted until the voltage reaches a threshold value determined by the dielectric properties of the insulator and phosphor films. Above this threshold a dissipative current flows, and light emission occurs. The brightness increases very steeply with the applied voltage but is finally saturated. The light output, or average brightness, is roughly proportional to the frequency up to at least 5 kHz, and also depends on the waveform of the applied voltage.

Traditional IEL displays are bright, very fast in video response time and highly tolerant of environmental extremes. However, the lack of full-color has limited their application for the mainstream consumer television market. In order to get efficient inorganic electroluminescence, we have proposed the layered optimization scheme. The correctness and reasonableness of this idea was fully realized [1].

2.2 Inject electroluminescence

Injection electroluminescence results when an inorganic semiconductor *pn* junction (LED) or a point contact is biased in the forward direction. This type of emission is the result of radiative recombination of injected minority carriers, with majority carriers in a material. Such emission has been observed in a large number of semiconductors. The wavelength of the emission corresponds to an energy equal, at most, to the forbidden band gap of the material. If a *pn* junction is biased in the reverse direction, so as to produce high internal electric fields, other types of emission can occur, but with very low efficiency. LEDs present many advantages over incandescent light sources including lower energy consumption, longer lifetime, improved robustness, smaller size, faster switching, and greater reliability. For room lighting LED are powerful enough and relatively expensive and require more precise current and heat management than compact fluorescent lamp sources of comparable output.

Electroluminescence in organic materials may be observed in a single organic layer sandwiched between two electrodes, but the efficiency is not high. In order to enhance the efficiency of OLED, multilayer structure devices are introduced. To explain the operation of a typical multilayer OLED, the physical processes have to be identified [8]. Firstly Charge carriers are injected from electrodes on the top and bottom of the OLED and transport in organic materials. A minimum layer thickness has to be retained in order to prevent minority carriers to migrate to the opposite contact. Opposite space charges extend from the two contacts into the device. The zone in which they overlap and where recombination takes place is only a few nanometers wide. The recombinative process is strongly influenced by traps. It results in a charge-transfer complex or an excited molecule, which is called a (Frenkel-)exciton. Due to the spin multiplicity, excited singlet and triplet states are formed with a ratio of one to three. The excitation energy can be transferred to neighboring or more distant molecules of the same type or to molecules with a lower excitation energy. While triplet excitons decay nonradiatively, luminescence is caused by the radiative decay of singlet excitons. Nonradiative decay may take place at contacts and impurities.

2.2.1 Integrated excitation of SSCL and OEL

SSCL of organic materials is resulted from the recombination emission and exciton emission. When the acceleration layer is an suitable material in which electrons can be accelerated and holes can inject to the organic layer, it is possible to detect the luminescence of organic materials due to the electron impact excitation(SSCL) and carrier injection (OEL). The integration of the excitation of SSCL and OEL was achieved [5].

2.2.2 Mixed luminescence of EL and OEL

Electric field induced luminescence provides the way of direct transformation of electric energy into light. According to the materials, there are two kinds of electroluminescence, inorganic electroluminescence and organic electroluminescence. The inorganic electroluminescence (IEL) suffers from the lack of bright blue luminescence, and the organic electroluminescence (OEL) suffers from its stability. But they have their own advantageous to be used in flat plate display. For example, OLED is widely considered as the potential display technology. Therefore we hope to combine their superiority and minimize their inferiority in order to improve their performance. To combine IEL and OEL, We tried to use organic materials to substitute the inorganic materials in the layered optimization scheme [2-3]. Firstly we used the primary property (simple electron conduction) of ZnO or ZnS [9] as the electron transport material for enhancing the electroluminescence of organic layer and achieved positive results. In both of these two kinds of electroluminescence the required electric field is almost of the same order of magnitude. We tried to combine ZnS:Mn and

PPV layers together and really obtained the simultaneous emission from both layers[9]. On the other hand, we successfully applied the layered optimization scheme to improve the overall performance of IEL.

3. Characterization of SSCL

The spectral characteristics of SSCL of organic luminescent materials are the appearance of two light emission peaks when the applied electric field increases. More detailed examination of these two peaks shows that the longer wavelength peak is originated from excitons and the shorter wavelength peak is attributed to transitions from LUMO to HOMO levels directly. When the applied electric field increases, the longer wavelength peak increases at first, then passes a maximum and monotonically drops down to zero. The shorter wavelength peak gradually increases monotonically after the diminishing of the longer wavelength peak. The reason of the variation lies in the ionization of excitons in an electric field.

3.1 When the applied voltage increases，the two bands varies according to ionization of excitons

We used a series of organic phosphors, such as Alq_3, PPV, MEH-PPV, C_9-PPVto substitute the inorganic layer in the layered optimization scheme. The device structure is Al/insulator layer/organic layer/insulator layer/ITO. The methods to prepare different organic layers were different. For example the sample of MEH-PPV with the structure Al/SiO_2/MEH-PPV/SiO_2/ITO was prepared by means of spin casting. The thickness of MEH-PPV is about 50nm. The thin film SiO_2 was obtained by electron beam evaporation (EVA450, alliance concept Co. Ltd., France) at the growth rate of 1Å/s under the high vacuum of 2×10^{-6} Torr with a thickness of about 200nm. The other sample of Alq_3 within the structure Al/SiO_2/ Alq_3 /SiO_2/ITO was prepared by means of thermal evaporation of Alq_3. Excited by AC voltage, these samples showed a similar spectral behavior of having dual spectral peaks and shifting to short wavelength when the applied voltage was increased.

The spectrum of ITO/SiO_2/Alq_3/SiO_2/Al is shown in Fig.1. It is a wide band emission which can be fitted by the Gaussian fitting to two peaks [6]. One peak is 510nm, another one is 451nm. In Fig.1, the electroluminescence of ITO/Alq_3/Al is observed also under DC bias. The peak locates at 510nm which is the emission from the singlet exciton recombination of Alq_3. In a typical OLED, electrons and holes inject to the fluorescent material from the cathode and anode respectively, then they meet each other and singlet and triplet excitons are formed. The recombination of singlet excitons gives the electroluminescence. In our study, SiO_2 is used in this new kind of devices. As we known SiO_2 is an insulator. So it is difficult to transfer electrons, especially holes. In such case, the mechanism of electroluminescence of ITO/SiO_2/Alq_3/SiO_2/Al is different to that of the typical OLED. This means that the electroluminescence of ITO/SiO_2/Alq_3/SiO_2/Al is not originated from the carrier injection recombination.

According to above simulation, electrons which enter to Alq_3 layer have been accelerated in SiO_2 of the device of ITO/SiO_2/Alq_3/SiO_2/Al under AC bias. These electrons with high energies should bombard Alq_3, then Alq_3 is excited by this electron bombardment as excited by light directly. In this case, more singlet excitons are formed. Along with the increasing of the electric field on Alq_3, singlet excitons should be dissociated into free electrons and holes to LUMO and HOMO respectively. If these free electrons and holes recombine, they emit 450nm light. Therefore, we observed a new emission peak in the electroluminescence spectrum of ITO/SiO_2/Alq_3/SiO_2/Al besides the singlet exciton emission of 510nm. Theoretically, in this kind of devices, the quantum efficiency is higher than that of typical OLED devices because the impact excitation likes the excitation by light. There are almost no triplet exciton formations under electron impact excitation. Therefore, the efficiency of this kind device is larger than that of the typical OLED.

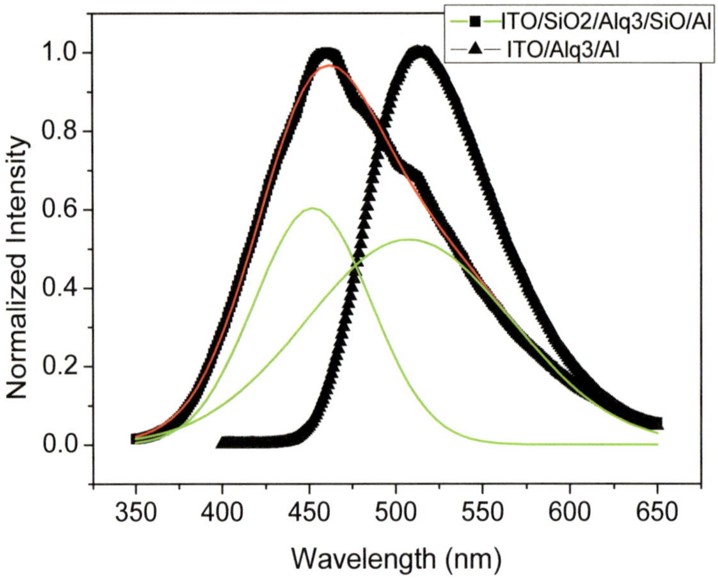

Fig. 1. The electroluminescence spectrum of ITO/SiO_2/Alq_3/SiO_2/Al (square) and ITO/Alq_3/ Al (triangle) under AC bias

The spectrum of Al/SiO_2/MEH-PPV/SiO_2/ITO consists of two peaks at 405 nm and 580 nm (Fig.2)[10]. In ordinary organic electroluminescence we have only the long wavelength peak. In SSCL, the spectrum shifts from the long wave length peak to blue one when the applied voltage increases. The apperance of short wave length peak is a characteristic phenomenon for SSCL due to the appearance of short wave length peak. The voltage range is divided into three different parts according to the spectrum: short wavelength peak at 405 nm only, a mixture of the two spectral components 405 nm and 580 nm, the long wavelength peak at 580 nm only. The 580 nm peak is attributed to the exciton emission, the 405nm peak corresponds to the radiation related to the recombination of carriers in conduction and valence band. When the applied voltage is increased, the spectrum shifts to blue.

We have also observed the solid state cathodoluminescence in organic phosphors, for example, $Ir(ppy)_3$ [7]. $Ir(ppy)_3$ is a well known organic phosphor. Its phosphorescence emission is a green emission with a peak at 520nm. We doped $Ir(ppy)_3$ in PVK, then prepared PVK:$Ir(ppy)_3$ thin film on SiO_2 by the method of spin coating. SiO_2 layers and Al cathode were prepared as before. Then the device of ITO/SiO_2/ PVK:$Ir(ppy)_3$/SiO_2/Al was realized. Under AC bias, the electroluminescence of this device which locates at

520nm was observed under different bias, as shown in Fig.3. This emission corresponds to the triplet excition emission of Ir(ppy)$_3$. As discussed above, the electroluminescence essentially is the solid state cathodoluminescence, that is, the emission from the bombardment of high energy electrons accelerated in SiO_2. It is very surprise that there is no the singlet exciton emission from PVK. It is possible that PVK is excited also by electrons and singlet excitons are formed. These singlet excitons do not give the radiative emission but transfer their energy to Ir(ppy)$_3$ directly. According to our previous study, there is the spectrum overlap between the emission spectrum of PVK and the excitation spectrum of Ir(ppy)$_3$. So the Forster energy transfer is possible between PVK and Ir(ppy)$_3$. Therefore, there are two ways for the excitation of Ir(ppy)$_3$. One is the directly excitation of the high energy electrons, another one is the energy transfer from PVK. Therefore, the efficiency should be very high theoretically.

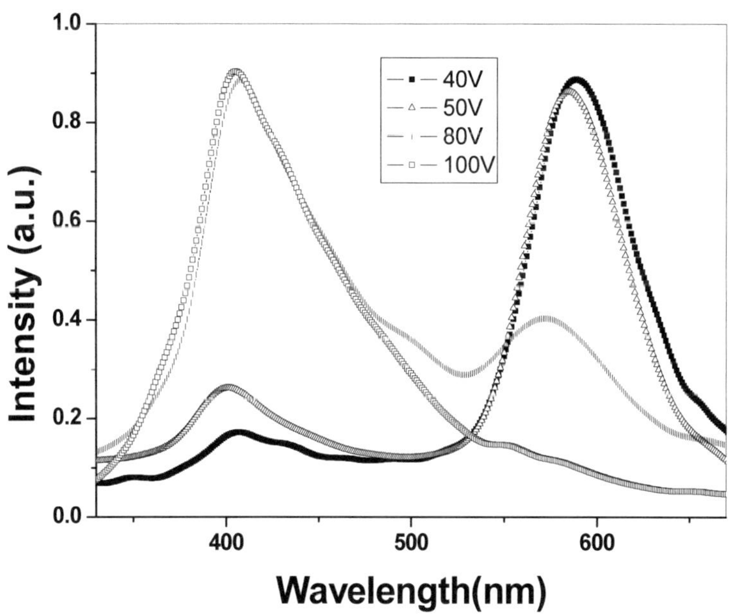

Fig. 2. The electroluminescence spectrum of ITO/SiO_2/MEH-PPV/SiO_2/Al under AC bias

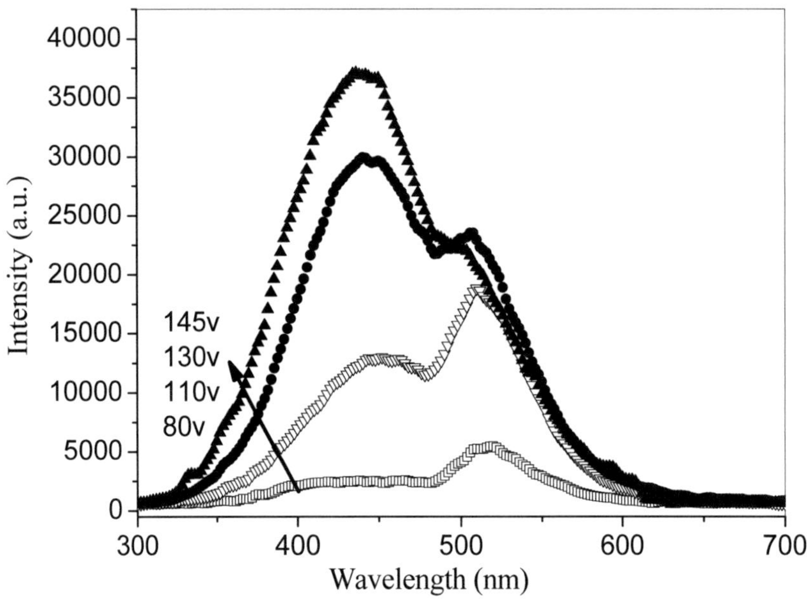

Fig. 3. The electroluminescence spectrum of ITO/SiO_2/PVK:Ir(ppy)$_3$/SiO_2/Al under AC bias

3.2 What is the nature of this emission? Cross proof of this emission being really a new kind of luminescence

We prepared a sample of the structure Al/SiO_2/MEH-PPV/ITO and discussed its emission mechanism under AC bias. If the emission is originated from gaseous discharge, the luminescence excited by it should be observed at both half periods of the applied voltage because the light may propagate ether in front direction, or in tail direction. This is simply because the amplitude of applied voltage remains the same in each of both half periods, the gaseous discharge might occur equally during each half period although their amplitude may be different. But experiment showed that emission occurs only when Al electrode is negative. Really during this half period the accelerated electrons may bombard MEH-PPV and in the opposite half period the electrons can not. Only the direct bombardment produces luminescence. This spoke of the impossibility of gaseous discharge. The same conclusion is true for symmetric structure.

This emission is not the injection luminescence as in p-n junction of LED and organic EL also. Because the energy barrier for holes on the interface of Al/SiO_2 is higher than 5.2eV and on the interface of ITO/ SiO_2 is 4.7eV, no holes can be injected into organic phosphor in a symmetric structure and consequently no recombination of electrons with holes, i.e., no LED or OEL is possible.

This emission is not also the self developed EL as in inorganic materials, in which the acceleration of hot electrons, collision excitation and luminescence proceed in the same layer, because the carrier mobility in organic phosphors is less than 10^{-2} $cm^2/v.s$ or in special case less than $1cm^2/v.s$. The kinetic energy of hot electrons accumulated in this material is lower than the value required for excitation. More than this, the ordinary IEL may be observed only in very limited class of semi-conductors, but the above mentioned new emission occurs not only in semiconductors but also in insulators if it is an ordinary luminescent phosphors.

All these exclusions of trivial phenomena from these aspects stick out the SSCL as a new type of luminescence.

3.3 In brightness waveforms the traditional method for determining phase relation fails to be effective[11]

In order to make the SSCL mechanism clear, we measured the temporary changes in the emission intensity along with the AC bias, which is the emission waveform, as shown in Fig 4. The line with the triangle symbol is the excitation AC bias with the amplitude of 60 V and frequency of 500 Hz. The lines with square and circle symbols are the waveforms of the emissions of 405 nm and 580 nm, respectively. It is found that the waveforms of these two emissions are different. The maximum intensity of 405 nm occurs under forward voltage, but the maximum intensity of 580 nm occurs under reverse voltage. According to the traditional phase detection method, the position of the maximum intensity is taken as the phase of the emission. Apparently 405 nm is emitted before 580 nm according to their waveforms if we take the position of the maximum brightness peak in the forward direction. When the bias is increased to another higher value, the order of these two SSCL peaks appears to be reversed. Certain important fact should be hidden behind this uncertainty of phase.

The detected intensities of 580 nm and 405 nm with increases in voltage are shown in Fig. 5. The threshold bias of this device is 20 V. When the AC voltage increases, 580 nm first enhances and then decreases. In the meantime, 405 nm is detected at the point which corresponds to the voltage of the maximum intensity of 580 nm. Then, 405 nm increases along with the increase in the voltage. We have concluded before that 405 nm emission is due to the direct recombination of electrons and holes which results from the decomposition of excitons. Their changes with changes in voltage can be described by the rate equation. The intensity of 405 nm is:

$$B^s = \sigma_1 \upsilon G^*[G + \alpha(E) n_0] \tag{1}$$

where $\alpha(E)$ is the exciton decomposition velocity under electric field E, σ_1 is the cross section of the recombination emission, υ is the average electron velocity in the conduction band, G is the excitation rate, and G^* is the number of holes. The simplest condition is G^* = G. They are functions of the electric field E.

The exciton formation rate is

$$\frac{dn_0}{dt} = \sigma_2 \upsilon G^*[G + \alpha(E) n_0] - \frac{1}{\tau} n_0 - \alpha(E) n_0 \tag{2}$$

Here, σ_2 is the cross section of the exciton formation. Because of the weak interaction between molecules, σ_1 and σ_2 can be considered as constants. But but according to the experiment, the recombination emission is difficult to detect in typical organic electroluminescence devices, so σ_1 is much less than σ_2. n_0 is the exciton concentration, and τ is the lifetime characteristic parameter of exciton emission.

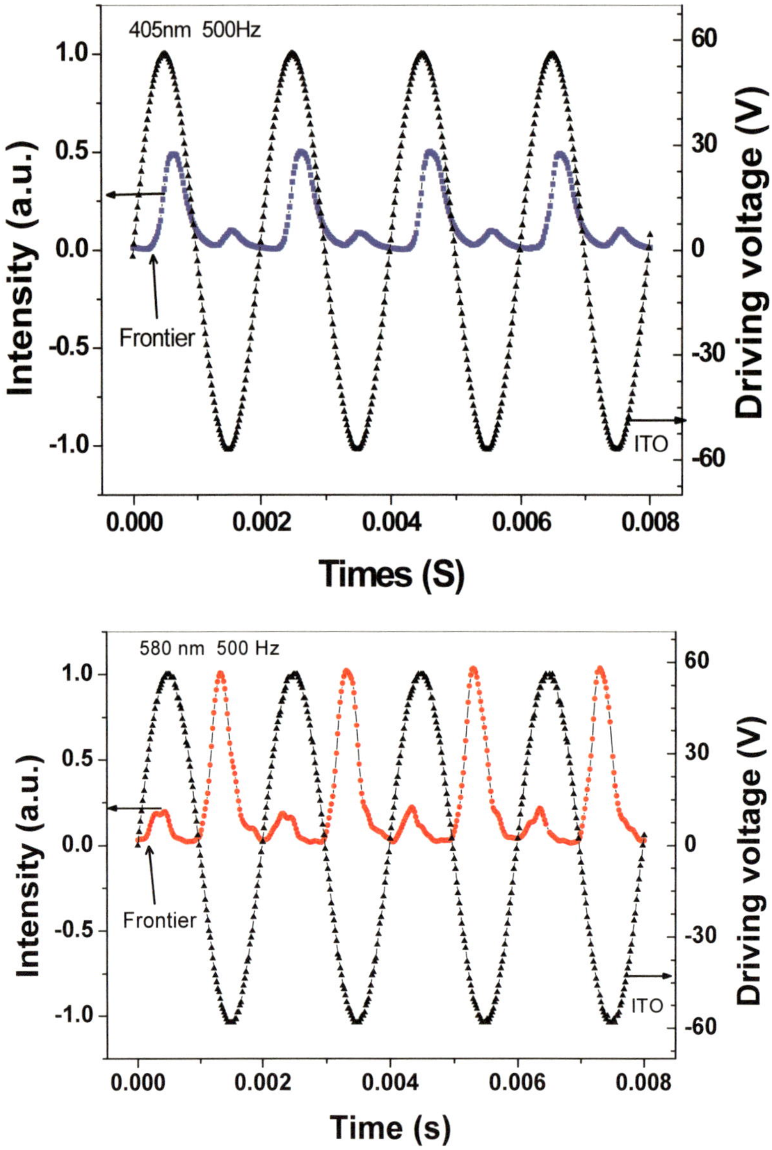

Fig. 4. The emission waveform of ITO/SiO_2/MEH-PPV/SiO_2/Al

The intensity of exciton emission is

$$B^{l} = \frac{1}{\tau} n_0 = \sigma_2 \upsilon G^{*}[G + \alpha(E) n_0] - \alpha(E) n_0 \tag{3}$$

Under forward bias, the electron injection from the cathode is very effective and electrons are accelerated in the SiO_2 layer. So the voltage on MEH-PPV is relatively high. But for reverse bias, the electron injection from ITO is not effective, and most of the voltage falls on the SiO_2

layer. The electric field on MEH-PPV is low. The simulation of the electric field on MEH-PPV is shown in Fig. 6. At low electric fields, $\alpha(E)$ is nearly equal to zero. E increases, and G and G* increase too. So the exciton emission increases. When E reaches a certain value, $\alpha(E)$ begins to increase along with E. As E increases, the exciton emission increases slowly, then decreases because of the decomposition of excitons. Meanwhile, the recombination emission is detected thanks to the decomposition of excitons. The bias on the device is sine wave, and should be changed from low to high. In such cases, 580 nm is detected firstly and then 405 nm is detected along with the increase in the bias. The AC bias put on the device is a changing bias. The voltage is not a constant but changes with time. We take V+ as the average voltage of the organic layer in forward bias, and V– as the average voltage in reverse bias. The actual bias ΔV+ and ΔV– on the organic layer is less than V+ and V– because those other layers take part of the bias, as shown in Fig. 6. The bias changes, and the injection current changes. Therefore, the bias ΔV+ and ΔV– changes and results in a changed electric field on the organic layer. This is critical for the exciton decomposition which relates to the exciton emission and band-band recombination.

Therefore, if we denote the phase of the emissions according to the fact shown in Fig.5, the phase corresponding to the maximum intensity in Fig. 4 is not the actual phase of the emissions. It is contrary to that found by the traditional phase detection method, which means that the traditional phase detection method is not suitable for the SSCL. It is found that the frontiers of their waveforms are consistent with their emission order. Thus if we use the phase of the frontier to denote the phase or the valley phase (minimum intensity) of the SSCL, we will obtain the right results. Therefore, the frontier phase instead of the maximum intensity phase is the true phase of the SSCL.

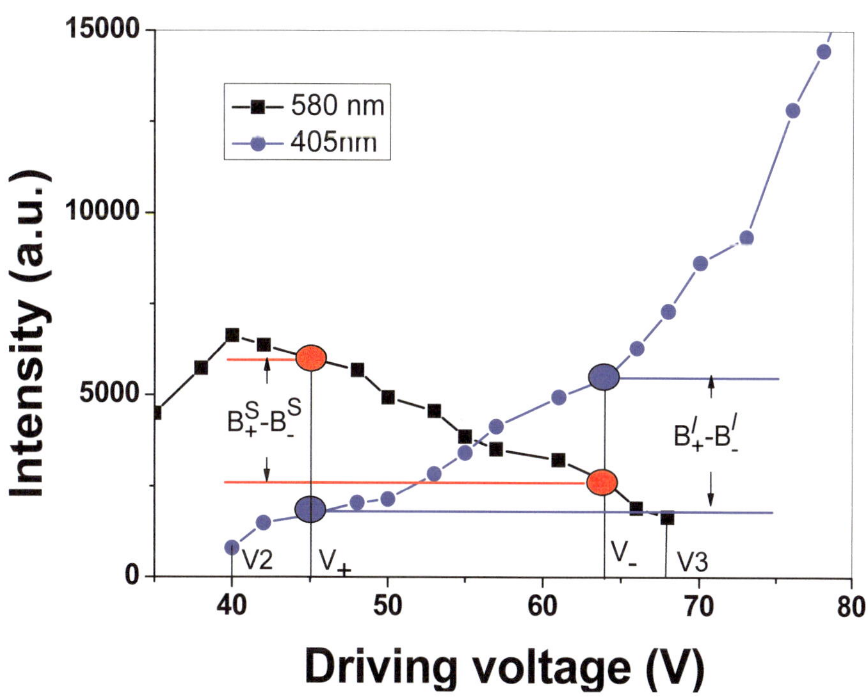

Fig. 5. The dependence of SSCL on the driving voltage

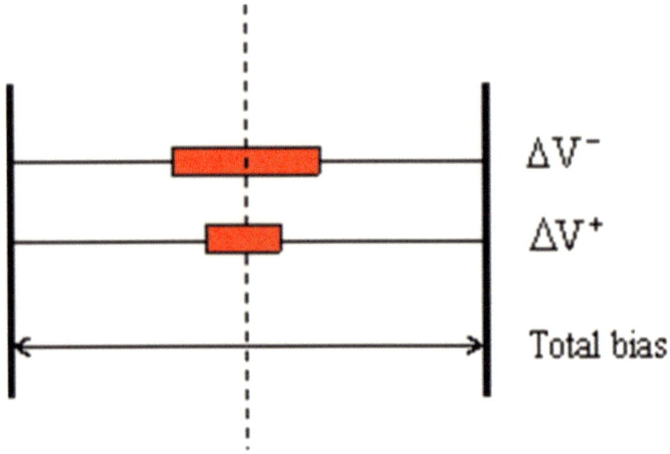

Fig. 6. The simulative bias on the organic layer

4. The special feature of SSCL lies in the accumulation of hot electrons energy

It is known that the SSCL are related to the electric field. The detected intensities of 580 nm and 405 nm of MEH-PPV with increases in voltage are shown in Fig. 5. The threshold bias of this device is 20 V. When the AC voltage increases, 580 nm first enhances and then decreases. In the meantime, 405 nm is detected at the point which corresponds to the voltage of the maximum intensity of 580 nm. Then, 405 nm increases along with the increase in the voltage. We have concluded that 405 nm emission is due to the direct recombination of electrons and holes which results from the decomposition of excitons. All these emissions are due to the electron impact excitation. In this way, the SSCL are depending on the hot electron energy.

4.1 Increment of hot electron energy in a stepwise way along the increase of electric field intensity

Since the excitation of electroluminescence originates from impact by hot electrons, the behavior of electroluminescence (EL) is chiefly determined by the energy of hot electrons. Thus the diverse phenomena in EL inspire the study of electron energy. In the scheme of layered optimization in which we used an acceleration layer located at the front of the luminescent layer in order to form a thrust for the acceleration of hot electrons in the electroluminescent layer, and hence enhance the energy of hot electrons. We observed really the obvious superiority of energy enhancement of this layered optimization scheme over the conventional sandwich structure. The intensity, efficiency, and percentage of short wave length emission are increased [12-14]. Moreover, a second jump of light intensity appears at higher electric field [15]. Here, three problems arise: 1) Does the additional acceleration layer plays the role of providing thrust to the acceleration of electrons in luminescent layer really? 2) How does the density of energy states changes with electric field strength? 3) how is the distribution of electron energy in dependence on electric field strength. Is it in accordance with the experimental result of second jump of luminescence? We limited our research to inorganic electroluminescence and investigated the behavior of electron energy (EE) to clarify the reason for these obvious improvements of performance. In order to find the energy distribution, the Boltzman equation should be solved. However, as it has no rigorous analytical solution we need to ask help of calculation techniques such as numerical calculation or Monte Carlo simulation. We chose the latter route. The initial electron energy and electric field intensity in electroluminescent layer were assumed to vary in a large scope. First, an analytical model [16] of the conduction band is used. It consists of three valleys (Γ, L, X) of the first conduction band, a Valley (X valley) of the second conduction band, and also a valley of still higher energy. For the latter band we proposed a higher conduction band for free electrons instead. The parameters for the free electron band are taken from reference [17].

Second, once the analytical band structure is constituted, we calculated the electron energy density state, as shown in Fig.7, which is defined as $N(E) = \lim_{\Delta E \to 0} \frac{\Delta Z}{\Delta E}$, where ΔZ is the number of energy states between E and $E + \Delta E$ [17]. In this process we took into account the anisotropy of dispersion and averaged the weighted result obtained along different directions.

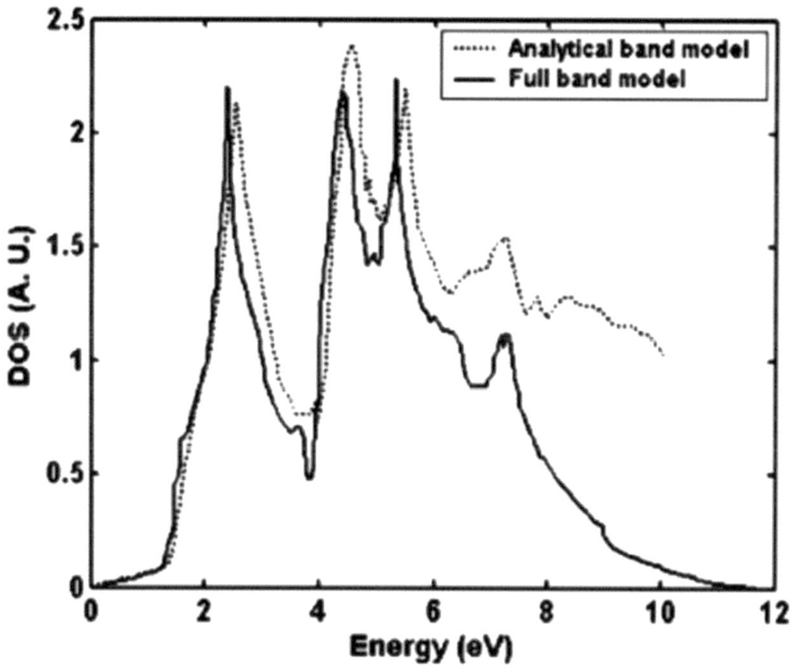

Fig. 7. Comparison of the overall density of states for our new analytical model of conduction band and for full band model

Third, we found out the different scattering rates:

1. The acoustic and optic phonon scattering rates for lower and higher electron energies.
2. The equivalent and non-equivalent inter-valley scattering rates were cited from reference[18].

3. The approximate scattering rate owing to the ionization of valence electron into the conduction band was cited from [19, 20].

$$\lambda_{ion} = R(E - E_{\hbar})^{\alpha}$$

Where $R = 5.14 \times 10^{10} s^{-1} (ev)^{-\alpha}$, $E_{\hbar} = 3.5ev$, and $\alpha = 5.183$

This process dominates the approach to energy stabilization and the beginning of electron multiplication. Based on these kinds of scattering rates, we carried out the Monte Carlo simulation. Electrons have a free flight time which may be determined stochastically according to the total scattering rate. From the total scattering rates we may work out a random number r (0<r<1) from the equation $\int_0^t S(\vec{k}.t')dt' = r$. $S(\vec{k}.t')$ represents the probability of the existence of $\vec{k}$ state within t. Because the variation of the total scattering rates with time is very complicated, the idea of self-scattering proposed by Rees is used. Then the free flight time is simplified to $t_f = -\frac{1}{\Gamma}\ln r$, where Γ is the total scattering rate (including self-scattering). During this free flight time the electron accumulates energy until this electron energy is balanced by scattering. At this time this flight is ended. The scattering which is active in this flight is chosen according to its relative magnitude. In the second flight we again chose a free flight time and new dominant scattering process.

The temporary variation of electron energy shows that after nearly 300fs the energy of hot electrons reaches a stationary value, which is responsible to the luminescence intensity. Thus, we calculated the stationary value of hot electron energy and get the following important results:

1. Initial electron energy might improve the stationary brightness only when the initial electron energy in ZnS attains 8eV in an electric field either lower than 0.6Mv/cm or greater than 4Mv/cm, as shown in Fig.8.
2. The distribution of hot electron energy under different electric fields is shown in Fig.9. The energy of hot electrons centralizes at higher energy valleys and the position of centralized energy consequently moves to the higher energy end when the electric field is increased owing to the increase in electron energy after the acceleration in the electric field and the balance with high energy scattering, including inter-valley scattering,. This tendency is very important and greatly increases the effectiveness of hot electrons in excitation.
3. The growth of hot electron energy proceeds in a stepwise way as shown in Fig. 10. From Fig. 10, we can see that if the initial electron energy ε_0 attains 8eV, the first plateau in EE appears at an average energy of hot electrons of more than 2.0eV, then after a sloping variation a second plateau in EE appears at an average energy of electrons above 4.8eV. The first and second plateaus constitute a stepwise growth in the electron energy. This behavior of EE is fully reflected in EL[15].

The complicated behavior of EE demonstrates the close dependence of electron transport on material properties, including conduction band structure, different scattering characteristics, the concrete structure of the device, and electric field strength. We obtained the satisfactory answers to the previous three problems.

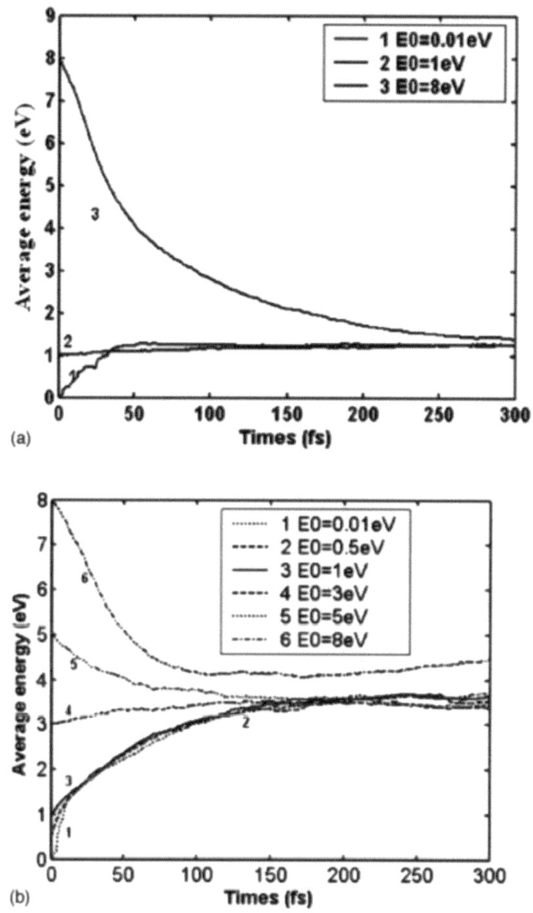

Fig. 8. Time evolution of electron energy at 0.6MV/cm (a) and 4MV/cm (b) field strength for different initial energies of electrons ε_0 in luminescent layer

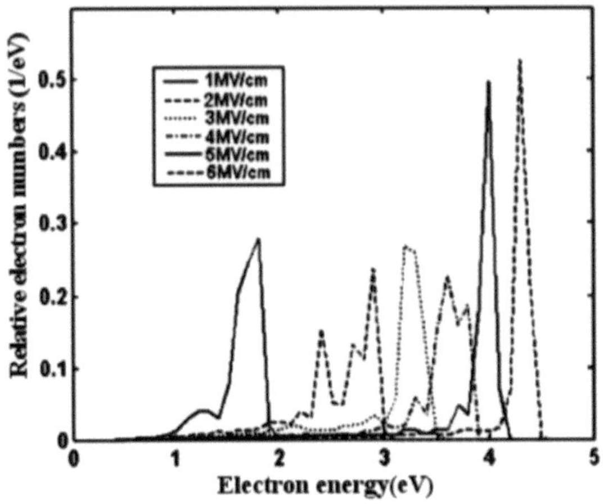

Fig. 9. Distribution of electron energy in different electric fields

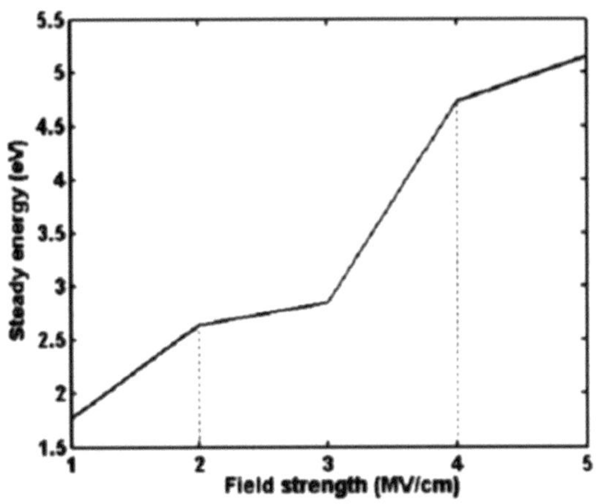

Fig. 10. Dependence of stationary electron energy on electric field strength when initial electron energy attains 8eV

The reason of all these behaviors or physical processes may be understood as follows:

1. When the electric field strength is less than 0.6MV/cm, the strength of scattering is not strong enough, and hence a part of the initial electron energy may be reserved and the stationary energy of hot electrons is raised. When the electric field strength is greater than 4MV/cm, although the strength of scattering is much increased, the free flight of electrons is also strengthened correspondingly. As only a part of the initial electron energy is required for energy balance, the stationary electron energy is thus increased. When the electric field is in between 0.6MV/cm and 4 MV/cm, no influence of 8eV

initial electron energy is detected. That means the sum of free flight energy and initial energy were balanced by total scattering.

2. In the process of electron transport the inter-valley scattering transfers the electrons from lower energy valleys to higher energy ones. This process leads to the concentration of hot electrons at a higher energy valley. During the further increase in the electric field, these hot electrons will be transferred to another even higher valley. Thus, the hot electron energy will be concentrated and shifted to another higher energy valley simultaneously. This tendency agrees with the distribution of density of states shown in Fig.7. When the electric field is high, most of electrons are evacuated from lower states and moved to higher energy states and the result appears in Fig.9.
3. When the electric field strength increases, the electron energy increases correspondingly. We observed a sloping energy increase, but when the field strength is high enough to start inter-valley scattering, the energy is concentrated and we observed a plateau. After the valley is fulfilled, the increase in the slope starts again. Thus we obtain a stepwise growth in hot electron energy. This phenomenon was observed in [15].

4.2 Choice of better acceleration material

The energy of hot electrons depends on the electric field and used materials. We have compared different materials as the acceleration materials and then compared the SSCL of a same emission layer. The acceleration materials used are Si_3N_4, SiO_2, ZnS or complex accelerated layer SiO_2/ZnS respectively [21-23]. The structure of all devices is same as ITO/accelerated layer/MEH-PPV/ accelerated layer/Al. In these devices, two luminescence peaks corresponding to the exciton emission and the extended band emission respectively are observed. The electron acceleration ability of SiO_2 is better than that of ZnS and Si_3N_4, but ZnS is better than SiO_2 when the ability of injection charges (as shown in Fig. 11 and 12) is compared. In Fig.11, under same bias, the intensity of short wavelength of MEH-PPV in the device of ITO/ SiO_2/MEH-PPV/SiO_2 /Al is stronger than that of ITO/ZnS/MEH-PPV/ ZnS/Al. So it is concluded that the acceleration ability of SiO_2 is better than that of ZnS. We investigated the complex accelerated layer SiO_2/ZnS. The results showed that under low voltage, the performance of the complex accelerated layer is superior than that of SiO_2 and inferior than that of ZnS. Under high voltage, the complex accelerated layer is the best among different devices. Thus, SiO_2 is the main accelerated layer, and ZnS is very useful to improve the performance of electrons injection of SSCL device.

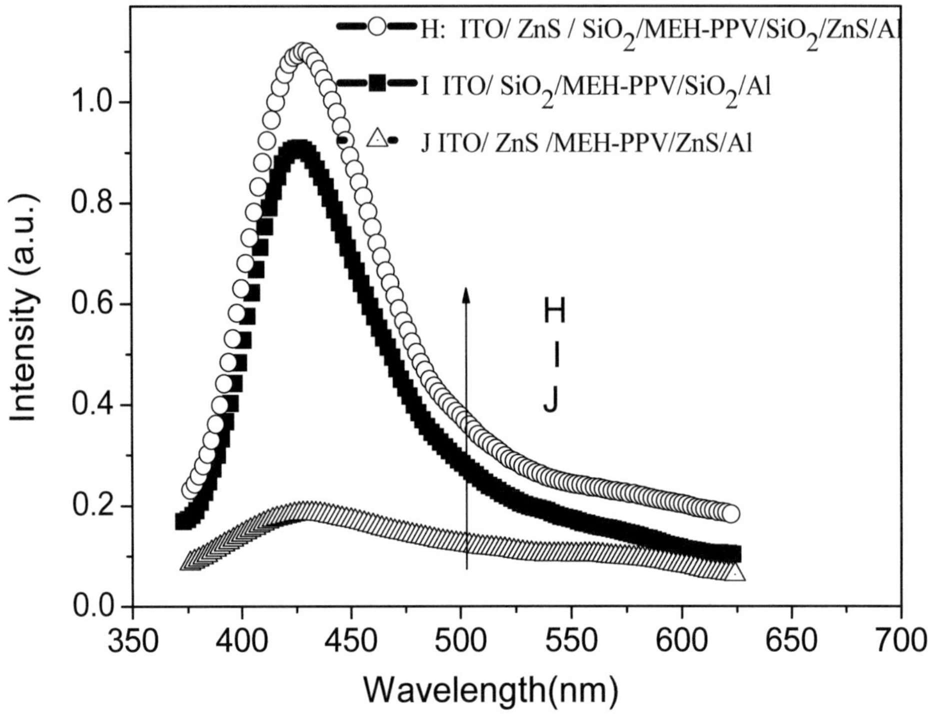

Fig. 11. The contrast of three devices on AC 65V

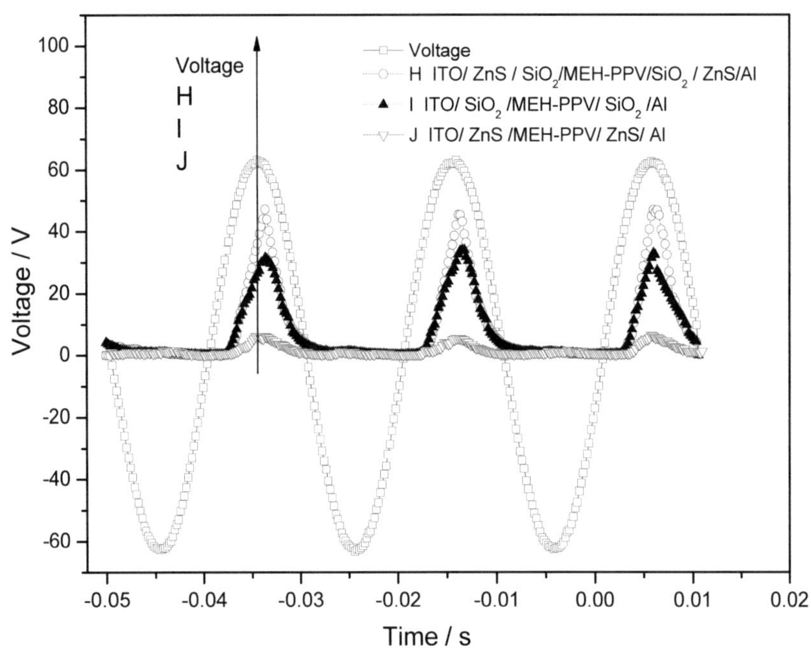

Fig. 12. The waveform of derivative luminescence (405nm) with applied voltage

5. The prospect of SSCL

Concerning above discussion, it is concluded that SSCL have more potential in the display field. The excitation mechanism of SSCL is superior to the organic electroluminescence. In organic electroluminescence, the maximum of the internal quantum efficiency is not more than 1/4 because of the limitation of the photon transition selection rule. After electrons and holes injected to the organic layer, only 1/4 number of electrons and holes can relax to form singlet excitons and give the light emission. But SSCL behaves nearly the photon excitation. In SSCL excitation, more singlet excitons are formed. There are almost no triplet excitons. Therefore, the efficiency of SSCL is better than that of pure OLED. The effort of raising luminescence efficiency is proposed to carry out in 4 aspects:

1. Examination of factors influencing the electron acceleration ability, e.g., crystal orientation of materials.

Using the Monte Carlo simulation method we have found out that the distribution of hot electrons concentrates and moves towards to higher energy end along the increase of energy. This situation is obviously benefit for raising the effectiveness of excitation. Another very important result is that the energy of hot electrons increases with electric field not harmonically but in stepwise way. This result coincides with experimental result of second jump in light intensity at higher electric field. The improvement of acceleration ability of the inorganic layer in the device should enhance SSCL. For example, if electrons transfer along the crystal orientation of the inorganic layer, under electric field, electrons should be accelerated easily.

2. Improvement of light intensity by enlarging the percentage of singlet exciton emission of organic materials.

From the consideration of excitation mechanism, it is beneficial to integrate of SSCL and OEL in the same device and enlarge singlet exciton formations.

3. Extension of complex display panel used in SSCL to FED

We published a paper on Optics Letters proposing the combination of electron source of FED and the complex screen in SSCL to form an improved variant of FED, in order to avoid the gas from the target under electron bombardment, to absorb the incoming high energy electrons b y injecting holes and consequently to hold the vacuum of the device.

4. Determination of trap depth of local electron source.

The release of electron from traps is the main source of the primary electron in SSCL devices. How to enhance the number of primary electrons? We investigated the trap depth of electrons. In luminescence the method of determining trap depth by analyzing glow curve with the assumption of kinetics order being 1 or 2 has been used more than 60 years, we pointed the large error of their results and proposed a quasi-equilibrium method which traces the real physical processes and provide an accurate result.

6. References

[1] Xu Xurong et al. Proc.of 7th international workshop on EL, Beijing, 1994: 42

[2] Zheng Xu, Chong Qu, Feng Teng, et al, Appl. Phys. Lett.,2005, 86(6): 061911

[3] Zheng Xu, Feng Teng, Chong Qu, et al , Physica B: Condensed Matter, 2004, 348(1-4): 231

[4] Suling Zhao, Zheng Xu, Fujun Zhang, et al., Opt. Lett. , 2007, 32:2094

[5] Xu XL, Xu Z, Hou YB, et al, J. Appl. Phys., 2001, 89 (2): 1082

[6] Xu XL, Chen XH, Hou YB, et al, Chem. Phys. Lett., 2000, 25 (4): 420

[7] Dong Qian, Zhao Suling, Xu Zheng, et al, Acta Physica Sinica, 2008, 57(12):7896

[8] K. Book, V.R. Nikitenko, H. Bässler , et al, Synthetic Metals, 2001, 122(1):135

[9] Xiaohui Yang1 and Xurong Xu, Appl. Phys. Lett. ,2000, 77: 797

[10] Fujun Zhang, Suling Zhao, Zheng Xu, et al, Chin. Phys., 2007,16(5): 1464

[11] Suling Zhao, Zheng Xu, Fujun Zhang, et al, J. Appl. Phys., 2009,106 ：023513

[12] X. Xu and G. Lei, Acta Polytech. Scand. , 1990, 170: 133

[13] X. Xu, G. Lei, and M. Shen, J. Cryst. Growth, 1990, 101: 1004

[14] X. Xu, G. Lei, and Z. Xu, J. Cryst. Growth, 1991, 117: 925

[15] W. J. Wang, Y. B. Hou, C. H. Wei, et al, Proceedings of the 1994 International Workshop on Electroluminescence Science, Beijing, 1994 : 206.

[16] Q. F. He, Z. Xu, F. Teng, et al, Chin. Phys. Lett. , 2006, 23: 701

[17] Q. F. He, Z. Xu, D. A. Liu, et al, Solid-State Electron., 2006, 50: 456

[18] G. Weireick, T. M. Sanders, and W. G. White, Phys. Rev. , 1959,114: 53

[19] M. Reigrotzki, K. F. Brennan, R. Wang, et al, J. Appl. Phys., 1995, 83: 1456

[20] M. Reigrotzki, R. Redmer, I. Lec, et al, SAP, 1996, 80: 5054

[21] Li Yuan, Zhao Suling, Xu Zheng, et al., Acta Physica Sinica, 2007, 56(09):5526

[22] Li Yuan, Zhao Suling, Xu Zheng, et al., Spectroscopy and spectral analysis, 2008, 28(9): 1974

[23] Kong Chao, Xu Zheng, Zhao Suling, et al, Acta Physica Sinica, 2008, 57(12): 7891

13

A Lighting Mechanism for Flat Electron Emission Lamp

Jung-Yu Li et al.*
Green Energy and Environment Research Laboratories,
Industrial Technology Research Institute, Hsinchu,
Taiwan

1. Introduction

A novel lighting device, flat electron emission lamp (FEEL), has been developed for generating planar light. The mechanism of FEEL is different from the mainstream techniques for lighting and display. The basic idea for FEEL is, for the first time, to integrate the discharge mechanisms of fluorescent lamp at cathode with the cathodoluminescent effect at anode. In that the electrons are induced by gas discharge, accelerated toward anode under electrical field, and finally impacting the phosphor-coated anode to emit visible light. The light spectrum radiated from FEEL is thus, dominated by the detailed composition of the phosphor coated on the anode. Unlike most of the fluorescent lamps, ultraviolet light is unnecessary in this case, and therefore there is no need to have mercury inside the device. By taking into account the human factors, such as safety and health, the uniform lighting obtained in FEEL further offers a prominent characteristic which prevents the eyes from the uncomfortable glare and drastically improves the persistence of vision. In this article we also demonstrate various prototype potential applications featured by FEEL, namely the double-side lighting, transparency, colored gray-scale image and heat insulation, etc. It is indicative that FEEL not only could have broad potential markets for normal lighting, but also might play a prominent role in power-saving building material and gray-scale ambiance lighting. More importantly, the unique mechanism prevailing in FEEL shows the high flexibility of expanding its applicability albeit it is still in the early stage of laboratory studies. We believed that it may become one of the competing candidates for the next generation green lighting in the near future.

2. Motivation of the technique development

In the nineteenth century, the new electrical lighting, incandescent lamp, was invented and had totally changed the human behavior and social environment in the early stage of electrical power generation. Hereafter many different kinds of lighting and display technologies were

---

*Ming-Chung Liu[1], Yi-Ping Lin[1], Shih-Pu Chen[1], Tai-Chiung Hsieh[1,2], Po-Hung Wang[1], Chang-Lin Chiang[1,2], Ming-Shan Jeng[1], Li-Ling Lee[1], Hui-Kai Zeng[3] and Jenh-Yih Juang[1,2]
[1]*Green Energy and Environment Research Laboratories, Industrial Technology Research Institute, Chutung, Taiwan*
[2]*Department of Electrophysics, National Chiao Tung University, Hsinchu, Taiwan*
[3]*Department of Electronic Engineering, Chung Yuan Christian University, Chung Li, Taiwan*

developed vigorously for illuminating every corner in the world. At present, the major branches of lighting techniques can be catalogued into four kinds of mechanisms: (1) Fluorescent Lamps (FL) (2) Cathodoluminescence (CL) (3) Solid-state lightings (SSL) (4) Gas-discharge lamps (GDL). All of these technologies have been developing for several decades and spread into broad range of daily applications. The maturity and wide-spread industrialization of these techniques have, consequently, led the manufacturing of most lighting and/or display equipments to be confined within the concepts derived from these mechanisms.

In this chapter, a novel lighting device, flat electron emission lamp (FEEL), for generating uniform planar light with competitive power efficiency in an environmentally recuperative manner is described. The lighting mechanism prevailing in FEEL is different from that operating in the mainstream techniques, such as FL, CL, SSL, and GDL. The originality is primarily based on the innovative integration of the working mechanisms of FL and CL. As depicted schematically in Fig. 1, the lighting model of FEEL can be clearly interpreted by dividing it into three steps: (1) Analogous to FL mechanism, the secondary electron emissions are generated by ion bombardment at the cathode surface. It is noted that, in FEEL, the gas pressure used to initiate the production of electrons is significantly lower than that used in FL, being in the range of 10^{-1} to 10^{-2} torr. (2) The electrons are subsequently accelerated by the electric field established between the cathode and the anode. During the course of flight, the gas excitation and ionization are induced by electron collision with gas. The probability of electron collision depends on the pressure, electron kinetic energy and gas type. (3) As an analogy to the anode structure of CL, electrons impact with CL phosphor-coated anode and transfer the kinetic energy to the emission of visible light. Evidently, over the whole process of lighting in FEEL, ultraviolet light is unnecessary and therefore there is no need to have mercury inside the device. Thus, the mechanism meets requirements for developing green lamps by completely avoiding the pollution issues from the involvement of mercury.

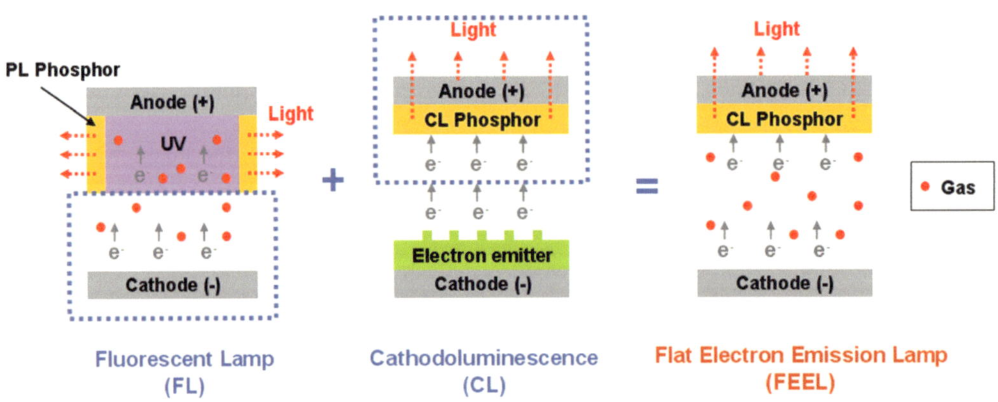

Fig. 1. The lighting mechanism of FEEL is proposed by integrating the mechanisms operating at the cathode of FL and the anode of CL. Briefly, the electrons are induced by discharge in the vicinity of cathode, accelerated by electrical field, and finally impacts on the phosphor-coated anode, which gives rise to the eventual light emission by transferring the electron kinetic energy into phosphorous radiations. The combination is advantageous in generating large area planar electron beams for obtaining uniform photon distributions over the entire CL phosphor screen.

So far FL is the most mature technology for general daily lighting, while CL has been ubiquitously used in the long history of the display technology. For instance, the fluorescent tubes and plasma display panels (PDP) belong to the field of FL mechanism, while the cathode ray tube (CRT), field emission display (FED), and field emission lamp (FEL) are all based on CL mechanism. It is interesting to note that over the long history of development in the two respective fields there was essentially no overlap between FL and CL technologies. In fact, they are working at very different gas pressure conditions. From the required conditions of FL, it is necessary to infuse a sufficient amount of gas (in the range of 100 torr) in order to sustain the stable glow discharge; it is also necessary to further produce high enough ultraviolet intensity during the discharge process to maintain photoluminescence effect for obtaining the visible light radiated from photoluminescence phosphor. On the contrary, in the case of CL, high vacuum environment ($< 10^{-6}$ torr) is essential for reducing energy dissipation from electron scattering. The suitable vacuum pressure conditions for FL and CL can be different by eight orders magnitude or even larger. As will be discussed in more details later, the working pressure in the current FEEL device is in a more compromised condition, which is in the middle range between the pressure of FL and CL, and, hence exhibiting some unique features that were not fully explored previously.

In contrast to the traditional line or point shaped lighting sources, the planar lighting sources with the uniform light have the important advantage of no glare, thus can create a comfortable lighting environment. The mercury-free field emission-like (FEL) lighting technology has been one of the potential candidates for the planar lighting or backlighting applications. Unfortunately, the random dark points constantly appearing in the lighting screen of FEL have severely hindered its market development. Usually, an additional diffuser covering the entire the device surface to obtain uniform light emission is required. However, the extra diffuser also reduces the effective lighting intensity of the lamp. Thus, either from the consideration of cost down or from the view point of energy-saving, the issue of dark points has become one of the technology bottlenecks for FEL. The primary source of such non-uniform light emission is attributed to the grown-in disorders during the growth of nanoscale electron emitters. For instance, one of the most favored materials for electron emitter, the carbon nanotubes (CNTs), turned out to be very difficult to control the uniformity of nanotube length and alignment over large area by the processes of screen printing or direct growth. To this respect, the gas discharge and emitter-free characteristic of FEEL provides an alternative solution for easily generating planar beam-like electrons from cathode. As an example, the lighting screen shown in Fig. 2 for a typical FEEL device evidently demonstrates the complete elimination of the dark points. In addition, it also avoids the unfavorable cost and process time spent in growing the electron emitter material. Nevertheless, it should be emphasized that making use of gas discharge in unconventionally low pressure regime is the first attempt for the development of cathodoluminescence technology and further understandings of the detailed mechanisms are certainly needed.

3. The lighting mechanism

In order to investigate the intrinsic discharge property of FEEL, an experimental setup was established for simultaneously measuring current-voltage (I-V) curves, Paschen curves, luminance and optical emission spectroscopy (OES) of FEEL device. As illustrated in Fig. 3, the experimental FEEL device consists of a cathode glass, a glass spacer, and a

CL phosphor-coated anode glass. The electrode glasses were deposited with a transparent conducting oxide film (fluorine-doped tin-oxide/FTO). The electrode and spacer were sealed by glass glue with infused gas as the working gas. The height of glass spacer was 10 mm and size of lighting screen (phosphor surface) was 30 mm x 30 mm. A glass tube connected the glass spacer and the vacuum system for evacuating and adjusting pressure inside the device. A DC voltage power (Keithley 248) and a picoammeter (Keithley 6485) were used for analyzing the discharge and electrical behavior; a Multi-function Color Analyzer (Ruyico Tech.) and an OES (optical emission spectroscopy) probe (StekkarNet EPP200V) were positioned on top of device for measuring the optical properties. The OES spectrum was measured to further reveal the correlated behaviors between glow excitations and the lighting mechanism of FEEL devices. To obtain the optimal conditions for FEEL lighting, the gas pressure inside the device was gradually decreased and the transitions of the lighting mode were simultaneously analyzed by observing the images of the lighting screen.

Fig. 2. The photo shows the lighting screen of a typical FEEL device demonstrating the uniform light radiation and absence of dark points ubiquitously encountered in traditional FED devices.

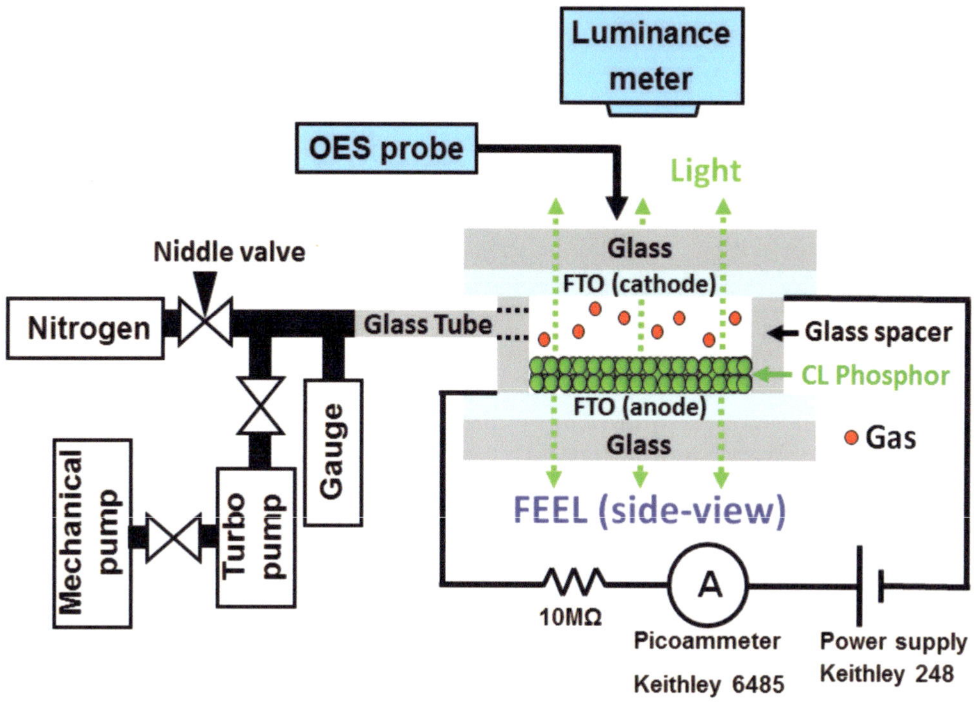

Fig. 3. The schematics of the FEEL device structure and the experimental setup for measuring the lighting and discharge behaviors. The device consists of cathode, glass spacer, gas, and CL phosphor-coated anode. The pressure inside the device can be adjusted via the vacuum system. The experimental setup for simultaneously measuring I-V curves, Paschen curves, luminance and OES spectrum enabled the revelation of correlations between discharge and lighting mechanism of FEEL.

Figure 4 depicts the prevailing mechanisms in different gas pressure regimes. With an applied DC voltage of 5 kV, which ensures the turn-on status for FEEL, the top view and side-view images of the device show that the glow discharge features (Fig. 4A) gradually transform to the FEEL mechanism (Fig. 4C) as the N_2 pressure is approaching to an optimized condition. To further confirm the predominant light source(s) in each gas pressure regime the OES spectra are also displayed in Fig. 4. It is clear that only glow excitation is active and the phosphor coated on the anode does not emit light when the N_2 pressure is higher than 0.2 torr (Fig. 4A), consistent with what's seen in the images. The four distinguished peaks identified in the OES spectrum with the wavelength respectively locating at 337, 358, 391.5 and 427.5 nm are originated from the nitrogen gas excitations belonging to the first negative system ($B^2\Sigma_u^+ - X^2\Sigma_g^+$) (358.2, 391.4 and 427.8 nm) and the second positive system ($C^3\Pi_u - B^3\Pi_g$) (337.1 nm). In this case, obviously the kinetic energy of electrons is too low to activate the emission from the phosphor, presumably due to the frequent electron scatterings by collidings with gas molecules in such high gas pressure environment. We note that the frequent electron-gas scatterings are exactly responsible for the discharge glow as what one would expect.

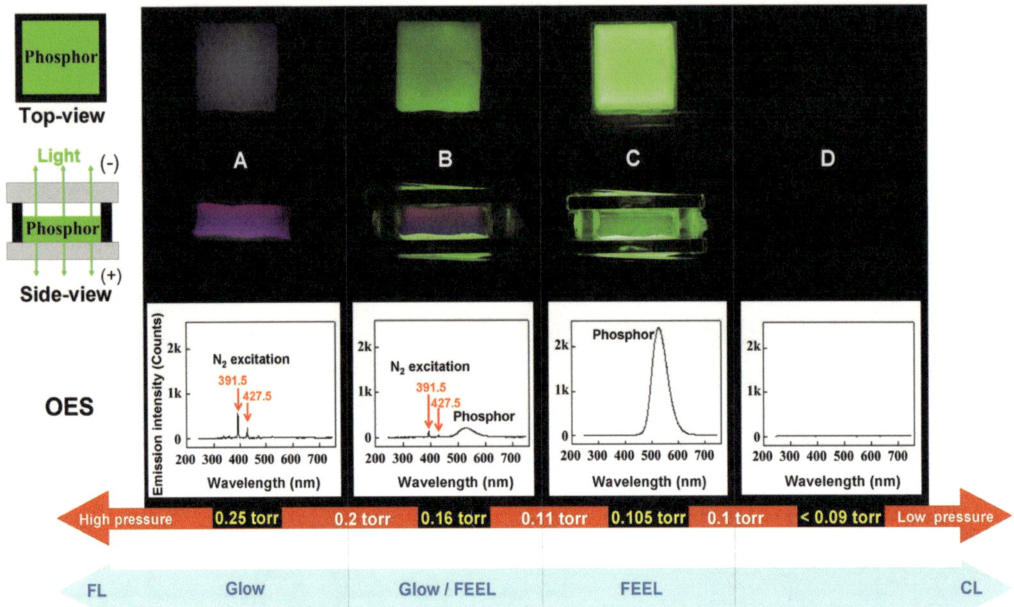

Fig. 4. The lighting mode transitions as a function of gas pressure inside the FEEL devices with a fixed applied voltage of 5 kV. Both the device images and OES spectra show the close interplay between the scattering-induced glow discharge and the prevailing of electron emission lighting (APL, Vol.94, 091501, 2009). Typically, four pressure regimes within which different dominant lighting mechanisms can be clearly identified:
(A) (P_{N2} > 0.2 torr): predominantly nitrogen excitation associated glow discharge.
(B) (0.2 torr> P_{N2} >0.11 torr): coexistence of nitrogen glow and phosphor lighting at the anode observed.
(C) (0.11 torr > P_{N2} > 0.1 torr): predominantly phosphor lighting due to the FEEL mechanism.
(D) (P_{N2} < 0.1 torr): no lighting mechanism is active.

As the N_2 pressure is decreased to within the range between 0.2 torr to 0.11 torr (Fig. 4B), it is clear from the side-view image that phosphor emission starts to emerge at the expense of the attenuated glow background. The OES spectrum evidently shows the weakened peak intensity from the N_2 excitation in this pressure range, with an apparent peak center at the wavelength of 530 nm, which is the characteristic wavelength of the phosphor used in this study. We believe that, in this lower pressure range, the dissipation of electron kinetic energy will be less than that in the previous case and there will be some population of electrons with residual kinetic energy higher than the threshold energy for activating phosphor emission. In this regime, nonetheless, the gas excitation glow and phosphor emission are coexistent, indicating that the predominant mechanisms operating in the system are strongly dependent on the gas pressure and, in fact, controllable. As the N_2 pressure is further decreased to the range of 0.11 torr to 0.1 torr (Fig. 4C), the OES spectrum displays only a strong peak at 530 nm, reflecting the green light emitted from the phosphor coated at the anode. Furthermore, the characteristic peaks of the glow excitation of N_2 gas disappear altogether under this condition. We refer this state as the FEEL state because it appears to be the dominant lighting mechanism. It is noticed that, although the pressure (0.11 torr) at which the system switches from the glow/FEEL coexistent state to the pure FEEL state (Fig. 4B and Fig. 4C) appears to be somewhat arbitrary, the existence of even a slight portion of glow can affect the luminance of the device dramatically. It suggests that the electron-gas scatterings during the flight journey from cathode to anode can have significant effect on the luminance efficiency.

Lastly, as the pressure of N_2 is below P = 0.1 torr, neither any lighting is seen in the images nor any peak is revealed in the OES spectrum. It indicates that the glow excitation of N_2 gas and lighting from phosphor emission are all not active under this condition (Fig. 4D). Intuitively, one would expect that at this low pressure condition, the device should be operating similarly to the CL mode prevailing in FED. However, we note that, for FED applications, usually the turn-on electrical field of cathode emitter is well-above order of V/μm as indicated by the Fowler-Nordheim (F-N) theory. Consequently, recent development of the cathode emitter for FED was mainly focused to lowering the turn-on electrical field. Indeed, the CNT with its advantageous high aspect ratio has successfully reduced the turn-on field to E = 0.5 ~ 1V/μm. In our case, the estimated electrical field is about 0.5 V/μm, near the turn-on field of CNT. It implies that by combining the glow discharge effect, the present FEEL device has the equivalent effect of CNT field emission. Nevertheless, without glow discharge, there will be no electron because the applied external field is not enough to trigger the FED mechanism, leading to virtually no response in the FEEL device when operated at low gas pressure regime (Fig. 4D).

In order to further clarify the correlations between the characteristics of luminance and glow excitation in a more quantitative manner, the current density and luminance were analyzed in the pressure range of FEEL and glow/FEEL states. Figure 5 shows the current density and the luminance as a function of gas pressure in the FEEL device operated at an applied dc voltage of 5 kV. It is evident that the maximum luminance is obtained at the same threshold pressure of 0.11 torr, in accordance with the glow excitation effect on the device luminance described above. In addition, the current densities of FEEL state are always lower than those obtained in the glow/FEEL state. It implies that, under the same luminance condition, the luminance contribution per unit current density in the FEEL state is always higher than that in the glow/FEEL state. It also indicates that high degree of glow excitation, which strongly correlates with electron scattering, in fact diminishes luminance performance of the FEEL device. The significant influence from gas excitation in reducing the device luminance is believed to arise from the dissipation of electron kinetic energy, which, in turn, leads to inefficient phosphor excitations.

Another set of systemic measurements that might further help in understanding the intrinsic properties of the FEEL devices are the current-voltage (I-V) and luminance-current characteristics of the devices. Figure 6A shows the I-V curves of the FEEL and glow/FEEL states operated over a wide range of gas pressures. The behaviors display nothing more than the usual electric breakdown behavior expected from the normal discharge theory. When breakdown occurs, it indicates that light is turning on and the static voltage drops. Since the output voltage of the power supply is higher than the breakdown voltage (V_b), the sustain voltage ($V_{sustain}$) of the device remains the nearly constant, albeit the power supply voltage increases continuously to enhance current density. Also, when the FEEL device is in the lighting state, the system shows features of self-sustain discharge and an apparent negative resistance. It is evident that FEEL works after breakdown occurs, a behavior akin to normal glow discharge. The results also indicate a tendency of lower V_b and $V_{sustain}$ for higher pressure devices. It suggests that the operation window for FEEL and glow/FEEL states are both located in the left hand side of the Paschen curve, which are sensitively dependent on parameters such as the gas type, cathode material, pressure condition and discharge gap width.

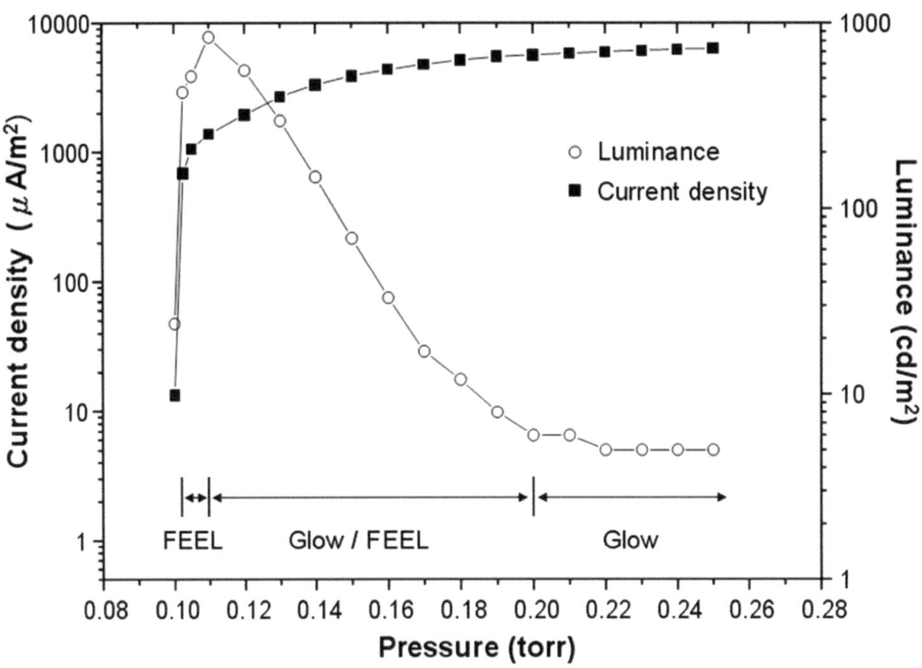

Fig. 5. The current density and luminance as a function of gas pressure inside the FEEL device. A close correlation between luminance efficiency and glow excitation is observed. The characteristic threshold pressure of 0.11 torr distinguishing the regimes of FEEL and glow/FEEL states is consistent with the observations shown in Figure 4.

Figure 6B shows the characteristics of luminance-current density for FEEL devices operated over the same gas pressure range. It is noted that, except the apparent saturation behaviors occurring for the two samples with the lowest gas pressures in FEEL state, there appears to exist a linear relation between luminance and current density in glow/FEEL state, albeit the slope is a function of gas pressure. To understand these, we consider the energy budget of the electrons under discussion. At anode, the maximum arrival kinetic energy for electrons emitted from the cathode by ion bombarding will be the summation of the initial kinetic energy and the acceleration energy from electrical field. However, over the course of flight, the electrons may collide with the gas molecules and loss their kinetic energies. The glow excitation observed within the pressure region of 0.11-0.25 torr shown in Fig. 4 is primarily due to these dynamical interactions. Also, as shown in Fig. 5, although the device current remains essentially constant after the luminance reaches its maximum at P_{N2} = 0.11 torr, further increase in device pressure does cause a reducing device luminance. Combining with the observation of the linear dependence of luminance on current density under a fixed device pressure (Fig. 6B), it indicates that, although the photon energy density (luminance) of the device is linearly related to the electron number density arriving at the anode per unit time, the internal quantum efficiency of the phosphor is very much dependent on the effective kinetic energy carried by the electrons. Consequently, a large amount of low kinetic energy electrons do not guarantee efficient luminance in the glow/FEEL state. On the other hand, as shown in Fig. 6B, for the low pressure cases even though the number of electrons (current density) is much lower with their high kinetic energy, high luminance can be achieved in the FEEL state. It appears that the slopes of luminance versus current density curves shown in Fig. 6B can be regarded as the effective kinetic energy of the electrons and the electron kinetic energy is the dominant factor to determine the final luminance in FEEL mechanism. Finally, the apparent saturation behaviors for the low pressure cases of the FEEL state might be due to the thermal quench inside the phosphor, which in turn prevents further increase of device luminance from electron kinetic energy transferring.

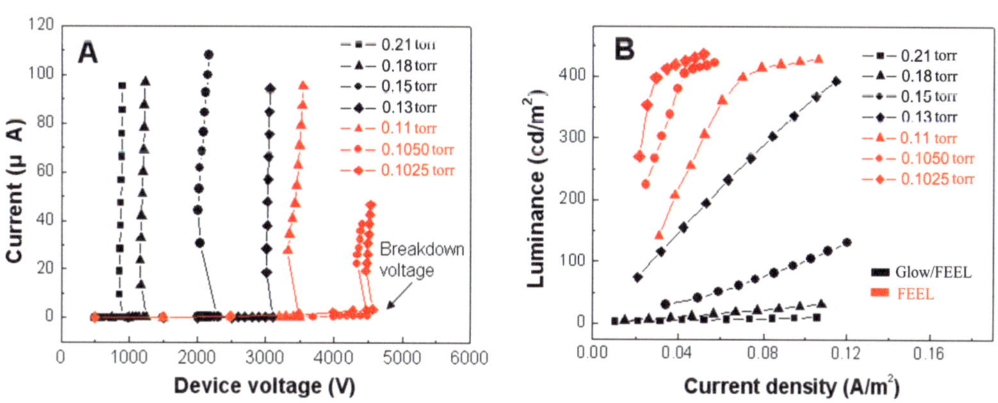

Fig. 6. (A) The I-V characteristics and (B) current density-luminance correlations in the FEEL and glow/FEEL states of the device (APL, Vol.94, 091501, 2009). The I-V curves indicate that FEEL emits light after electric breakdown. As power supply voltage is higher than breakdown voltage, the device voltage remains nearly constant although the power supply voltage continuously increases to enhance current density. (B) Luminance versus current reveals that luminance of FEEL is linearly related to current density after electric breakdown.

Finally, we discuss the variations of the breakdown voltage V_b near the operation conditions of FEEL devices by measuring the Paschen curve of the device. As shown in Fig. 7, the Paschen curve demonstrates that regimes of FEEL and glow/FEEL are indeed operating in the lower pressure regime locating on the left hand side of the Paschen curve. As a result, the steep change in V_b within a slim region of pressure change is currently a major hurdle to be dealt with. Improvements by using different working gases as well as cathode materials with better secondary electron emissions are currently under extensive investigations.

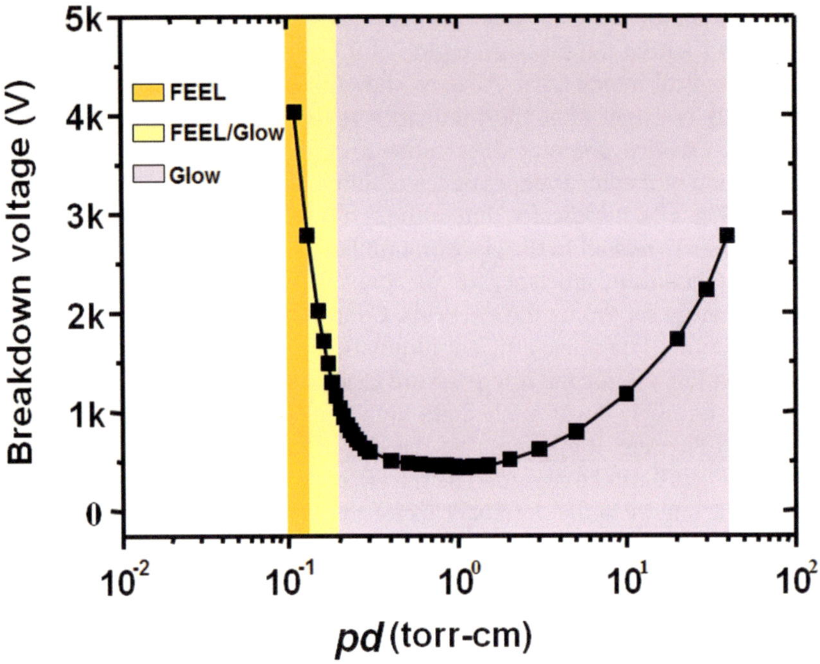

Fig. 7. The Paschen curve of FEEL devices with N_2 as the working gas. The areas highlighted by the dark-yellow and light-yellow colors respectively indicate the operation regime of the FEEL and glow/FEEL states in the devices.

4. Energy-saving issues and applicability of FEEL

As described above, the lighting mechanism operating in FEEL devices can be conceptually regarded as a combination of gas discharge working for FL devices with CL phosphor lighting. Such combination, nevertheless, has been proven to be feasible in generating large area planar electron beams which successfully eliminated the dark spots on the phosphor screen by providing uniformly distributed excitations over the entire anode surface. Other equally important and extremely crucial issues in general lighting are the considerations of health and safety, namely how to protect human eyes and skin from long-time exposure to the lighting environment. To take into account these human factors, the planar light source which provides the uniform light emission, can naturally support the sufficient amount of illumination with reduced lighting intensity to avoid of the glare and persistence of vision from lamp. Thus, the unique planar feature of FEEL nicely meets the requirement of healthy human factors. To further compare with the traditional planar lamps with diffusers, which are composed of spot or line shaped light sources inside the lamp structure, FEEL doesn't need extra diffuser to obtain uniform planar lighting from originally concentrated radiation sources. Hence, without energy dissipation from diffuser, FEEL doesn't need to increase the power to obtain higher radiated intensity for compensating the energy loss from diffuser. It means that even total illumination and input power of the FEEL are reduced for energy saving, it still simultaneously matches the requirements of specification and lighting environment. Obviously, lower material cost and power consumption become the production advantages for FEEL. Besides, to consider the heat extraction issue, FEEL is similar to the planar heat source with low heat density. Without heat concentration, it also doesn't need special device design for heat extraction.

In addition to planar lighting market, FEEL has high potential to expand its applicability by modifying the device design. We find that, especially in the FEEL anode, there are several methods to adjust the styles of light emission through modifying the phosphor pattern. In the following demonstrations, we illustrate that FEEL not only can be applied to planar lamp, but also can be designed as many kinds of unique lighting productions. For instance, the prototypes of transparent lighting glass, gray-scale image lighting, and colorful image lighting have all been developed.

Figure 8 individually reveals the power turned-off and turned-on status of transparent FEEL device. It shows that the FEEL is transparent when its power is turned off (Fig. 8A). In Fig. 8B, the phosphor coating on the FEEL anode radiates the visible light when it is turned on the power. Due to the influence of the light emitted from phosphor, at the turn-on status the transparency of the device is reduced to a lower level. For transparency purpose, in this case transparent FTO films were used as the electrodes of the device. The only opaque material, the phosphor coating on the anode surface, was designed as regular dot-matrix pattern for light transparency. The area of the dot pattern is around 100 μm, so the detailed structure of the pattern cannot be differentiated by human eyes. The device transparency could be evaluated by the coverage of phosphor on the whole anode surface. The maximum device transparency is about 60%, which is limited by the transparency of transparent electrodes and total coverage of phosphor. It means that FEEL can be a lighting window to receive the extra natural light in the daytime-indoor, and provides the indoor lighting at night. Additionally, we found that the isolated glass structure and a very small amount of gas inside the 4 inch FEEL device naturally give rise to the property of low heat conductivity (0.07 W/mK), which is about 20% of the heat conductivity of a 4 inch normal glass (1.4 W/mK). The thermal conductivity of the device was measured by thermal conductivity meter (EKO HC-072). It is interesting to note that FEEL is the only lighting technology to have the property of heat isolation. It is very useful for the windows of passive house to prevent the heat from transferring through the FEEL window. In other words, it can preserve the thermal energy inside the house in winter and avoid the outside heat from transferring into the indoor in summer.

It is clear from the above description that FEEL does have the potential of being simultaneously used as the energy-saving thermal insulation window, daylighting system, and indoor lighting in sustainable passive house for saving the power consumption of air-condition in summer, indoor heating in winter or daytime-indoor lighting (Fig. 9). In particular, with the unique features of transparency and heat insulation, it can also meet the demand of the sustainable building material. It may be anticipated that the building's

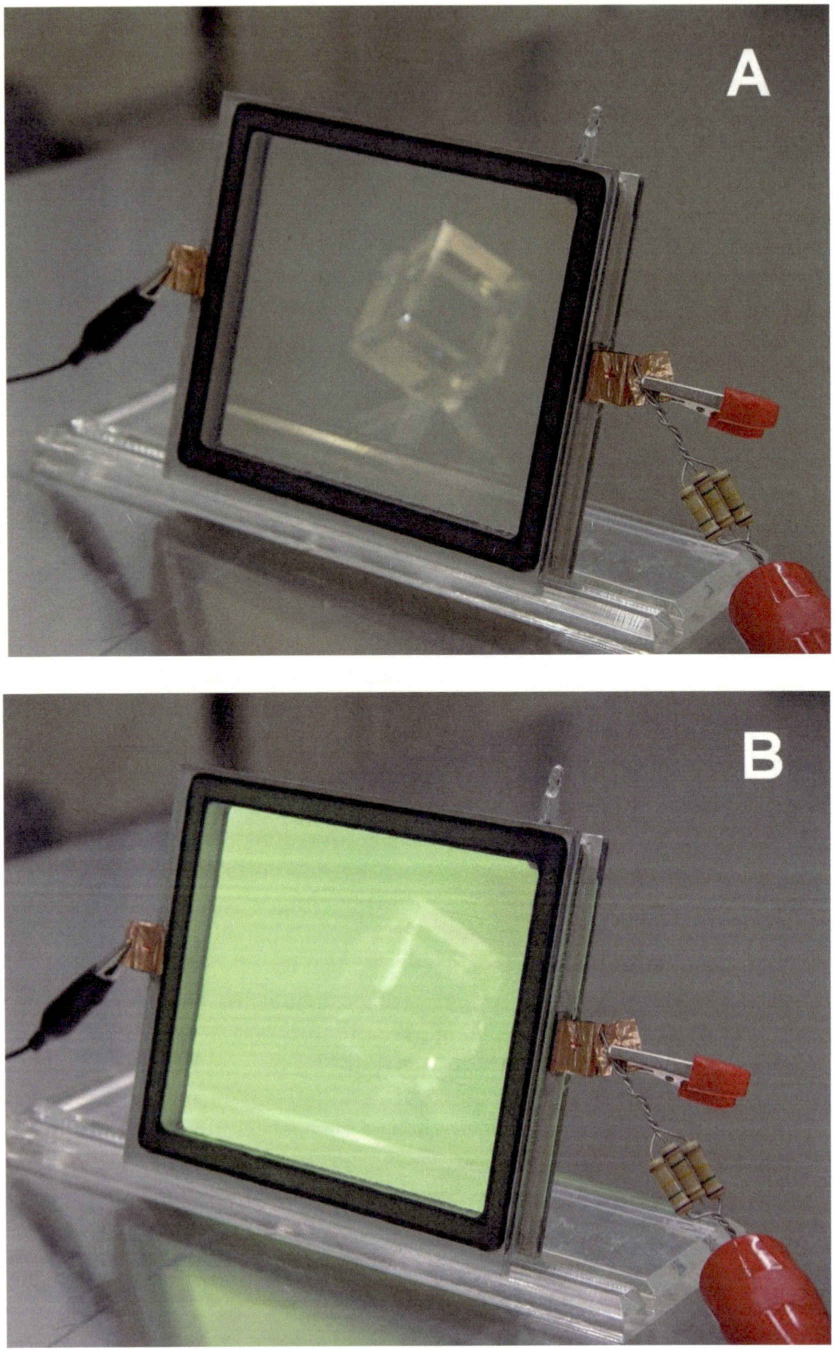

Fig. 8. The photo of a transparent FEEL. (A) It shows that the lamp is transparent when power is turned off. The phosphor dotted matrix can not be differentiated because of the size of dotted patterns is only around 100 μm. (B) The phosphor coating on the anode of FEEL radiates the visible light when its power is turned on. It also reveals the partial transparent effect while seeing through the device.

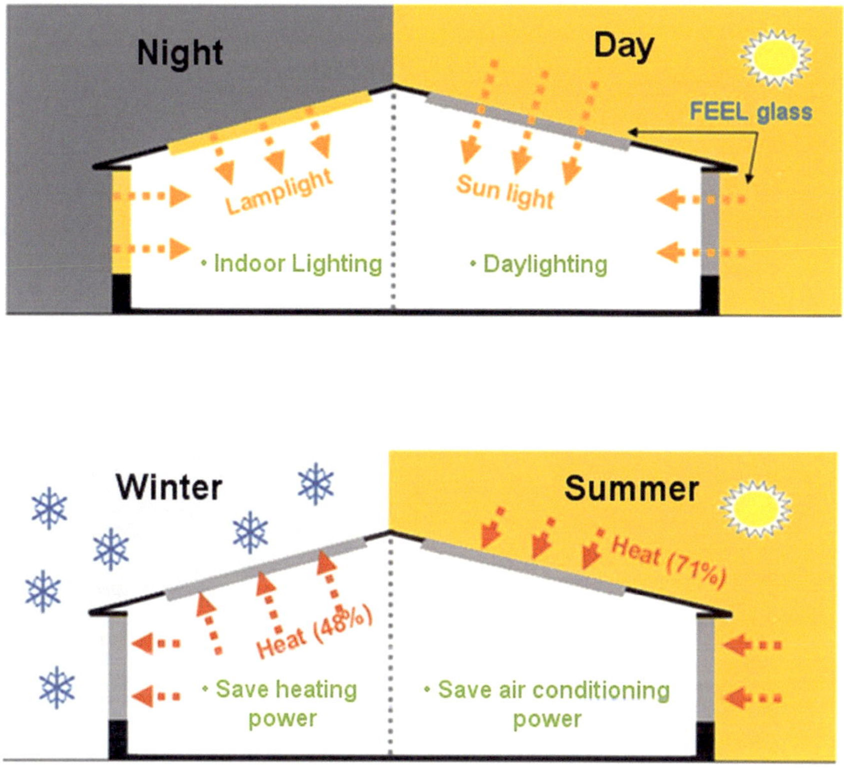

Fig. 9. Schematics demonstrating that it is possible to use FEEL to simultaneously accomplish the energy-saving thermal insulation glass, daylighting system, and indoor lighting in sustainable passive house for saving the power consumption of air-condition in summer, indoor heating in winter or day time indoor lighting.

skylights and elevation windows are the potential applications for green building markets. Furthermore, the building's owners are able to independently control the turn-on status for every FEEL window on the building elevation. If the elevated walls of the building are analogy to the display screens, then the FEEL glasses can be represented as the display pixels, which are named as window pixels in this case (Fig. 10). We believe that, the linkage between lighting and green building material, will change the business model and the user's habits in the lighting market. However, we expect that the installation and material cost for FEEL will be higher than the normal glass, but the extra cost can be refunded from the power consumption saving generated from daily operations as described above. In addition, at night, it is possible to utilize the FEEL windows in commercial mansions to be rewarded for the advertisement applications such as show of the company's logo and instantaneous information.

Fig. 10. FEEL can be applied to building skylights and building elevation windows. The building elevation looks like a large size display when every FEEL glass is independently controlled the turn-on status.

The rectangle shape is not the limit for FEEL technology. As shown in Fig. 11, the flat FEEL bulb, which challenges the arbitrary shape capability of FEEL, subverts the normal impressions of ball shaped incandescent bulb. It also shows the special lighting effect after phosphor is printed with designed pattern on anode surface.

Fig. 11. The flat FEEL bulb subverts the normal impressions of ball shaped incandescent bulb.

As illustrated in Fig. 12, the radiated gray-scale image on FEEL anode screen is another unusual demonstration by special screen printing design for phosphor. The way for revealing the static gray-scale image by FEEL is making a big difference from display panel. On normal display, the screen is composed of dotted matrix pixels. Every individual pixel radiates its own correspondence intensity for revealing the gray-scale effect. On FEEL lighting surface, however, there are no regular pixel matrix and external control circuit. Instead, it utilizes the area of individual dotted phosphor pattern to form the corresponding radiated intensity under the same power condition. It means that the radiated intensities are linear relation to the areas of dotted phosphor patterns. Similar to the case of transparent device, the areas of the phosphor patterns are too small to differentiate the pattern's structure by human eyes. By this design rule, any digital picture or document content can be transformed to the phosphor patterns with correspondent areas for reproducing the radiated image. Thus, FEEL should be able to radiate the high resolution image with gray-scale. It is interesting to note that the static lighting image from FEEL is always confused with display image. Although it is only a lamp, it indeed provides unlimited imagination for display applications.

In Fig. 13, we further extend the gray-scale function to show the possibility of using FEEL for colorful image demonstrations. Basically, colorful radiation is linear superposition of three original colors (Red, Green and Blue). It means that any color image can be decomposed into three individual images with original color. As described before, these individual RGB images are further transformed into dotted phosphor patterns in correspondent areas to produce the gray-scale image with single color. At last, RGB phosphor patterns are overlapped (multi-layers) printing on the same anode of FEEL to reveal the color image. Once again, its image resolution looks like the static color picture in display, although there is no display driver and pixel structure inside the device.

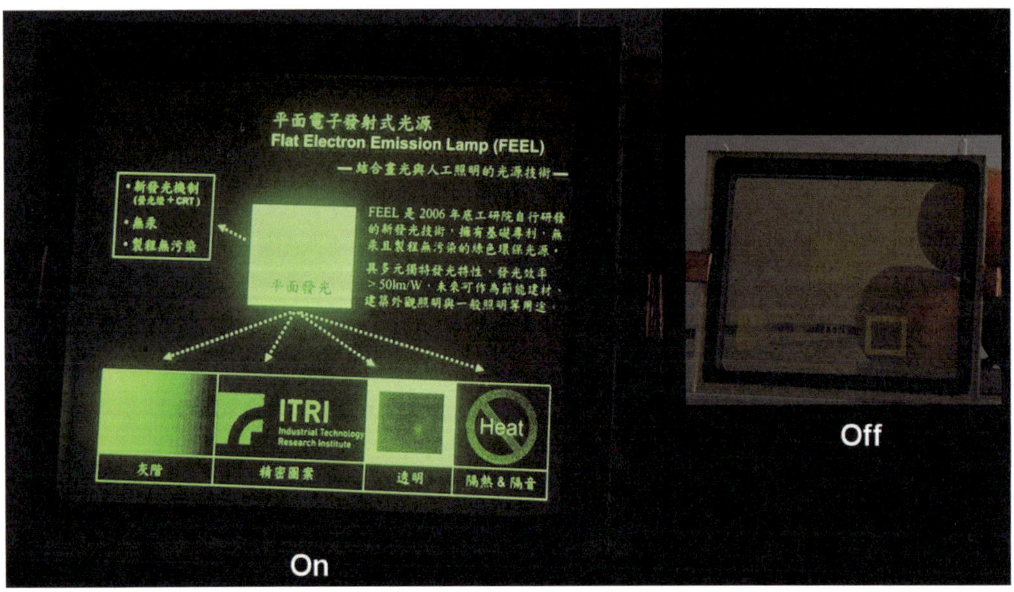

Fig. 12. FEEL radiates the precise image with gray-scale. It provides the imagination for display applications.

Fig. 13. FEEL shows the capability of colorful image demonstration by the overlapped printing of RGB phosphor patterns.

5. Conclusion and remarks

The lighting mechanism for FEEL is based on the innovative integration from mature theories, such as gas discharge and cathodoluminescence. It has been indicated that the working range for FEEL may obey the rule of normal glow discharge, with the features of voltage drop and negative resistance. The measurements of I-V curves, Paschen curve, luminance and OES show that the electron kinetic energy is the dominant factor to determine the final luminance efficiency. The first priority to enhance the luminance efficiency of FEEL is to reduce the electron scattering dissipation by decreasing gas pressure. However, it conflicts with the requirement for stable discharge. In general, it needs sufficient amount of gas for maintaining self-sustain discharge, or higher gas pressure means lower breakdown voltage. The solution to solve such dilemma between low gas pressure and efficient discharge is to improve the secondary electron emission efficiency of cathode material under the low gas pressure environment. It is noticed that the suitable gas type should be carefully selected to match with cathode material for getting higher discharge efficiency. Besides, the optimized phosphor composition and printing process need to be further investigated for obtaining higher quantum efficiency of phosphor. So far, FEEL mechanism shows the high flexibility and potential for applicability in the early stage. It already presents the potential to expand applicability for normal lighting, power-saving building material and gray-scale ambiance lighting. We believe that FEEL could be the candidates of next generation green lighting for providing the comfortable lighting, power-saving, and ambiance applications.

6. Acknowledgments

The authors would like to acknowledge the support from the Energy Fund of Ministry of Economics Affairs, Taiwan. The authors would also like to acknowledge the support from the Energy Foundation. JYJ is supported partially by National Science Council of Taiwan and by MOE-ATU program operated at National Chiao Tung University (NCTU).

7. References

Jung-Yu Li, Shih-Pu Chen, Chia-Hung Li, Yi-Ping Lin, Yen-I Chou, Ming-Chung Liu, Po-Hung Wang, Hui-Kai Zeng, Tai-Chiung Hsieh, and Jenh-Yih Juang (2009). A lighting mechanism for flat electron emission lamp, *APPLIED PHYSICS LETTERS*, Vol.94, pp.091501-1-091501-3, Taiwan

Chia-Hung Li, Ming-Chung Liu, Chang-Lin Chiang, Jung-Yu Li, Shih-Pu Chen, Tai-Chiung Hsieh, Yen-I Chou, Yi-Ping Lin, Po-Hung Wang, Ming-Shin Chun, Hui-Kai Zeng, and Jenh-Yih Juang (2011). Discharge and photo-luminance properties of a parallel plates electron emission lighting device, *Optics Express*, Vol.19, pp.A51-A56, Taiwan

T. Jüstel, H. Nikol, C. Ronda (1998). New Developments in the Field of Luminescent Materials for Lighting and Displays, *Angew. Chem. Int. Ed*, Vol.37, pp. 3084-3103, Germany

M. Ilmer, R. Lecheler, H. Schweizer, M. Seibold (2000). Hg-free Flat Panel Light Source PLANON - a Promising Candidate for Future LCD Backlights, *SID International Symposium, Digest of Technical Papers*, XXXI, pp. 931-933, Germany

A. A. Talin, K. A. Dean, J. E. Jaskie (2001). Field emission from carbon nanotubes: the first five years, *Solid-State Electronics*, Vol.45, pp. 893-914, Switzerland

U. Kogelschatz (2004), Excimer Lamps: History, *Discharge Physics, and industrial Applications Proceedings of SPIE*, Vol.5483, pp. 272-286, Switzerland

G. G. Lister, J. E. Lawler, W. P. Lapatovich, V. A. Godyak (2004). *The physics of discharge lamps, Rev. Mod. Phys.*, Vol. 76, pp. 541-598, USA

J. P. Boeuf (2003). Plasma display panels: physics, recent developments and key issues, *J. Phys. D: Appl. Phys.*, Vol.36, R53-79, France

M. Itoh, L. Ozawa (2006). Cathodoluminescent phosphors, *Annu. Rep. Prog. Chem., Sect. C*, Vol.102, pp. 12-42, Japan

Y. X. Liu, J. H. Liu, C. C. Zhu (2009). Flame synthesis of carbon nanotubes for panel field emission lamp, *Applied surface science*, Vol.255, pp.7985-7989, China

R. W. B. Pearse and A. G. Gaydon (1965). The Identification of Molecular Spectra,3rd ed., Chapman and Hall, pp. 209–220, London

Y. Saito and S. Uemura (2000). Field emission from carbon nanotubes and its application to electron sources, *Carbon*, Vol.38, pp.169-182, Japan

J.-M. Bonard, H. Kind, T. Stöckli, and L.-O. Nilsson (2001). Field emission from carbon nanotubes: the first five years, *Solid-State Electron*, Vol.45, pp.893-914, Switzerland

M. J. Druyvesteyn, F. M. Penning (1940). The mechanism of electrical discharges in gases of low pressure , *Reviews of Modern Physics*, Vol.12, pp.87-174, Holland

M. M Pejović, Goran S Ristić, J. P Karamarković (2002). Electrical breakdown in low pressure gases, *J. Phys. D: Appl. Phys.*, Vol.35, R91-103, Yugoslavia

G. Cho, J. Y. Lee, Dae H. Lee, S. B. Kim, H. S. Song, J. Koo, B. S. Kim, J. G. Kang, E. H. Choi, U. W. Lee, S. C. Yang, J. P. Verboncoeur (2005). Glow Discharge in the External Electrode Fluorescent Lamp, *IEEE Transactions On Plasma Science*, Vol.33, pp.1410-1415, Korea

J. Reece Roth (1995). Industrial Plasma Engineering-V.1 : Principle, *Institute of Physics*, p.353, London

Yuri P. Raizer (1991). Gas discharge physics, *Springer-Verlag Berlin Heidelberg*, pp.134-135, New York

A. Miller (1988). Conflguration-independent minimum voltage for the Townsend model of breakdown in gases, *J. Appl. Phys.*, Vol.63, pp.665-667, New Jersey

Nikolai N. Chubun, Andrei G. Chakhovskoi, Charles E. Hunt (2003). Efficiency of cathodoluminescent phosphors for a field-emission light source application, *J. Vac. Sci. Technol. B*, Vol.21, pp.1618-1621, California

J. J. Rocca, J. D. Meyer, M. R. Farrell, G. J. Collins (1984). Glow-discharge-created electron beams: Cathode materials, electron gun disigns, and technological applications, *J. Appl. Phys.*, Vol.56, pp.790-797, Colorado

A.P. Bokhan, P.A. Bokhan and D.E. Zakrevsky (2005). Peculiarities of electron emission from the cathode in an abnormal glow discharge, *Appl. Phys. Lett.*, Vol.86, pp.1-3, Russia

P. Hartmann, H. Matsuo, Y. Ohtsuka, M. Fukao, M. Kando, and Z. Donko (2003). Heavy-particle hybrid simulation of a high-voltage glow discharge in helium, *Jpn. J. Appl. Phys.*, Vol.42, pp.3633-3640, Hungary